Methods for Reliability Improvement and Risk Reduction

Methods for Reliability Improvement and Risk Reduction

Michael Todinov
Oxford Brookes University
UK

WILEY

This edition first published 2019

Registered Offices
John Wiley & Sons, Inc., 111 River Street, Hoboken, NJ 07030, USA
John Wiley & Sons Ltd, The Atrium, Southern Gate, Chichester, West Sussex, PO19 8SQ, UK

Editorial Office
The Atrium, Southern Gate, Chichester, West Sussex, PO19 8SQ, UK

For details of our global editorial offices, customer services, and more information about Wiley products visit us at www.wiley.com.

Wiley also publishes its books in a variety of electronic formats and by print-on-demand. Some content that appears in standard print versions of this book may not be available in other formats.

Library of Congress Cataloging-in-Publication Data

Names: Todinov, M. T., author.
Title: Methods for reliability improvement and risk reduction / Michael Todinov, Oxford Brookes University, UK.
Description: Hoboken, NJ, USA : Wiley, 2019. | Includes bibliographical references and index. |
Identifiers: LCCN 2018033729 (print) | LCCN 2018036493 (ebook) | ISBN 9781119477310 (Adobe PDF) | ISBN 9781119477594 (ePub) | ISBN 9781119477587 (hardcover)
Subjects: LCSH: Reliability (Engineering) | Risk management. | System failures (Engineering)
Classification: LCC TA169 (ebook) | LCC TA169 .T649 2019 (print) | DDC 620/.00452–dc23
LC record available at https://lccn.loc.gov/2018033729

Cover Design: Wiley
Cover Illustration: © Michael Todinov

Set in 10/12pt TimesLTStd by SPi Global, Chennai, India

Printed and bound by CPI Group (UK) Ltd, Croydon, CR0 4YY

10 9 8 7 6 5 4 3 2 1

To the bright memory of my mother

Contents

Preface

The common approach to risk reduction is domain-specific and relies exclusively on detailed knowledge from a specific domain. Measures specific to the domain are selected for reducing the risk and risk reduction is conducted exclusively by the experts in the domain. The underlying argument is simple yet powerful. Why should, for example, a welding specialist or automotive engineer listen to and seek advice from a general risk expert on how to improve the reliability of the welds or the reliability of a car? After all, the risk expert is not normally familiar with the welding or automotive technology.

This argument contributed to creating the illusion that efficient risk reduction can be delivered successfully solely by using methods offered by the specific domain without resorting to general methods for risk reduction. This led to a situation that in many domains, even the existence of a general risk science has been forgotten. In textbooks on design of machine components, for example, there is hardly any mention of general methods for improving reliability and reducing the risk of failure.

The price for this illusion is that many industries have been deprived of effective risk reducing strategy and solutions. The same mistakes are made again and again, resulting in numerous accidents and inferior products and processes, associated with high risk of failure.

An important contributing reason for this highly undesirable situation is the absence of a framework of domain-independent methods that could provide vital methodological knowledge in reliability improvement and risk reduction.

With the exception of a very few simple and well-known domain-independent methods for risk reduction, such as implementing redundancy, strengthening weak links, upgrading with more reliable components, simplification of components, systems and operations, and condition monitoring, the framework of domain-independent methods for risk reduction is missing. The absence of a framework of domain-independent risk reduction methods diminishes significantly the importance of risk science and poses serious questions about whether it actually adds value to various areas of human activity.

Consequently, proposing a framework of domain-independent methods for improving reliability and reduce risk was the primary motivation behind writing this book.

In this book, methods and principles related to improving reliability and reducing risk that can be classified as domain-independent are first reviewed and their limitations discussed. Next, new domain-independent principles and methods for reliability improvement and risk reduction are introduced, with a detailed discussion of the mechanisms through which they reduce risk.

The methods of reliability improvement and risk reduction presented in this book are based on a large number of available solutions, most of which came from mechanical engineering. Each of the available solutions was analysed for recurring reliability-enhancing patterns and invariants. A certain level of abstraction was used to strip available solutions from the specific mechanical engineering context and uncover the underlying patterns governing the reliability improvement and risk reduction.

From the analysis of available solutions, various patterns and invariants emerged which were captured and distilled into categories, classes, and individual techniques. The application of the distilled new methods and principles has been illustrated with numerous real-life application examples and case studies. Many of the domain-independent methods reduce risk at no extra cost. This is a significant advantage to many traditional methods for reducing risk (e.g. redundancy, upgrading components, condition monitoring) which are associated with substantial investment.

The proposed framework of domain-independent risk reducing methods is not a substitute for domain-specific methods. It rather serves as a powerful enhancement of the domain-specific risk reduction and helps to obtain superior solutions.

The framework of domain-independent methods and principles for risk reduction proposed in this book prompts design engineers not to limit themselves to a few familiar domain-specific methods for improving reliability and reducing risk which often lead to solutions that are far from optimal. Using appropriate combinations of domain-independent and domain-specific methods brings superior results.

Consequently, the proposed domain-independent methods form an important part of risk science and firmly support the design for reliability and the decision making in the presence of uncertainty.

The proposed framework will enhance the reliability of products and operations for any company and organisation. To any company, reliability is one of the most important attributes of its products. High product reliability means high company reputation, high customer satisfaction, low warranty costs and repeat business. For a company, this translates into a big competitive advantage and secure market position. Low product reliability means financial losses, human injuries and fatalities, and damaged environment and infrastructure. For companies, this translates into a loss of reputation, loss of market share and, ultimately, a loss of business.

The framework of domain-independent methods for risk reduction proposed in this book provides the basis for a strong interdisciplinary research. Researchers and engineers, after receiving training in domain-independent risk reduction methods, are capable of solving complex reliability improvement problems in their specific industries/domains. In turn, the problems encountered in the specific industry/domain stimulate further development of the domain-independent reliability improvement and risk reduction methods. This creates a positive self-reinforcing feedback loop which benefits both the industry and the reliability and risk science.

The proposed framework of domain-independent reliability improving methods works particularly well in improving the reliability of existing designs. In this respect, the author's experience in teaching these methods to engineering students shows that after minimal initial training in domain-independent methods, students are capable of significantly improving the reliability of engineering designs of their choice by original and effective solutions based on these methods.

The domain-independent methods for reliability improvement and risk reduction, are analogous to the proof techniques in mathematics. Proofs consist of a finite number of ideas and techniques (e.g. proof by contradiction, proof by contrapositive, mathematical induction, the extreme principle, the pigeonhole principle, proof based on parity, proof based on continuity of functions, probabilistic proof, etc.). Just as studying general proof techniques ensures success in building sound mathematical proofs, studying the domain-independent methods for improving reliability and reducing risk ensures success in designing capable and reliable products.

It is the author's firm belief that the domain-independent methods for reliability improvement and risk reduction should not only be an integral part of the reliability and risk science, they should also be an integral part of the education of every design engineer.

In conclusion, I thank the acquisition editor, Anne Hunt, the project editor, Jemima-India Kingsly, the copyeditor Wendy Harvey and the production editor Sathishwaran Pathbanabhan at John Wiley & Sons for their excellent work and cooperation. My thanks also go to my many colleagues from universities and the industry for their useful suggestions and comments.

Finally, I acknowledge the immense help and support I received from my wife, Prolet, during the writing of this book.

Oxford, 2018

Michael Todinov

1

Domain-Independent Methods for Reliability Improvement and Risk Reduction

1.1 The Domain-Specific Methods for Risk Reduction

A systematic classification of generic methods for reducing technical risk is crucial to risk management, safe operation, engineering designs, and software. However, this very important topic has not been covered with sufficient depth in the reliability and risk literature. For many decades, the focus of the reliability research has been primarily on identifying risks, risk assessment, and reliability prediction rather than methods for reliability improvement and risk reduction. The as low as reasonably practicable (ALARP) approach to risk management (Cullen 1990; HSE 1992; Melchers 2001), for example, advocates that risks should be reduced ALARP. This is commonly interpreted in the sense that risks have to be reduced to a level at which the cost associated with further risk reduction outweighs the benefits arising from further reduction (HSE 1992; Melchers 2001). While a decision about implementation of risk-reducing measures can be taken by implementing cost–benefit analysis, the focus of the ALARP approach is whether risk-reducing measures should be implemented or not. There is little clarity on the risk-reducing methods that can be used to achieve the risk reduction.

Reliability improvement and risk reduction also relied for a long time on the feedback provided from reliability testing or on feedback from customers. Once the feedback about a particular failure mode is available, the component is redesigned to strengthen it against that failure mode. The problem with this approach is that the feedback always comes late, after the product has been manufactured. Therefore, all changes consisting of redesign to avoid the discovered failure modes will be costly or impossible. In addition, conducting a reliability testing programme to precipitate failure modes is expensive and adds significant extra cost to the product.

General guidelines on risk management do exist. Risk management, according to a recent review (Aven 2016) can be summarised to (i) establish the purpose of the risk management activity, (ii) identify adverse events, (iii) conduct cause and consequence analysis, (iv) make judgement about the likelihood of the adverse events and their impact and establish risk description and characterisation, and (v) risk treatment.

While a great deal of agreement exists about the necessary common steps of risk assessment, there is profound lack of understanding and insight about the general methods for reducing risk that can be used. The common approach to risk reduction is the domain-specific approach which relies heavily on *root cause analysis* and detailed knowledge from the specific domain. Measures specific to a particular domain are selected for reducing the likelihood of failure or the consequences from failure and the risk

Methods for Reliability Improvement and Risk Reduction, First Edition. Michael Todinov.

reduction is conducted exclusively by experts in the specific domain. The risk reduction is effectively fragmented into risk reduction in numerous specific domains: nuclear industry, aviation, construction industry, food storage and food processing, banking, oil and gas industry, road transportation, railway transportation, marine transportation, financial industry, cyber security, environmental sciences, etc.

As a result, the domain-specific approach to risk reduction created an illusion: *that efficient risk reduction can be delivered successfully solely by using methods offered by the specific domain without resorting to general methods for risk reduction.*

The direct consequence of this illusion is that many industries have been deprived from effective risk-reducing strategy and reliability improvement solutions. The same mistakes are made again and again, resulting in numerous accidents and inferior products and processes, associated with high risk of failure. Examples of such repeating mistakes are:

- insufficient reliability built in products with very high cost of failure;
- designing components with homogeneous properties where the stresses are clearly not uniform;
- creating systems with vulnerabilities where a single failure causes the collapse of the system;
- redundancy compromised by a common cause.

At the same time, excellent opportunities to improve reliability and reduce risk are constantly missed. Examples of such missed opportunities are:

- failure to increase reliability of systems and components at no extra cost (e.g. by a simple permutation of the same type of components in the system);
- failure to increase the reliability of components and systems by a separation of properties and functions;
- failure to reduce by orders of magnitude the probability of erroneous conclusion from imperfect tests;
- failure to increase by orders of magnitude the fault tolerance of components;
- failure to reduce risk by including deliberate weaknesses.

The weaknesses of the risk management in many specific domains were exposed by a string of costly failures and disasters (e.g. catastrophic oil spills, financial crises, serious industrial accidents, transport accidents, power blackouts, etc.).

In some cases, correct solutions were indeed found by 'reinventing the wheel', after a series of costly and time-consuming trials and errors.

An important contributing reason for this highly undesirable situation is *the absence of a framework of domain-independent methods for reliability improvement and risk reduction that could provide vital methodological knowledge to many unrelated domains.*

With the exception of a few simple and well-known domain independent methods such as *implementing redundancy, strengthening weak links, upgrading with more reliable components, simplification of components, systems and operations*, and *condition monitoring*, the framework of domain-independent methods for reliability improvement and risk reduction is missing.

Thompson (1999) stressed the importance of effective integration of maintainability and reliability considerations in the design process and the importance of failure mode analysis in design. Thompson (1999) correctly identified that knowledge of the principles of risk are important aids to achieving good reliability, however, no domain-independent principles for improving reliability and reducing risk have been formulated.

Samuel and Weir (1999) covered problem solving strategies in engineering design and stressed the importance of satisfying design inequalities in defining the domain of acceptable designs. However, no domain-independent methods for improving reliability have been discussed.

French (1999) formulated a number of general principles to be followed in conceptual design, but they were not oriented towards improving reliability and reducing technical risk. General principles to be followed in engineering design have also been discussed in Pahl et al. (2007). Most of the discussed principles, however, are either not related to reducing the risk of failure or are too specific (e.g. the principle of thermal design), with no general validity. Collins (2003) discussed engineering design with failure prevention perspective. However, no risk-reducing methods and principles with general validity were formulated.

Taguchi's experimental method for robust design through testing (Phadke 1989) achieves designs where the performance characteristics are insensitive to variations of control (design) variables. This method can be considered to be a step towards formulating the domain-independent risk reduction principle of robust design for which the performance characteristics are insensitive to variations of design parameters.

1.2 The Statistical, Data-Driven Approach

A common approach to reliability improvement is to select a statistical-based, data-driven approach. This approach relies on critical pieces of data: *failure frequencies, load distribution, strength distribution*, etc., in order to make predictions about the reliability of components and systems.

To describe the reliability on demand, which is essentially the probability that strength will exceed load, data covering the variation range of the load and the variation range of the strength are needed. These data are necessary to fit an appropriate model for the strength, an appropriate model for the load, and to estimate the parameters of the models fitting the load distribution and strength distribution. Next, a direct integration of the load–strength interference integral or a Monte Carlo simulation can be used to estimate the probability that, on demand, strength will be greater than the load (Todinov 2016a).

To calculate the time to failure of a system, the time-to-failure models of the components are needed. For each component, from the past times to failure, an appropriate time-to-failure model must be fitted and subsequently used to evaluate the reliability of the system built with the components (Todinov 2016a). However, the time-to-failure models of the components depend strongly on the environmental stresses. For example, increasing temperature accelerates material degradation and shortens the time to failure. Because of this, the time to failure of a seal working at elevated temperatures is significantly shorter than the time to failure of a seal working at room temperature. Reducing temperature also gives rise to dangerous failure modes (e.g. brittle fracture) which reduce the time to failure. The time to failure in the presence of a corrosive environment, high humidity and vibrations is shorter than the time to failure in the absence of such environmental stresses.

It is a well-established fact that time-to-failure models built on past reliability data collected for a particular environment yield poor predictions if applied automatically to another environment. To fit a reliable and robust model, the past failure data need to cover all possible duty cycles and different environmental conditions (temperature, humidity, vibrations, pressure, corrosion stresses, etc.) the product could encounter. This is a near to impossible task. As a result, the time-to-failure models built on past failure data, not covering all possible operating conditions and environmental stresses, lead to increased levels of uncertainty in the model parameters and poor predictive power.

Finally, the data-driven approach provides feedback about the product performance after it has been released in the field. The time delay in receiving this feedback, long after the design stage, makes it expensive to improve reliability.

Calculating the reliability built in a product by using past data is often a difficult task because reliability-critical data (failure frequencies, strength distribution of the flaws, failure mechanisms, and repair times) may not be available, particularly for new designs, with no failure history. This does not permit meaningful analysis and correct prediction of the reliability of components and systems, for different environments or uses.

The pure statistical approach does not normally address the physical principles controlling the operation and failure of engineering systems and components.

Even if all critical pieces of information were available, in some cases, a meaningful reliability prediction would still be a problem. Consider, for example, a simple assembly including a large number of identical components from the same batch. Suppose that the assembly fails whenever any of the identical components fail. The reliability of the assembly is then estimated by raising the estimated reliability R of a single component to a power equal to the number of the components in the assembly. Increasing the number of tests will reduce the uncertainty associated with the estimated reliability R of a single component but will never eliminate it.

A small error in estimating the reliability R of the identical components would result in an unacceptably large error (uncertainty) in the estimated reliability of the assembly, which renders the reliability prediction meaningless.

This can be illustrated by a simple analytical argument. Consider a simple assembly built on n identical components from the same batch, each characterised by reliability R. The reliability of the system then becomes $R_{sys} = R^n$. An error ΔR in the reliability of a single component leads to a relative error $\Delta R_{sys}/R_{sys} = n\,(\Delta R/R)$ in the predicted reliability for the system. For a system composed of 50 capacitors, logically arranged in series, from the same production batch, a mere 1% relative error in the estimated reliability of a single capacitor will result in $\Delta R_{sys}/R_{sys} = 50 \times 0.01 = 0.5$ (50%) error in the estimated system reliability, which makes the reliability prediction meaningless.

Furthermore, while the time to failure due to a particular wearout failure mode can, in some cases, be predicted with a great deal of confidence, no reliable prediction of the time to failure of software components is possible. It depends on a particular combination of input data, user choice, and other conditions, which cannot normally be predicted with confidence.

Some of these difficulties led some authors to question the appropriateness of reliability prediction based on past failure rates.

1.3 The Physics-of-Failure Approach

The development of the physics-of-failure approach to reliability improvement (Pecht et al. 1990; Pecht 1996) has been prompted by the major deficiency of the data-driven approach discussed earlier – the critical dependency on the availability of past failure rates. The physics-of-failure approach created a widespread view among many reliability practitioners that only developing physics-of-failure models can deliver reliability improvement and developing physics-of-failure models is the only proper method to assess the reliability of a product. This view has also been fuelled by the failure of some statistical models to predict correctly the life of engineering components.

According to the physics-of-failure approach, failures and decline in performance of components and systems occur due to known underlying failure mechanisms. Unlike the data-driven approach, the physics-of-failure approach addresses the underlying causes of failure. Many failure mechanisms lead to accumulation of damage. Failure is initiated when the amount of accumulated damage exceeds the endurance limit and the time to failure of components can be physically modelled.

Despite these advantages, building accurate physics-of-failure models of the time to failure is not always possible because of the complexity of the physical mechanisms underlying the failure modes, the complex nature of the environment and the operational stresses. There is usually a great deal of uncertainty associated with the parameters of the physics-of-failure models. If the goal, for example, is to increase strength, the physics-of-failure modelling can help increase strength by conducting research on the link between microstructure and mechanical properties. This approach requires arduous and time-consuming research, special equipment and human resources while a positive outcome from the research is not guaranteed.

Furthermore, in many failure events, several failure mechanisms are often involved, interacting in a very complex fashion. Such is, for example, the corrosion fatigue where two very complex, interdependent, and synergistic failure mechanisms ('corrosion' and 'fatigue') contribute to failure. Corrosion increases the rate of fatigue damage accumulation and the progression of the fatigue crack increases the extent of corrosion. This complex interaction and synergistic behaviour cannot be captured and modelled successfully if limited research is done on corrosion, fatigue and their interaction.

Often, limited experimental evidence is available because of cost limitations. The experimental evidence necessary to build a correct model can be limited not only in terms of quantity but also in terms of quality. There is a difference between observed phenomena and driving force. The experimental evidence usually captures the visual component of damage (the observed phenomena), which may not necessarily reflect the driving force behind the damage accumulation and the total existing damage precipitating failure. As a result, only the visual manifestation of the damage is captured and quantified as opposed to the driving force behind the damage accumulation and the total damage reaching the damage endurance limit. Acquiring the necessary knowledge and data related to the failure mechanisms, capturing and quantifying all types of uncertainty, necessary for a reliable prediction of the time to failure, is a formidable task.

Despite their success and popularity, physics-of-failure models cannot transcend the initial narrow domain they serve and cannot normally be used to improve reliability and reduce risk in unrelated domains.

Implementing measures aimed at eliminating hydrogen embrittlement in welds, for example, is a very important step towards improving the reliability of welds and reducing the risk of failure. These measures, however, cannot transcend the narrow domain of the welding technology. They cannot, for example, be applied to reduce risk in computer networks, finance, economics, etc.

For this reason, physics-of-failure methods cannot normally be generalised as domain-independent reliability improvement methods.

1.4 Reliability Improvement and TRIZ

The need for increasing efficiency and reducing the weight of components and systems while maintaining high reliability is a constant source of technical and physical contradictions. Hence, it is no surprise that several principles for resolving technical contradictions formulated by Altshuller in the development of TRIZ methodology for inventive problem solving (Altshuller 1984, 1996, 1999) can also be used for reducing technical risk. Eliminating harmful factors and influences is the purpose of many inventions and Altshuller's TRIZ system captured a number of useful general principles which could be used to eliminate harm.

Here, it needs to be pointed out that many of the principles for technical risk reduction with general validity are rooted in the reliability and risk theory and cannot possibly be deduced from the general inventive principles formulated in TRIZ, which serve as a general guide in developing inventive solutions, as an alternative to the trial-and-error approach. For example, the domain-independent principle which states that the reliability built-in a system should be proportional to its cost of failure is rooted in the risk theory and cannot be deduced from general inventive principles.

Another limitation of TRIZ is the lack of coverage of the mechanisms through which the inventive methods could achieve reliability improvement and risk reduction.

Another weakness preventing the effective use of TRIZ for reliability improvement is that the TRIZ methods are not backed with mathematical models or algorithms which, in a number of cases, are absolutely necessary to unlock the reliability improvement resource. Indeed, by providing a succinct description of the system, a mathematical model or algorithm could deliver significant benefits:

- The system can be described by taking into consideration a very complex interaction of risk-critical factors which could not be intuitively contemplated by design engineers. In many cases, the only way to extract risk-reduction benefit is to build and analyse a mathematical model or algorithm. Such is the case, for example, of reducing the risk of overlap of risk-critical events which requires a relevant mathematical model to be built.
- A mathematical model or an algorithm provides a way of tracking the impact of the risk-critical factors on the level of risk. In this respect, the mathematical model/algorithm provides an insight into which control variables are essential and which seemingly important variables have actually no practical impact on the reliability and risk level.
- A mathematical model or algorithm provides insight into which factor needs to be altered and by how much in order to extract the maximum risk-reduction benefit.

1.5 The Domain-Independent Methods for Reliability Improvement and Risk Reduction

The reliability and risk literature (Bazovsky 1961; Barlow and Proschan 1965, 1975; Ang and Tang 1975; Dillinton and Allan 1992; Ramakumar 1993; Ebeling 1997; Meeker and

Escobar 1998; Vose 2000; Booker et al. 2001; Bedford and Cooke 2001; Kuo et al. 2001; Trivedi 2002; Andrews and Moss 2002; Aven 2003) is oriented towards risk modelling, risk assessment, risk management and decision making and there is very little discussion related to domain-independent principles for reducing technical risk.

Well-known methods for improving reliability and reducing risk such as *simplification of components and systems, introducing redundancy, strengthening weak links* or *condition monitoring* have the potential to reduce risk in many unrelated domains (e.g. computer networks, finance, economics, etc.). They are domain-independent and do not rely on reliability data or knowledge of physical mechanisms underlying possible failure modes. As a result, they are very well suited for developing new designs, with no failure history and unknown failure mechanisms.

The domain-independent risk reducing methods are not a substitute for domain-specific methods. Rather, they are a powerful enhancement of the domain-specific risk reduction methods and help to obtain superior solutions. Consequently, the domain-independent methods form an important part of risk science.

The systematic distilling, formulating and classifying of domain-independent methods and principles for improving reliability and reducing technical risk started in a 2007 book (Todinov 2007), continued in a subsequent book (Todinov 2016a) and was recently accelerated in a series of papers introducing a number of new domain-independent methods for reliability improvement and risk reduction (Todinov 2017a,b,c). In the 2007 book (Todinov 2007), the domain-independent principles for risk reduction have been broadly divided into: 'preventive' – reducing mainly the likelihood of failure; 'protective' – reducing mainly the consequences from failure; and 'dual' – oriented towards reducing both the likelihood of failure and the consequences from failure.

The recently proposed new domain-independent reliability improvement and risk reduction principles and methods transcend the area of mechanical engineering where they originated and can be applied in diverse areas of human activity. For example, the new risk reduction methods (*separation, stochastic separation, segmentation, self-reinforcement, inversion, reducing the rate of accumulation of damage, introducing deliberate weaknesses, permutation and limiting the space and time exposure*) can be applied in various unrelated domains.

Without sound methodological knowledge of domain-independent methods, opportunities for decreasing risk by separating functions and properties, by segmentation, inversion or by introducing deliberate weaknesses cannot be seen and are often missed. Many of the domain-independent methods for risk reduction presented in this book reduce risk at no extra cost. This is a big advantage over many traditional methods for reducing risk (redundancy, upgrading components, condition monitoring) which are associated with substantial investment.

Consider the problem of increasing the reliability of a cylindrical pressure vessel with a specified volume. Typical domain-specific solutions are to increase the strength of the material of the pressure vessel or to increase the thickness of the shell. The first solution is costly because it requires, research, human resource, time, special equipment, and expensive alloys. The second solution is associated with increasing the weight and the cost of the pressure vessel. These undesirable consequences associated with both solutions can be avoided if the domain-independent method of *inversion by maintaining an invariant* (Todinov 2017b) is used. This method alters the shape of the pressure vessel by reducing its diameter, increasing the length, and decreasing the thickness of the shell while preserving the volume. This reduces the maximum stress in the shell and improves reliability with a simultaneous reduction in weight and cost (see Chapter 8).

Improving the reliability of a seal on a pressure vessel by inverting the relative location of the cover or improving the reliability of drilling by inverting the orientation of the component are also improving reliability at no extra cost. By using separation in geometry, the reliability of a cantilever beam can be improved at no extra cost (Todinov 2017a). The logical separation, for example, leads to low-cost yet very efficient designs eliminating safety risks (Todinov 2017a). By using permutation of components of the same type and different age, the reliability of common parallel series systems can be improved at no extra cost (Todinov 2015). Using the domain-independent principle of introducing deliberate weaknesses towards which failure is channelled in the case of overloading, also leads to low-cost risk reduction solutions. The alternative is expensive upgrading of the reliability of components.

For an expert in a specific domain it is not obvious that the risk of an erroneous conclusion from an imperfect test can be decreased by orders of magnitudes by segmenting the test. The domain-independent method of segmentation leads to a low-cost risk reduction solution because no investment is made into more precise test equipment, associated with a smaller measurement error (Todinov 2017c).

In a number of cases, the reliability improvement by capturing a self-reinforcing response, by self-anchoring, and self-alignment is also achieved at no extra cost (Todinov 2017b).

Without understanding the reliability improvement mechanisms of the segmentation, it is counter-intuitive and difficult to see why a monolithic column loaded in compression is less fault-tolerant compared with a column built with bricks of the same material that are not linked.

For many domain experts, for example, it is not obvious that the reliability of a piece of wire loaded in tension can be increased significantly simply by decreasing the loaded length without altering the material of the wire or its cross-sectional area. The reliability improvement is achieved at a reduced cost because the amount of material used is reduced.

Without the prompt from the domain-independent method of separation, designers often select material with uniform properties even though the loading and the stresses are not uniform. This leads to premature failures and unreliable products. Applying the method of separation also leads to reliability improvement at no extra cost. Instead of selecting expensive material with the ability to resist all environmental stresses, a combination of cheap materials can be used, optimised to resist individual environmental stresses.

The application examples and case studies featured in this book cover *mechanical engineering, civil engineering* and *construction, electronics, software engineering, chemical engineering, financial control, management, project management, environmental sciences, logistics supply, economics*, etc.

By providing the theoretical foundation for solving risk management problems in diverse areas of human activity, the domain-independent methods change radically the existing risk reduction paradigm based exclusively on domain-specific methods.

The domain-independent methods for reducing risk are so powerful that some of them (e.g. *the method of segmentation, the method of separation*, and *the method of inversion*) can effectively be used as self-contained tools for solving problems unrelated to risk reduction.

2

Basic Concepts

2.1 Likelihood of Failure, Consequences from Failure, Potential Loss, and Risk of Failure

The purpose of risk analysis is to provide support in making correct management decisions. By evaluating the risk associated with a set of decision alternatives, the risk analysis helps to identify the alternative which maximises the expected utility for the stakeholders while complying with a set of specified criteria and constraints. Key concepts in risk management are *probability of failure*, *consequences from failure*, *potential loss*, and *conditional loss*. Risk is associated with an adverse event (event with negative impact).

Despite the large number of definitions and interpretations of risk (e.g. risk as uncertainty about the performance of a system (Aven 2003) or risk as a measure of how variable our losses are (Crouhy et al. 2006)), risk seems to incorporate three fundamental characteristics (Kaplan and Garrick 1981): (i) a loss-generating factor (event with a negative impact); (ii) the likelihood of an exposure to the loss-generating factor; and (iii) the consequences associated with the exposure and the uncertainty associated with them.

Hazards are situations, conditions, or objects with the potential of causing harm. Hazardous sources are, for example, objects storing a significant amount of electrical, chemical, heat, kinetic, or potential energy. Examples of hazardous sources are: particular actions, flammable materials, highly reactive chemical compounds, toxic wastes and chemicals, pressurised gas, explosive materials, materials with very high or very low temperature, high voltage, electromagnetic fields and high-intensity electromagnetic radiation, high-intensity sound and vibration, fast moving or falling objects, sharp moving parts, and suspended heavy objects.

Hazards, under certain triggering conditions, cause adverse events with negative impact.

Failures also are adverse events and appear as frequent loss-generating factors.

Failure modes express the way a system, component, process, or operation fails to perform as intended. An essential part of the risk management process is identifying, preventing, or mitigating potential failure modes.

Most of the losses associated with the materialisation of an event with negative impact can be classified in several major categories:

- *Loss of life or damage to health.*
- *Losses associated with damage to the environment and the community infrastructure.*

Methods for Reliability Improvement and Risk Reduction, First Edition. Michael Todinov.

- *Financial losses* including warranty costs, loss of production, loss of capital assets, cost of intervention and repair, compensation payments, insurance costs, medical costs, penalty payments, and legal costs.
- *Loss of reputation* including loss of market share, loss of customers, loss of contracts, impact on share value, loss of confidence in the business, etc.

Depending on the category, the losses can be expressed in monetary units, number of fatalities, lost time, volume of lost production, volume of pollutants released into the environment, number of lost customers, amount of lost sales, etc. Often losses from failures are expressed in monetary units and are frequently referred to as *cost of failure*.

A specific domain brings its unique set of failure modes. Common failure modes in mechanical engineering, for example, are: fracture, corrosion, excessive wear, overheating, congestion, deformation, stress relaxation, fracture, buckling, and jamming. In the area of project management, the following failure modes may appear: unavailability or delay in receiving a key project resource (materials, equipment, software, trained personnel, etc.); loss of key people on the project; a third party failing to deliver a key task to the project; change of project scope; delay in receiving training for a key task; breakdown of equipment; denied access to a data base or software; poor quality measurements due to faulty equipment, etc. The loss (consequences) from the realisation of any of these failure modes is failure to achieve project objectives, inferior quality, missed deadlines, and excessive project costs.

Risk is a unity of an adverse event, probability of its occurrence and loss, given that the adverse event has occurred. Risk is not synonymous with the magnitude of the loss, irrespective of whether this magnitude is constant or variable. Without exposure to an adverse event the loss will never materialise. According to a classical definition (Henley and Kumamoto 1981; Vose 2000), the risk of failure is measured by the product of the probability of failure and the cost given failure:

$$K = p_f C \tag{2.1}$$

where p_f is the probability of failure and C is the cost given failure. To an operator of production equipment, for example, the cost given failure C may include several components: cost of lost production, cost of cleaning up a polluted environment, medical costs, insurance costs, legal costs, costs of mobilisation of emergency resources, lost sales due to loss of reputation and low customer confidence, etc. The cost of failure to the manufacturer of production equipment may include: warranty payment if the equipment fails before the agreed warranty time, loss of sales, penalty payments, compensation, and legal costs.

The theoretical justification of Eq. (2.1) can be made on the basis of the following thought experiment. Suppose that a non-repairable piece of equipment is put in operation for a length of time a, which can be the warranty period for the equipment. If the equipment fails before the specified warranty time a, its failure is associated with a constant loss C due to warranty replacement. Suppose that the experiment involves N identical pieces of equipment, of which N_f fail before time a. Since failure before time a is associated with losses, the total loss generated by failures among N pieces of equipment is $N_f \times C$. The average (expected) loss K is then $K = (N_f \times C)/N$. This expression can be rearranged as

$$K = (N_f \times C)/N = (N_f/N) \times C \tag{2.2}$$

If the number of pieces of equipment N is sufficiently large, $p_f \approx N_f/N$ gives an estimate of the probability of failure of the equipment before time a and Eq. (2.1) is obtained.

Usually, a relatively small number for N gives a sufficiently accurate estimate of the true probability of failure. As a result, Eq. (2.1) describes the average (expected) warranty loss per single piece of equipment. This is one of the reasons why, in the engineering context, the risk is often treated as *expected loss from failure*.

It needs to be pointed out that the likelihood of the exposure to a loss-generating factor is conditional on the existing knowledge about the adverse event. Indeed, consider two identical sensors (sensor A and sensor B) working independently in a high-temperature environment and participating in controlling a chemical process. Each of the sensors survives a period of operation in high-temperature conditions of one week, with probability r. If $P(A)$ and $P(B)$ denote the probabilities of the sensors working and $P(\overline{A})$ and $P(\overline{B})$ denote the probabilities that the sensors will be in a failed state at the end of the week: $P(A) = P(B) = r$ and $P(\overline{A}) = P(\overline{B}) = p = 1 - r$.

The correct operation of at least one of the sensors is needed to control the chemical process. The probability that both sensors will fail at the end of the week (which constitutes a system failure), in the absence of any knowledge about the general state of the sensors is the product of the probability that the first sensor will be in a failed state and the probability that the second sensor will be in a failed state: $P(\overline{A} \cap \overline{B}) = P(\overline{A}) \times P(\overline{B}) = p^2$. Suppose that at the end of the week, a condition monitoring test circuit indicates that at least one sensor has failed. This additional piece of knowledge now alters the probability that both sensors will be in a failed state. The probability that at least one sensor will be in a failed state at the end of the week is

$$P(\overline{A} \cup \overline{B}) = P(\overline{A}) + P(\overline{B}) - P(\overline{A}) \times P(\overline{B}) = p + p - p^2 = p(2 - p)$$

The probability that both sensors have failed given that at least one sensor has failed is

$$P(\overline{A} \cap \overline{B} \mid \textit{at least one sensor failed}) = \frac{P(\overline{A} \cap \overline{B})}{P(\overline{A} \cup \overline{B})} = \frac{p^2}{p(2-p)} = \frac{p}{2-p}$$

For a probability of failure $p = 0.3$ at the end of the week, characterising each sensor, the probability that both sensors will be in a failed state at the end of the week is $P(\overline{A} \cap \overline{B}) = p^2 = 0.09$. The probability that both sensors will be in a failed state at the end of the week, given that one of the sensors is in a failed state at the end of the week, is

$$P(\overline{A} \cap \overline{B} \mid \textit{at least one sensor failed}) = \frac{p}{2-p} = \frac{0.3}{2-0.3} = 0.176$$

As can be seen, the presence of an additional piece of information from the condition monitoring circuit altered the probability of system failure significantly: the conditional probability that both sensors will be in a failed state given that at least one sensor has failed is almost twice the absolute probability that both sensors will be in a failed state at the end of the week.

Because risk is conditional on pre-existing knowledge, risk is relative (Kaplan and Garrick 1981). The same action of a particular subject, from the point of view of the subject and the observer can be a low-risk action or a high-risk action depending on the possession of additional information by the subject or by the observer.

There are numerous misconceptions about risk. Some authors equated risk with uncertainty about the occurrence of an event. Uncertainty about the occurrence of an event does not constitute risk unless some of the outcomes of the event have a negative impact (are associated with a loss) (Kaplan and Garrick 1981). Consider a case where an investor bets \$100 on a particular prospect. With probability 0.5 the investor receives a total of \$200

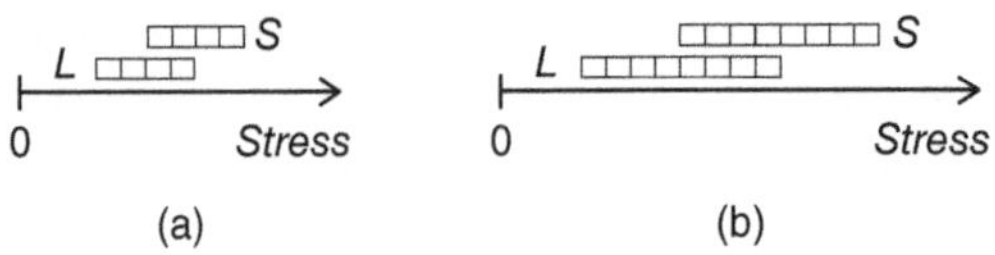

Figure 2.1 Larger variability does not always translate into a larger risk.

or with probability 0.5 the investor receives \$400. Despite the uncertainty related to the outcome of the investment, there is no negative impact (there is no loss) therefore there is no risk in this investment.

Some authors equated risk with variability by stating that the higher the variability the higher the risk. This is not always correct. While often, variability is a major source of risk, larger variability does not always translate into larger risk. Consider the two intervals in Figure 2.1a depicting the independent variation of load (interval L) and strength (interval S). Each interval has a length of four units and load and strength are uniformly distributed along their corresponding intervals. Failure is present if load is greater than strength. The probability of failure is equal to the probability of the event that load will be greater than strength. In order for this event to occur, both load and strength must belong to the intersection region (the probability of this is $0.5 \times 0.5 = 0.25$) and given that both load and strength are within the intersection region, load must exceed strength (the probability of this event is 0.5 because, due to the symmetry, in half of the load–strength random configurations, load is greater than strength, and in half of the configurations, strength is greater than load). The probability of the compound event that both load and strength are within the intersection region and load exceeds strength is therefore $0.25 \times 0.5 = 0.125$. In Figure 2.1b, the variation intervals of both load and strength have been increased twice. Although the variability has been increased, the probability that load will exceed strength remains the same, equal to 0.125. *Increasing variability had no effect on the risk of failure.*

Another counter-example contradicting the common view that increased variability is equivalent to increased risk is given in Chapter 3, with a load–strength interference of asymmetric load and strength distributions.

Another misconception about risk is related to a sequence of statistically independent repeated bets. These were the focus of a paper by Samuelson (1963). In this paper, the author uses the following argument, through a story in which he offered his colleague a good bet (with a positive expected value): 50–50 chance of winning 200 or losing 100. The colleague refused by saying that he would feel the 100 loss more than the 200 gain. He said that he would agree to participate if he was offered to make 100 such bets. Samuelson criticised the reasoning of his colleague and went on to propose and prove a 'theorem' which stated that if a single good bet is unacceptable, then any finite sequence of such bets is unacceptable too. Samuelson claimed that increasing the number of unacceptable bets does not reduce the risk of a net loss and termed accepting a sequence of individually unacceptable bets 'a fallacy of large numbers'. Samuelson's 'theorem' has been reproduced in several other papers (Ross 1999). This 'theorem' spawned several related papers where researchers have extended Samuelson's condition to assure that they would not allow the 'fallacy of large numbers'. Contrary to Samuelson's theory, increasing the number of unacceptable bets does reduce the risk of a net loss, and this can be demonstrated by using Samuelson's own example.

Consider the good bet where, with probability 0.5, the person receives \$200, or with probability 0.5 loses \$100. If a single bet is made, this investment is risky: with probability 0.5 the person will lose \$100. If 10 independent bets are made sequentially, the probability

Table 2.1 Expected profit, variance, and probability of a net loss with increasing the number of good bets.

Number of good bets	Expected profit	Standard deviation	Probability of a net loss
1	50	150	0.5
10	500	474.3	0.17
20	1000	670.8	0.057
30	1500	821.6	0.049
50	2500	1060.7	0.0077
80	4000	1341.6	0.0012
90	4500	1423	0.0010
100	5000	1500	0.00044
130	6500	1710.3	0.00007
150	7500	1837	0.00003

of no win is equal to the probability of no win in all bets plus the probability of exactly one win from all 10 bets plus the probability of exactly 2 wins from all 10 bets. The probability of this combined event is 0.17.

Table 2.1 lists the results from the calculations using equations derived in the literature (Todinov 2013c) and confirmed by a Monte Carlo simulation.

As can be verified from the table, even though with increasing the number of good bets, the variance of the net profit increases, the probability of a net loss has decreased significantly. In other words, although selecting an individual bet is not acceptable because of the high probability of a loss, selecting repeated bets is beneficial, because a longer sequence of repeated bets is characterised by an increased expected profit, a small probability of a net loss and a small risk of a net loss. Contrary to the view expressed by Samuelson, which has also been adopted in related papers (e.g. Ross 1999), repeating the unacceptable bet reduced significantly the risk of a net loss.

With increasing the number of good bets, the variance of the profit increases significantly (Table 2.1). This seeming anomaly perhaps has led some researchers to conclude that because of the increased variance of the profit, the risk will also increase. However, the analysis shows that the variance of the profit can increase without a simultaneous increase in the risk of a loss. In this case, the common belief that a larger variance translates into a larger risk does not hold. In the case of repeated opportunity bets, a larger variance actually coexists with a smaller risk and the variance of the profit cannot serve as a risk measure.

There are several basic ways of reducing risk. Risk can be reduced either by reducing the consequences (the loss) given failure or by reducing the probability of failure or by reducing both. Depending on how the risk is reduced, the risk reducing methods can be broadly divided into three major categories: *preventive risk reducing methods*, reducing the probability of failure; *protective risk-reducing methods*, reducing the loss given failure; or *dual methods*, reducing both.

In cases where there is little control over the hazards and in cases where there is a great deal of uncertainty about the failure modes, hazards, working environment, and environmental stresses, the likelihood of failure is difficult or impossible to estimate. In this case, the principle of the risk-based design should be applied (Todinov 2006b) which states that the reliability of a system/component should be proportional to the cost of its failure. Protective measures also work well. They should be targeted at improving the resilience of the system to adverse events.

2.2 Drawbacks of the Expected Loss as a Measure of the Potential Loss from Failure

The risk Eq. (2.1) only estimates the average value of the potential loss from failure. A decision criterion based on the expected loss would prefer the design solution characterised by the smallest expected potential losses. What is often of primary importance *is not* the expected (average) loss, but the deviation from the expected loss (the unexpected loss). This is, for example, the case where a company estimates the probability that its potential loss will exceed a particular critical value after which the company will effectively be insolvent. Although the expected loss gives the long-term average of the loss, there is no guarantee that the loss will revert quickly to such an average (Bessis 2002). This is particularly true for short time intervals where the variation of the number of failures is significant.

Let us consider a real-life example where a selection needs to be made between two competing identical systems which differ only by the time to repair. A critical failure of the first system is associated with a time for repair which follows a normal distribution. As a consequence, the lost production due to the critical failure also follows a normal distribution. Suppose that this distribution is characterised by mean $\overline{C}_1$ and variance $\sigma_1{}^2$. The second system is associated with a constant time for repair and constant cost of lost production $\overline{C}_2 > \overline{C}_1$. The two systems are characterised by the same probability of failure $p_{1f} = p_{2f} = p_f$. Equation (2.1) yields $K_1 = p_f \overline{C}_1$ for the risk of failure characterising the first system and $K_2 = p_f \overline{C}_2$ for the risk of failure characterising the second system. Clearly, $K_1 < K_2$ because $\overline{C}_1 < \overline{C}_2$. However, the probability that the loss given failure will exceed a critical maximum acceptable value x_{max} is zero for the system characterised by the larger risk K_2, and non-zero for the system characterised by the smaller risk K_1 (Figure 2.2).

In other words, a smaller expected loss does not necessarily mean a smaller risk that the loss will exceed a critical limit.

If the expected value of the loss given failure was selected as a utility function of the consequences from failure, the first system would be selected by a decision criterion based on minimising the expected loss.

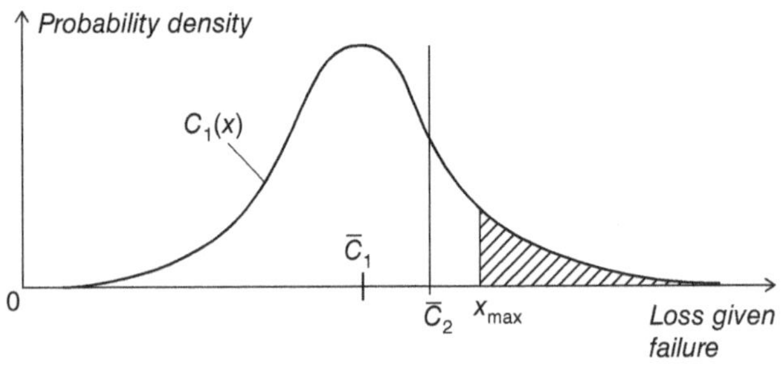

Figure 2.2 Distributions of the loss given failure for two systems.

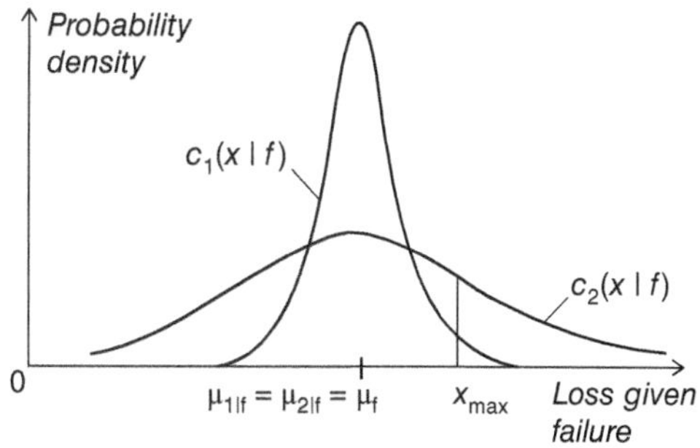

Figure 2.3 The variance of the loss given failure is strongly correlated with the probability that the loss will exceed a specified quantity.

Suppose that x_{max} is the maximum amount of reserves available for covering the loss from critical failure (Figure 2.2). No recovery can be made from a loss exceeding the amount of x_{max} and production cannot be resumed. With respect to whether a recovery from a critical failure can be made, the first system is associated with risk while the second system is not.

In order to make a correct selection of a system minimising the risk of exceeding a maximum acceptable limit the utility function should reflect whether the loss exceeds the critical limit x_{max} or not.

Increasing the variance of the loss given failure, increases the risk that the loss will exceed a specified maximum tolerable level. This is illustrated in Figure 2.3 by the probability density distributions of the loss given failure $c_1(x \mid f)$ and $c_2(x \mid f)$ of two systems characterised by different variances ($\sigma_1^2 < \sigma_2^2$) of the loss.

Despite the drawbacks of the expected values in risk quantification, one should not go to the extreme to wipe out the expected value as a measure of risk. The authors who advocate this extreme view should remember that the system reliability measure is nothing but an expected value. The system reliability can be defined as the expected fraction of times a system is in a working state at the end of a specified time interval. Alternatively, the system reliability can be defined as *the expected fraction of stochastically pruned reliability networks in which paths through working components exist from the start node to all terminal nodes* (Todinov 2016b). Production availability is also an expected value. Production availability is the ratio of the expected system output in the presence of component failures and the total output in the absence of component failures (Todinov 2016b). Risk can also be measured by the expected time fraction of simultaneous presence of critical events (see Section 6.2).

In another example, applying the expected value for calculating the risk associated with a limited number of repeated opportunity bets is a bad idea (Todinov 2013c). However, for a large number of repeated opportunity bets, applying expected values to assess the risk works. The problem is not in the expected value used as a measure of risk. The problem is with selecting the most appropriate measure of risk in each particular case for the particular conditions and constraints.

A new measure of the loss from failure which avoids the limitations of the traditional risk measure (Eq. (2.1)) is the *cumulative distribution of the potential loss*.

2.3 Potential Loss, Conditional Loss, and Risk of Failure

The concepts of *potential loss* and *conditional loss* are central to understanding risk. The quantity *a loss given failure* is a conditional quantity because it is defined *given that failure has occurred*. This is in sharp contrast with the *potential loss* which is an *unconditional quantity* and is defined *before failure occurrence*. While the conditional distribution of the loss given failure can be used to determine the probability that given failure, the loss will be larger than a specified limit, the distribution of the potential loss combines the probability that there will be failure and the probability that the loss associated with it will be larger than a specified limit. In other words, the measure 'potential loss' incorporates the uncertainty associated with the exposure to losses and the uncertainty in the consequences given exposure.

Historical data related to the losses from failures can only be used to determine the distribution of the conditional loss. Building the distribution of the potential losses however, requires also an estimate of the probability of failure.

Both the conditional loss and the potential loss are random variables. Thus, in the failure event leading to a loss of containment of a reservoir or a pipeline transporting fluids, the conditional loss depends on how severely the reservoir/pipeline is damaged.

Since the potential loss is a random variable, it is characterised by a cumulative distribution function $C(x)$ and a probability density function $c(x)$. The probability density function $c(x)$ gives the probability $c(x)\,dx$ (before failure occurs) that the potential loss X will be in the infinitesimal interval x and $x+dx$ $(P(x<X\leq x+dx)=c(x)dx)$.

Accordingly, the conditional loss (the loss *given* failure) is characterised by a cumulative distribution function $C(x\,|\,f)$ and a probability density function $c(x\,|\,f)$. The conditional probability density function $c(x\,|\,f)$ gives the probability $c(x\,|\,f)\,dx$ that the loss X will be in the infinitesimal interval x and $x+dx$ given that failure has occurred $(P(x<X\leq x+dx\,|\,f)=c(x\,|\,f)dx)$.

Let S be a non-repairable system composed of M components, logically arranged in series, which fails whenever any of the components fails. It is assumed that the components' failures are mutually exclusive; that is, no two components can fail at the same time. The reasoning below and the derived equations are also valid if instead of a set of components, a set of M mutually exclusive system failure modes are considered, i.e. no two failure modes can initiate failure at the same time. Because the system is non-repairable, the losses are associated with the first and only failure of the system. However, the reasoning presented next, is also valid for a repairable system if the focus is on the loss from the first failure only.

The cumulative distribution function $C(x)$ of the potential loss gives the probability that the potential loss X will not be greater than a specified value x. A loss is present only if failure is present. Consequently, the unconditional probability $C(x)\equiv P(X\leq x)$ that the potential loss X will not be greater than a specified value x is equal to the sum of the probabilities of two mutually exclusive events: (i) failure will not occur and the loss will be not be greater than x; and (ii) failure will occur and the loss will not be greater than x. The probability of the first compound event is $(1-p_f)\times H(x)$, where p_f is the probability of failure and $H(x)$ is the conditional probability that the loss will not be greater than x given that no failure has occurred. This conditional probability can be presented by the Heaviside unit step function $H(x) = \{^{0,x<0}_{1,x\geq 0}$. The probability of the second compound event is $p_f\,C(x\,|\,f)$ where $C(x\,|\,f)$ is the conditional probability that *given* failure, the loss will not be greater than x. Consequently, the probability $C(x)$ that the potential loss X will not be greater than x is given by the distribution mixture:

$$C(x) = P(X \leq x) = (1-p_f)\times H(x) + p_f \times C(x\,|\,f) \tag{2.3}$$

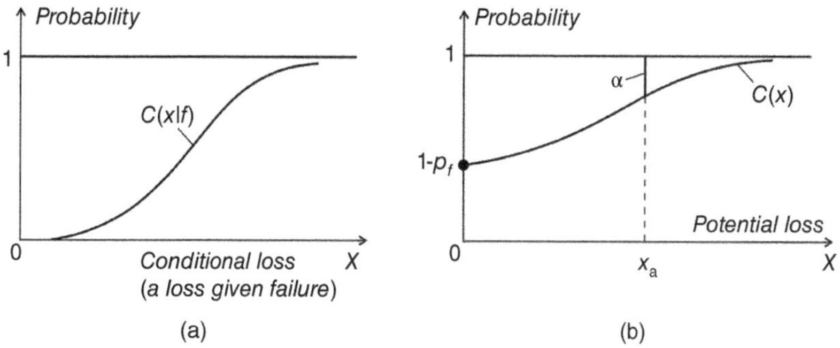

Figure 2.4 (a) A conditional loss (a loss given failure); (b) a potential loss and maximum potential loss x_α at a pre-set level α.

The difference between a potential and conditional loss is well illustrated by their distributions in Figure 2.4. A characteristic feature of the cumulative distribution of the potential loss is the concentration of probability mass with magnitude $1 - p_f$ at point $X = 0$ (Figure 2.4b) because with probability $1 - p_f$ failure will not occur and the potential loss will be zero.

If a level α for the probability of obtaining an extreme or more extreme loss is specified, a maximum potential loss x_a can be determined which corresponds to the specified level. α is the probability that the potential loss will exceed this maximum specified loss x_α ($\alpha = P(X > x_a)$, Figure 2.4b). Then, the maximum potential losses $x_{\alpha,i}$ characterising different design solutions can be compared.

The maximum potential loss at a pre-set level is a risk measure that specifies the loss limit whose probability of exceedence is equal to the pre-set level.

The maximum potential loss at a pre-set level can be used to determine the risk-based capital required to absorb the loss associated with failure. The expected loss is not sufficient to define the necessary resource of capital because the actual loss varies randomly around it.

Let $C_k(x \mid f)$ be the conditional cumulative distribution of the loss (the loss *given* failure) characterising the kth failure mode, and $p_{k|f}$ be the conditional probability that *given* failure, the kth component/failure mode has initiated it first ($\sum_{k=1}^{M} p_{k|f} = 1$). The conditional probability distribution $C(x \mid f) \equiv P(X \leq x \mid f)$ that the loss X *given* that failure has occurred will not be greater than a specified value x, can be presented by the union of the following mutually exclusive and exhaustive events: (i) it is the first component/failure mode that has initiated the failure and the loss X is not greater than x (the probability of which is $p_{1|f}C_1(x \mid f)$); (ii) it is the second component/failure mode that has initiated the failure and the loss X is not greater than x (the probability of which is $p_{2|f}C_2(x \mid f)$).... The final compound event is: it is the Mth component/failure mode that has initiated the failure and the loss X is not greater than x (the probability of which is $p_{M|f}C_M(x \mid f)$). The probability of a union of mutually exclusive events equals the sum of the probabilities of the separate events. As a result, the conditional distribution of the loss given failure (the conditional loss) becomes:

$$C(x \mid f) = \sum_{k=1}^{M} p_{k|f} C_k(x \mid f) \tag{2.4}$$

The distribution of the conditional loss $C(x\,|\,f)$ is a mixture of the distributions of the conditional losses $C_k(x\,|\,f)$ characterising the individual failure modes, scaled by the conditional probabilities $p_{k\,|\,f}$ of initiating failure first *given* that failure has occurred ($\sum_{i=1}^{M} p_{k\,|\,f} = 1$). Finally, Eq. (2.3) regarding the cumulative distribution of the potential loss becomes

$$C(x) = (1 - p_f)\,H(x) + p_f \sum_{k=1}^{M} p_{k\,|\,f} C_k(x\,|\,f) \tag{2.5}$$

The product of the probability of failure p_f and the probability $p_{k\,|\,f}$ that given failure, the kth component/failure mode has initiated it is simply equal to the probability p_k that the kth component/failure mode will initiate failure first ($p_f p_{k\,|\,f} = p_k$). Considering this relationship and also the relationship $\sum_{i=1}^{M} p_k = p_f$, Eq. (2.5) can also be presented as

$$C(x) = (1 - p_f)\,H(x) + \sum_{k=1}^{M} p_k C_k(x\,|\,f) \tag{2.6}$$

Equations (2.5) and (2.6) are fundamental and give the cumulative distribution of the potential loss associated with mutually exclusive failure modes.

From Eq. (2.6), it can be seen that *risk is measured by the sum of the products of the likelihood of each failure mode and the distribution of the consequences from the corresponding failure mode.*

Differentiating Eq. (2.6) with respect to x results in

$$c(x) = (1 - p_f)\,\delta(x) + \sum_{k=1}^{M} p_k\, c_k(x\,|\,f) \tag{2.7}$$

where $c(x) \equiv dC(x)/dx$ is the probability density distribution of the potential loss and $c_k(x\,|\,f) \equiv dC_k(x\,|\,f)/dx$ are the conditional probability density distributions of the loss *given* that failure has occurred, associated with the separate failure modes.

In Eq. (2.7), $\delta(x)$ is the Dirac's delta function which is the derivative of the Heaviside function $dH(x)/dx$. The expected value of the potential loss from failures $\overline{C}$ is obtained by multiplying Eq. (2.7) by x and integrating it ($\int x\delta(x)dx = 0$):

$$\overline{C} = \int x\, c(x)\, dx = \sum_{k=1}^{M} p_k \int x\, c_k(x\,|\,f)\, dx = \sum_{k=1}^{M} p_k\, \overline{C}_{k\,|\,f} \tag{2.8}$$

where $\overline{C}_{k\,|\,f} = \int x\, c_k(x\,|\,f)\, dx$ is the expected value of the loss given that failure has been initiated by the kth failure mode.

For a single failure mode, Eq. (2.8) transforms into

$$\overline{C} = p_f\, \overline{C}_f \tag{2.9}$$

Clearly, the measure of the risk of failure K in Eq. (2.1) can be defined as the expected value of the potential loss. However, while taking the expected values of the potential loss expressed by Eq. (2.7), the very important dynamic dimension of risk is lost. Consequently, Eq. (2.1) cannot be a true measure of risk. The true measure of risk, which also captures its dynamic dimension is the potential loss given by Eqs. (2.6) and (2.7).

Given that an accident/failure has occurred, for each identified set of initiating events, an assessment of the possible damage is made. In the case of loss of containment, for example, depending on the *release rate* and the *dispersion rate*, the consequences can vary significantly. In the case of a leak to the environment, the consequences are a function of the magnitude of the leak and the dispersion rate.

Where possible, a distribution of the conditional losses (consequences given failure) should be produced. This distribution gives the likelihood that the consequences given failure will exceed a specified critical threshold.

In the case of n mutually exclusive failure scenarios, the conditional cumulative distribution $C(x\,|\,f)$ of the loss given failure or accident is described by the equation

$$C(x\,|\,f) = p_{1|f}\, C_1(x\,|\,f) + p_{2|f} C_2(x\,|\,f) + \ldots + p_{n|f} C_n(x\,|\,f) \tag{2.10}$$

where $p_{i|f}$ is the conditional probability that the ith failure scenario will occur first $(\sum_i p_{i|f} = 1)$.

For the expected value of the conditional loss (consequences) given failure, the following equation holds:

$$\overline{C} = p_{1|f}\, \overline{C}_1 + p_{2|f} \overline{C}_2 + \ldots + p_{n|f} \overline{C}_n \tag{2.11}$$

The full spectrum of possible failure scenarios should be analysed. Event trees are often employed to map all possible failure scenarios. The conditional probabilities $p_{i|f}$ of the separate scenarios are calculated by multiplying the probabilities of the branches composing the paths leading to them. Usually, in the case of a large number of different failure scenarios or complex interrelationships between them, special Monte Carlo simulation software is needed to determine the distribution of the conditional losses. Consequence modelling tools help to evaluate the consequences from dispersion of toxic gases, smoke, fires, explosions, etc.

2.4 Improving Reliability and Reducing Risk

According to a commonly accepted definition (IEC 1991) reliability is 'the ability of an entity to perform required functions in a specified manner under stated conditions, for a stated period of time'. A system or component is said to have a failure, if the service it delivers to the user deviates from the specified one, for example, if the system stops production. System failures or component failures usually require immediate corrective action (e.g. intervention for repair or replacement), in order to return the system or component into operating condition. Each failure is associated with losses due to the cost of damage, cost of intervention, cost of repair and the cost of lost production.

The ability to perform required functions for a stated period of time is not necessarily measured with the probability of performing the required functions for a stated period of time. This ability can be assessed in a number of alternative ways. Thus, the tendency of random demands to overlap puts strain on limited sources servicing the demands and constitutes a common failure mode. This tendency can be measured not only with the probability of clustering the random events; it can also be measured with the expected (average) fraction of time corresponding to unsatisfied demand.

Avoidance of failure modes is an important part of the reliability improvement effort (Clausing 2004; Davis 2006). However, reliability cannot always be reduced to 'avoidance of failure modes'. Despite the fact that defining reliability as 'avoidance of failure modes' is parsimonious and correctly reflects the reliability improvement effort in some domains, this definition is limited. In some cases, a failure mode cannot be avoided. For example, a failure mode consisting of unsatisfied demand, due to overlapping of random demands competing for a single resource, cannot be avoided. This failure mode is intrinsic to the core function of the system and can only be made less likely, by a stochastic separation. This can be achieved by reducing the demand times, by increasing the time interval over which the random demands arrive, and by decreasing the number density of the random demands.

In another example, failures due to accumulation of damage (e.g. fatigue failure of a spring experiencing cyclic loading) cannot be avoided. They are intrinsic to the core function of the component and materialise inevitably as soon as the accumulated damage reaches a critical level. In the case of a compression spring experiencing cyclic loading, the fatigue failure mode can be delayed by reducing the rate of damage accumulation. The rate of damage accumulation can be reduced by reducing the stress range on the spring surface by shot peening, by improving the fatigue resistance of the material through micro-alloying and appropriate heat treatment, by selecting cleaner material with fewer inclusions and homogeneous microstructure, etc.

In some cases, to improve the reliability of critical components, failure modes are even deliberately created. Such is the case of the retrofitable deliberate weak link placed on a cable loaded in tension. In normal operation conditions, the load is carried only by the deliberate weak link. When the load exceeds a critical level, the deliberate weak link fails, and load is transferred to the cable. The deliberate failure provides ample warning to the operator that the permitted operating load has been exceeded. The load is reduced, and the reliability of the cable is improved. This is an example where reliability is improved by creating a deliberate failure.

Finally, it needs to be pointed out that improving reliability *is not synonymous with reducing risk*. Higher reliability does not necessarily mean low losses from failure. To demonstrate that selecting the more reliable system does not necessarily mean selecting the system with the smaller losses from failures, consider two very simple systems consisting of two components only, logically arranged in series (Figure 2.5).

Both systems consist of an electronic control module (EC) and mechanical device (M). The components are logically arranged in series and each system fails whenever the electronic control module fails, or the mechanical device fails.

For the first system (Figure 2.5a), suppose that the failure rate of the electronic control module EC1 is $\lambda_{A1} = 4$ year^{-1} and its replacement after failure costs $C_{EC1} = \$300$, while the failure rate of the mechanical device M1 is $\lambda_{M1} = 1$ year^{-1} and its failure is associated with $C_{M1} = \$200\,000$ cost for replacement. Suppose now that for an alternative system consisting of the same type of electronic control module and mechanical device (Figure 2.5b), the

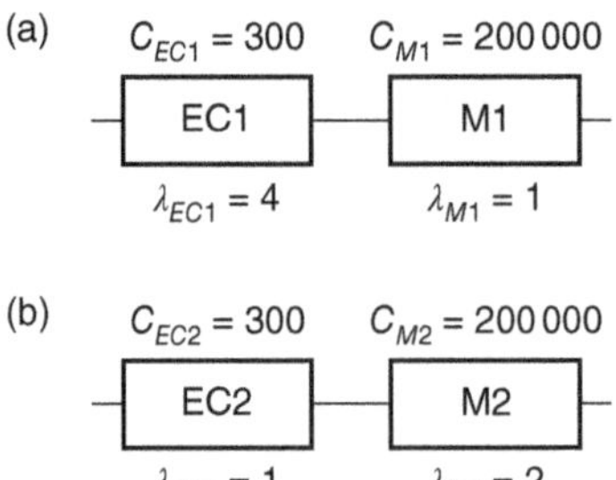

Figure 2.5 Systems composed of two components only, demonstrating that the more reliable system is not necessarily the one associated with the smaller losses from failures.

losses associated with failure of the separate components are the same but the failure rates are different: The electronic control module EC2 is characterised by a failure rate $\lambda_{EC2} = 1$ year^{-1} and the mechanical device M2 by a failure rate $\lambda_{M2} = 2$ year^{-1}. Clearly, the second system (b) is more reliable than the first system (a) because it is characterised by a failure rate $\lambda_2 = \lambda_{EC2} + \lambda_{M2} = 3$ year^{-1} whereas the first system is characterised by a failure rate $\lambda_1 = \lambda_{EC1} + \lambda_{M1} = 5$ year^{-1}. Because the first system fails whenever either component EC1 or component M1 fails, the expected (average) losses from failures for the system, during one year of operation are

$$\overline{L}_1 = \lambda_{EC1} C_{EC1} + \lambda_{M1} C_{M1} = 201\ 200 \tag{2.12}$$

For the second system, the expected losses from failures during one year of operation are

$$\overline{L}_2 = \lambda_{EC2} C_{EC2} + \lambda_{M2} C_{M2} = 400\ 300 \tag{2.13}$$

As can be seen, the more reliable system (the second system) is associated with the larger losses from failures.

This simple example shows that a selection of a system solely based on its reliability can be very misleading. Improving reliability and reducing risk do not carry the same meaning. In the case of a system whose component failures are associated with similar cost, a system with larger reliability does mean a system with smaller expected losses from failures. In the common case of systems whose component failures are associated with different costs, the system with the largest reliability is not necessarily the system with the smallest losses from failures. The reliability analysis of complex production systems should be based on the losses from failures (Todinov 2006b,c).

2.5 Resilience

Resilience is a relatively new concept in the reliability and risk literature and broadly means the ability of an entity to withstand a disruption from an adverse event, with acceptable level of degradation, and recover within a suitable time and reasonable costs (Haimes 2009; Hosseini et al. 2016). Vugrin et al. (2010, 2011) define resilience as the system's ability to reduce efficiently both the magnitude and duration of the deviation from targeted system performance levels, following a particular disruptive event.

Most generally, resilience reflects the capability of a system: (i) to resist disturbance from adverse event; (ii) to absorb disturbance from adverse event; (iii) to adapt to adverse events; and (iv) to recover its required functions following the adverse event.

Resilience does not coincide with *robustness*, *fault-tolerance*, *vulnerability*, or *survivability*. Resilience is a complex characteristic whose building blocks are in turn complex, with diverse properties.

Indeed, the capability of a system to resist disturbance is a complex characteristic and reflects the reliability of the system, the vulnerability, the built-in capacity and safety margins, the capability of the system to dampen an impact from adverse events, the capability of the system to provide a self-reinforcing response to external overload, etc.

In turn, the capability of a system to absorb shocks from adverse events is a complex characteristic and reflects the fault tolerance of the system, the capability of delaying the spread of damage once damage has been initiated, the levels of redundancy, the lack of dependencies between components and sub-systems, the safety margins, the capability of the system to dampen impacts from adverse events, the capability of the system to limit the losses below a specified limit in case of adversity, etc.

The capability of a system to adapt temporarily to new conditions immediately following the disruption is also a complex characteristic. It reflects the capability of the system: to maintain its vital functions after disruption or in degraded condition, to rearrange and reconfigure in order to limit the consequences of the adverse event, to disable or dispose of components and assemblies which are a potential hazard in case of accident, and to select a safe course of action among a number of possible alternative options, etc.

The capability of the system to recover its performance following the adverse event reflects both the speed with which efficient restoration measures are taken and the capability of the system to rearrange and reconfigure following a shock. Reconfigurable repairable flow networks are examples of a system with the capability to reconfigure after a failure of an edge which disrupts the flow. Non-reconfigurable flow networks do not have this capability (Todinov 2011a).

The resilience property is focused on the system's recovery following an adverse event. Thus, reducing the vulnerability of a system to adverse events by erecting additional barriers may not necessarily translate into an improved resilience if, following an adverse event, these barriers have been compromised (Haimes 2009).

3

Overview of Methods and Principles for Improving Reliability and Reducing Risk That Can Be Classified as Domain-Independent

A number of reliability improvement and risk reduction principles have already been introduced in the reliability and risk literature that can be classified as domain-independent. Among these are improving reliability by active and standby 'redundancy', by strengthening weak links and upgrading to more reliable components, by eliminating a common cause and by reducing variability (Barlow and Proschan 1975; Lewis 1996; Ebeling 1997; O'Connor 2002). A number of domain-independent methods for increasing reliability and reducing risk have also been introduced and discussed in Todinov (2007, 2016a). In what follows, existing domain-independent methods for reliability improvement and risk reduction will be reviewed and their advantages and shortcomings discussed.

3.1 Improving Reliability and Reducing Risk by Preventing Failure Modes

3.1.1 Techniques for Identifying and Assessing Failure Modes

The risk of failure is reduced by removing potential failure modes and the design for reliability effort is largely about preventing failure modes from occurring during the specified lifetime of the product.

These techniques ensure that as many as possible potential failure modes have been identified and their effect on the system performance assessed. The objective is to identify critical areas where design modifications can reduce the probability of failure or the consequences of failure. In this way, potential failure modes and weak spots which need attention are highlighted and limited resources for risk reduction are focused there.

The *failure mode* is the way a system or component fails to function as intended. It is the effect by which failure is observed. The physical process leading to a particular failure mode will be referred to as the *failure mechanism.* It is important to understand that the same failure mode (e.g. fracture of a component) can be associated with different failure mechanisms. Most commonly, the fracture of components could be the result of a *brittle fracture* mechanism, *ductile fracture mechanism,* or *fatigue fracture mechanism.* At high temperature, *creep fracture mechanism* is often present.

The brittle fracture mechanism is characterised by little or no plastic deformation. Once initiated, the crack extends at a high speed without the need for an increased stress. The component fails quickly, without any warning.

Methods for Reliability Improvement and Risk Reduction, First Edition. Michael Todinov.

In contrast, the ductile fracture mechanism is accompanied by a considerable amount of plastic deformation, redistribution, and relaxation of the high stresses in the zones of stress concentration. Ductile fracture requires more strain energy in order to develop. In the process of plastic deformation, the strength of the material is enhanced through strain hardening. The crack will not normally extend unless the load is increased. The component deforms before fracture, which gives early warning and sufficient time for intervention and repair. This is why, in engineering applications, where safety concerns are involved, materials with ductile behaviour are the obvious choice.

The fatigue failure mechanism involves nucleation and slow propagation of a fatigue crack under varying (cyclic) load. The rate of fatigue crack propagation is strongly affected by the stress range during loading, not by the maximum stress. After the crack reaches a particular critical size, the component fails due to brittle or ductile fracture.

Creep is a failure mechanism which consists of a very slow deformation of a component exposed to elevated temperatures and constant load. Creep deformation occurs primarily by grain-boundary sliding. The rate of creep deformation depends on the material, exposure time, exposure temperature, and the loading stress. Failure occurs when the deformation reaches a critical level at which the component can no longer perform its function.

In each particular case, the failure mechanism behind the failure mode 'fracture' is very different.

Apart from fracture, other examples of failure modes are: 'short circuit', 'open circuit', 'overheating of an electrical or mechanical component', excessive noise and vibration, leakage from a seal, excessive deformation, excessive wear, misalignment contamination, etc.

An essential part of the design for reliability is preventing failure modes from occurring during the specified lifetime of the product or reducing the likelihood of unavoidable failure modes to a tolerable level. Suppose that the permitted space within which the design variables vary is denoted by Ω (Figure 3.1). The permitted space of the design variables is obtained after considering the constraints associated with the variation of the design variables. Suppose that *n* distinct failure modes have been identified.

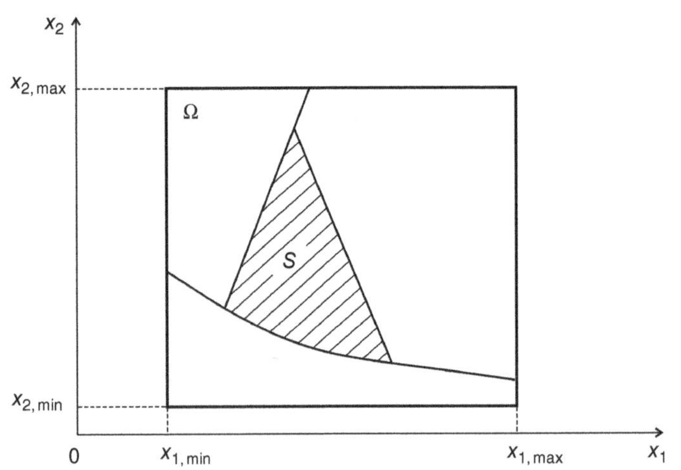

Figure 3.1 Specifying a domain for the design variables which prevents the identified failure modes from occurring or reduces their likelihood to a tolerable level.

The safe domain S of the design variables prevents all failure modes from occurring or reduces their likelihood to a tolerable level. An important objective of the design for reliability is to specify the design variables so that they belong to the safe domain preventing the identified failure modes from materialising or reducing their likelihood to a tolerable level. The benefits from identifying and eliminating failure modes are: improved reliability of the product/process, improved safety, reduced warranty claims and other potential losses.

It is vital, that identifying the failure modes and the required design modifications for reducing their likelihood or their full elimination are made *during the early stages of the design, before the failure modes have escaped into the field*. Design modifications during the early stages of the design are much less costly compared with design modifications executed during the late stages of the design or after the product has been released to customers. Failure modes during the early stages of the design are easy to rectify, even though their identification may be difficult. Failure modes after the product has been released to customers are difficult to rectify despite the fact that their detection is easy at this stage.

Consequently, in order to reduce the risk of failure of a product or a process, it is important to recognise their failure modes as early as possible in order to enable the execution of the necessary design modifications and specific actions reducing the risk of failure.

Systematic procedures for identifying possible failure modes in a system and evaluating their impact have already been developed. The best-known method is the failure mode and effects analysis (FMEA), developed in 1963 by the National Aeronautics and Space Administration (NASA) for the Apollo project. FMEA is a very useful method for discovering potential failure modes and has subsequently been applied in aerospace and aeronautical engineering, the nuclear industry, the electronics industry, the automotive industry, and in software development. Most literary resources concerning this method stem from the American Military Standard MIL-STD-1629A, 1977. The fundamental idea behind FMEA is to discover as many as possible potential failure modes early in the design stage, evaluate their impact, identify failure causes, and outline controls and actions limiting the risks associated with the identified failure modes. The extension of FMEA which includes *criticality analysis* is known as failure mode, effects and criticality analysis (FMECA). In the identification of failure modes, the following approaches and techniques are useful:

- The inductive approach is an important basic technique for identifying possible failure modes at a system level. It considers sequentially the failure modes of all parts and components building the system and tracks their effect on the system's performance.
- The deductive approach is another basic technique which helps to identify new failure modes. It considers an already identified failure mode at a system level and investigates what else could cause this failure mode.

Other techniques for identifying potential failure modes are:

- A systematic analysis by using check lists. An example of a simple check list which helps to identify a number of potential failure modes in mechanical assemblies is the following:
  - Are components sensitive to variations of load?
  - Are components resistant against variations of temperature?
  - Are components resistant against vibrations?
  - Are components resistant to corrosion?

- – Are components/assemblies robust against variation in tolerances?
  - – Are parts sensitive to precise alignment?
  - – Are parts prone to misassembly?
  - – Are parts resistant to contamination?
  - – Are components resistant against stress relaxation?
- Using past failures in similar cases. In many domains, a big weight is given to databases of the type 'lessons learned' which help to avoid failure modes causing problems in the past. Lessons learned from past failures have been particularly useful to prevent failure modes in the oil and gas industry, aerospace industry, and nuclear industry.
- Gathering information through interviews with specialists in the assessed area.
- Playing devil's advocate. Probing what could possibly go wrong. Asking lots of 'what if' questions.
- Root cause analysis often uncovers a number of unsuspected failure modes. It reveals processes and conditions leading to failures. Root cause analysis is also a very important method for revealing the genesis of failure modes.
- Assumption analysis. This consists of challenging and testing common assumptions about the followed design procedures, manufacturing operations, usage of the product, working conditions, and environment.
- Analysis of the constraints. The analysis of the technical constraints of the system, the work conditions and the environment often helps to discover new failure modes.
- Asking not only questions about what could possibly go wrong but also questions on how to make the system go wrong. This is a useful technique for discovering rare and unexpected failure modes.
- Using creativity methods and tools for identifying failure modes in new products and processes (e.g. brainstorming, TRIZ, lateral thinking, etc.). In the Anticipatory Failure Determination, for example (Kaplan et al. 1999), the focus is on how to precipitate failure by using available resources (factors) characterising the system and its environment. This approach often yields failure modes that are very difficult to discover by other methods.

Before an identification of failure modes is attempted, it is vital to understand the basic processes in the system and how the system works. In this respect, building a functional block diagram and specifying the required functions of the system is very important.

The functional diagram shows how the components or process steps are interrelated.

A common limitation of the traditional FMEA methodology is that there is no guarantee that all failure modes have been identified. Another limitation of the FMEA methodology is that failure modes of components are treated independently. The 'independence' assumption makes it very difficult to discover complex failure modes at a system level which materialise only if a combination of several failure modes and conditions at a component level is present.

FMEA cannot discover failure modes which result from unfavourable stack up of values of reliability-critical parameters. Consider a press-fit assembly. If elevated temperature is combined with a small coefficient of friction and a small value of the interference at a room temperature, the press-fit assembly may lose its capability to carry the specified torque and axial force. This type of failure mode can be discovered easily by using virtual testing based on a Monte Carlo simulation (Todinov 2009c, 2016a).

Another limitation of FMEA is its inability to discover failure modes dependent on the timing (or clustering) of conditions and causes. If a number of production units demand randomly a specified quantity of a particular resource (e.g. water steam), for a specified time, the failure mode 'insufficient resource supply' depends exclusively on the clustering of demands during the time interval and the capacity of the generator supplying the resource.

3.1.2 Effective Risk Reduction Procedure Related to Preventing Failure Modes from Occurring

The goal of effective risk management is to ensure that technical risks are correctly identified, assessed, prioritised, and reduced where possible. A basic step of the risk management is the identification of as many failure scenarios as possible, assessing their likelihood and impact. If there are problems in the risk management process, problems in design and inspection would be difficult to discover, which would give rise to dangerous and costly failure modes.

Deciding upon and selecting particular risk reduction measures may not necessarily reduce the total risk. Indeed, a common situation in designing complex systems is when design modifications introduced to eliminate a particular failure mode often create other failure modes. In order to reduce the possibility of introducing new failure modes, each time after deciding upon and selecting appropriate risk reduction measures (e.g. design modifications), possible new failure modes are identified and assessed again. Furthermore, risks are often interrelated. Decreasing the risk of a particular failure mode may increase the risk of other failure modes. Thus, in upgrading a new software system, reducing the risk of a security breach by removing a particular failure mode may increase the risk of loss of functionality or compatibility. Reducing the risk of wearout failures by substituting a mechanical assembly with an electronic assembly, may increase vulnerability to vibration and increased temperature and introduce new failure modes. Building a tourist attraction on a remote place with sunny weather reduces the risk of a reduced number of customers due to bad weather but simultaneously increases the risk of a reduced number of customers due to higher transportation expenses (Pickford 2001).

A good protection against interrelated failure modes is the *integrated risk management* which includes repeated assessment of all potential failure modes after deciding upon the implementation of each risk reduction measure.

Next, a rigorous objective method should be implemented for assessing the identified failure modes. Wider use of proven statistical methods must be used where possible in assessing the failure modes. The very popular risk assessment based on a qualitative risk matrix has drawbacks and often results in costly mistakes in major decisions. This method of risk assessment is subjective and often yields distorted risk assessments, particularly when probability and consequence values are negatively correlated (Cox 2008).

In the very common case of multiple failure modes, the traditional approach based on a risk matrix, reveals another weakness. Often, each individual failure mode is in the low-risk region of the risk matrix (therefore acceptable) which creates a dangerous perception of safety. In some cases however, the aggregated risk from all low risks cannot be tolerated. Although all individual risks may have low scores and for none of them a response plan is required, the total aggregated risk may not be acceptable. A large aggregated risk from multiple small risks, each below the tolerable level, can be just as damaging as a large single risk above the tolerable level. The qualitative risk matrix does not account for this common case.

Finally, an effective risk management process must make an optimal allocation of the risk reduction budget across identified failure modes. Errors in the risk allocation process result in waste of resources and insignificant risk reduction. Risk matrices do not take into consideration the costs of the risk reduction options and the magnitude of risk reduction achieved by the risk reduction options (Cox 2008). More sophisticated dynamic programming methods are necessary to account for the limited risk reduction budget, the amount of risk the risk reduction options achieve, and the cost of the risk reduction options (Todinov 2014).

A risk management approach based on simple scoring schemes, no matter how popular and simple, is the least supported by a serious theoretical or empirical analysis. To make it work, risk management should be backed whenever possible, by proven risk assessment methods such as *Monte Carlo simulation*.

3.1.3 Reliability Improvement and Risk Reduction by Root Cause Analysis

Root cause analysis is a proven technique for reliability improvement and risk reduction. Knowledge regarding the circumstances and processes which contribute to the failure events is the starting point for a real reliability improvement. The main purpose of root cause analysis is to identify factors promoting the failure modes and determine whether they are present in other parts of the system.

Identifying the root causes initiates a process of preventing a failure mode from occurring or reducing its likelihood by appropriate modifications of the design, the manufacturing process, or the operating procedures.

3.1.3.1 Case Study: Improving the Reliability of Automotive Suspension Springs by Root Cause Analysis

An example of improving reliability by root cause analysis can be given with hot-coiled Si-Mn automotive suspension springs suffering from premature fatigue failure. Typically, automotive suspension springs are manufactured by hot winding. The cut-to-length cold-drawn spring rods are austenitised, wound into springs, quenched, and tempered. This is followed by warm pre-setting, shot peening, cold pre-setting, and painting (Heitmann et al. 1996).

The initial step of the root cause analysis is conducting rig tests inducing fatigue failures to a large batch of suspension springs under various conditions. Fracture surfaces are then preserved and scanning electron microscopy is used to investigate the fatigue crack initiation sites. If large-size inclusions are discovered at the fatigue crack origin, a possible risk reduction measure is to switch to a supplier of cleaner spring steel.

Optical metallography of the failed suspension springs must also be carried out, in order to make sure that there is no excessive decarburisation of the surface layers of the spring wire. If the depth of the decarburised layer is significant, its fatigue resistance is low, and care must be taken to control the carbon potential of the furnace atmosphere, in order to avoid excessive decarburisation. Alternatively, the chemical composition of the steel can be altered by selecting a micro-alloyed steel (e.g. Si-Mn + vanadium), in order to make it less susceptible to decarburisation. The grain size at the surface of the spring wire must also be examined because microstructures with excessively large grain size are characterised by reduced fracture toughness and fatigue resistance. Correspondingly, the austenitisation temperature and the duration of the austenitisation process must guarantee that the grain size remains relatively small.

The spring surface after quenching must also be examined in order to make sure that no micro-cracks or excessive tensile residual stresses are present. Tempering must guarantee an optimal hardness and yield strength which maximise fatigue life. After tempering, the hardness and yield strength of the spring material must be tested in order to ensure that the optimal levels have been reached.

Finally, after shot peening, the compressive residual stress at the surface of the spring wire must be measured (e.g. by an X-ray diffractometer) to make sure that it is of sufficient magnitude and uniformly distributed over the circumference of the spring wire. If, for example, the compressive residual stress is highly non-uniform or of small magnitude, it would offer little resistance against fatigue crack initiation and propagation. An alteration of the shot-peening operation must then be implemented in order to guarantee a sufficient magnitude and uniformity of the compressive residual stress.

Root cause analysis is an excellent method for improving reliability and reducing risk but it is associated with difficulties. Revealing the root causes of failure modes often requires conducting experimental research involving specialised laboratories, equipment, and highly qualified researchers. This often makes root cause analysis an expensive and time consuming reliability improvement method.

Although the root cause analysis approach has been very successful in reliability improvement and risk reduction, it has a very narrow validity; it cannot normally transcend the narrow application domain it serves and be used to improve reliability in other, unrelated domains.

3.1.4 Preventing Failure Modes by Removing Latent Faults

An efficient way of preventing failure modes is the removal of latent faults from products, systems, and operations. A *fault* is an incorrect state, or a defect resulting from errors during material processing, design, manufacturing, assembly, or operation, with the potential to cause failure or accelerate the occurrence of failure under particular conditions. A software fault is synonymous with a bug and is, in effect, a defect in the code that can cause software failure.

A fault is not the same as a failure. Failure is an adverse event after which the service delivered by the system no longer complies with the specified requirements. A fault is a condition (a state). A latent fault is often necessary but not always sufficient for initiating failure. A system with faults, may continue to provide its service and comply with the specifications until some triggering input condition is encountered which triggers failure. A latent fault may lay dormant for years before the conditions which trigger failure appear. Thus, an incorrectly installed rupture disc on a pressurised vessel is a fault that could lay dormant for years before over-pressurisation is actually present and explosion occurs (Hedlund et al. 2016). The latent fault associated with the incorrectly installed rupture disc could have been easily discovered had safety checks been carried out.

In another example, a software bug allowing the system to act on signals from faulty sensors is a latent fault. The system will continue to operate as long as there is no failure of the sensor. This fault will materialise into a failure only when the sensor fails but the system continues to read it and act on its false indications as if the sensor was working correctly. In aviation, such a fault may result in an unexpected behaviour of an aircraft and cause an accident.

The presence of a software fault/bug does not necessarily result in immediate software failure. Failure will be present only when the logical branch containing the faulty piece of code is executed, which happens only when certain conditions are present.

Typical latent faults in electronic devices include poor solder joints, defective wire bonds, semiconductor impurities, semiconductor defects, and component drift. A fault could lead to a faster accumulation of damage. A common example is the presence of a large defect (e.g. pore) in a stressed component. The component may operate for some time, but the fatigue crack growing from the defect often causes a premature fatigue failure.

A large building can have as a latent fault improper foundation support which will only materialise into a catastrophic failure during an earthquake of certain magnitude.

A deviation of a parameter from its safe range is also a latent fault which could lead to failure. For example, a deviation of a clearance from its prescribed value could cause jamming and failure if temperature varies beyond a certain critical level.

Testing and thorough inspection for latent faults are key measures for their removal. In developing software applications, a thorough debugging and operational testing is the key to minimising the number of latent faults. Proper quality management processes must be in place in order to eliminate or minimise the latent faults in the released products.

The objective of *environmental stress screening* (ESS) is to simulate expected worst-case service environments. The stress levels used for ESS are aimed at eliminating (screening) the part of the population with faults. This population causes a heavy lower tail of the strength distribution and is the primary reason for many early-life failures.

ESS improves the strength distribution by removing the weak population. By trapping faulty and substandard items before they are released to the customer, the operation reduces early-life failures caused by items with substandard strength. ESS also helps to discover and eliminate sources of faults and weaknesses during design, manufacturing, and assembly.

During environment stress screening (or *burn-in*), it is important to find operating and environmental test conditions which permit efficient screening without consuming a substantial part of the life of the tested components that have passed the screening test. Thermal cycling of integrated circuits for example often reveals poor wire bonds, improperly cured plastic packages, poor die bonds, etc.

Particularly useful tests, which reveal a large number of failure modes are *Highly Accelerated Life Testing* (HALT) and *Highly Accelerated Stress Screens* (HASS) (Hobbs 2000). The purpose of these tests is to expose (precipitate) faults and weaknesses in the design, manufacturing, and assembly in order to provide a basis for reliability improvement. The purpose is not to simulate the service environment. To precipitate a latent fault means to change its state from undetectable to a detectable state. A latent fault 'poor solder joint' is usually undetectable unless it is extremely poor. Applying vibration, thermal or electrical stress helps to precipitate the fault under mild testing conditions and perform appropriate corrective action. The precipitated faults and weaknesses are used as opportunities for improvement of the design and manufacturing, in order to avoid expensive failures during service. The stresses used during HALT and HASS are extreme, applied for a brief period of time. They include all-axis simultaneous vibration, high-rate broad-range temperature cycling, power cycling, voltage, frequency and humidity variation, etc. (Hobbs 2000). During HALT and HASS, faults are often exposed with a different type of stress or a stress level than the ones that would be used during service. This is why the focus is not on the test conditions which precipitate the faults but on the faults themselves.

Some of the factors contributing to the genesis of latent faults are:

1. Combination of values for the controlling parameters leading to degraded strength or weak resistance to deterioration.
2. Inappropriate material or defects in the material.
3. Human errors.
4. Failures of the manufacturing equipment.
5. Failures of the control equipment.
6. Poor design methods.
7. Design errors.
8. Poor control of the production process.

Quality control checks and tests aimed at removing as many as possible latent faults should be placed in the key stages of the production process.

3.2 Improving Reliability and Reducing Risk by a Fault-Tolerant System Design and Fail-Safe Design

3.2.1 Building in Redundancy

Incorporating redundancy in the design is particularly effective in cases where random failures are present. *Redundancy* is a technique by which one or more components of a system are replicated in order to increase reliability (Blischke and Murthy 2000).

As it will be shown in Chapter 4, *redundancy is effectively an act of application of a more general reliability improvement method – the method of separation.* From the perspective of the method of separation, redundancy is an act of separating a single function into many components carrying the function. For active redundancy, all redundant units are in operation and share the load with the main unit from the time the system is put in operation (Figure 3.2a). Full active redundancy is present in cases where the assembly is operational if at least one of the units is operational.

Suppose that n components, working in parallel, have reliabilities $r_1, r_2,\ldots,r_n$. The probability that at least one of the components is operational is equal to 1 – 'the probability that none of the components is operational'. The probability that none of the components in Figure 3.2a is operational is $(1 - r_1)(1 - r_2)\ldots(1 - r_n)$. Consequently, the reliability of the system in Figure 3.2a is given by

$$R_{sys} = 1 - (1 - r_1)(1 - r_2)\ldots(1 - r_n) \tag{3.1}$$

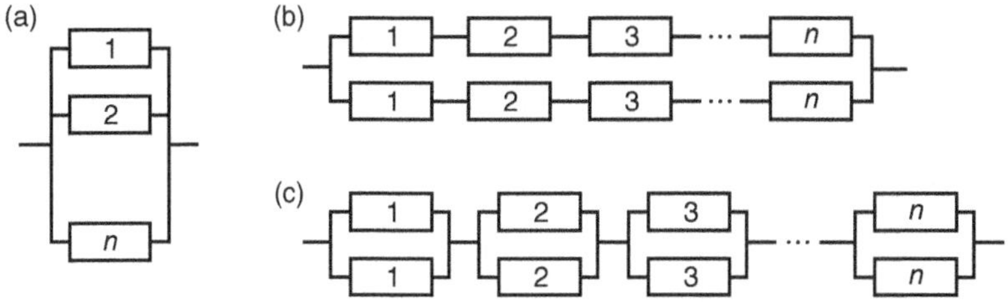

Figure 3.2 (a) Active redundancy; (b) redundancy at a system level; and (c) redundancy at a component level.

Active redundancy at a component level (where each component is replicated) yields significantly higher reliability compared with active redundancy at a system level (where the entire system is replicated).

Active redundancy at a system level (Figure 3.2b) where the entire system is replicated, yields lower reliability compared with active redundancy at a component level (Figure 3.2c) where each component is replicated (Hoyland and Rausand, 1994). The reliability network topology in Figure 3.2c is characterised by a significantly greater connectivity compared with the reliability network topology in Figure 3.2b, which makes it much less sensitive to failures in both branches. While failures of two components, each in a separate branch, always cause a system failure for the arrangement in Figure 3.2b, the system in Figure 3.2c fails only if the two failed components have the same index. There are n^2 different ways of having a single failure in each branch. While each of these ways invariably means failure for the system in Figure 3.2b, only n of these combinations fail the system in Figure 3.2c. The system in Figure 3.2c is therefore more reliable than the system in Figure 3.2b.

Furthermore, with increasing the number of components in the system, the active redundancy at a component level maintains significantly higher system reliability levels compared with active redundancy at a system level (Figure 3.3). This is caused by the higher degree of connectivity associated with redundancy at a component level.

This is confirmed if the reliabilities of both arrangements are compared. Assume for the sake of simplicity, that all components have the same reliability $r_1 = r_2 = \ldots = r_n = r = 0.9$. The reliability of arrangement 'b', with redundancy at a system level, is given by

$$R_b = 1 - (1 - r^n)^2 \tag{3.2}$$

while the reliability of arrangement 'c', with redundancy at a component level is given by

$$R_c = (1 - (1 - r)^2)^n \tag{3.3}$$

These two dependencies have been plotted in Figure 3.3 for different number of components in a single branch. As can be verified, the arrangement with redundancy at a component level (Figure 3.2c) is more reliable compared with the arrangement with redundancy at a system level (Figure 3.2b). Moreover, with increasing the number of components, the system from Figure 3.2c retains a significantly higher reliability level compared with the system from Figure 3.2b.

While for *active redundancy*, no switching is required to make redundant components available, using *passive* (*standby*) *redundancy* requires a switching operation to make redundant components available. In cases of passive redundancy, the redundant components do not share any load with the operating component.

The redundant components are put in use one at a time, after failure of the currently operating component and the remaining components are kept in reserve (Figure 3.4).

Standby components do not operate until they are switched in. The components in the standby system operate one after another which explains why a cold standby redundancy with perfect switching (Figure 3.4a) provides higher reliability levels compared with an active redundancy (Figure 3.4b).

If the switch is perfect, the time to failure of the standby system in Figure 3.4a is the sum of the times to failure of the components ($t_a = t_{c1} + t_{c2} + \ldots + t_{cn}$). The time to failure of the active redundant system in Figure 3.4b is equal to the maximal of the times to failure of the components ($t_b = \max\{t_{c1}, t_{c2}, \ldots, t_{cn}\}$). The cold standby system is characterised by a higher reliability ($t_a > t_b$) compared with the system with active redundancy.

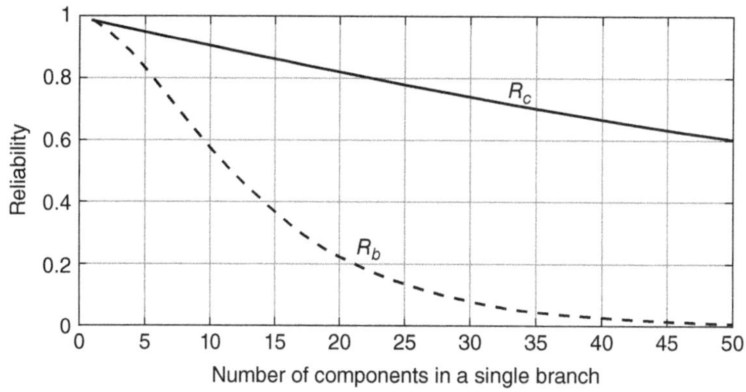

Figure 3.3 Improving the reliability of an active redundant section by transforming the redundancy at a system level into redundancy at a component level.

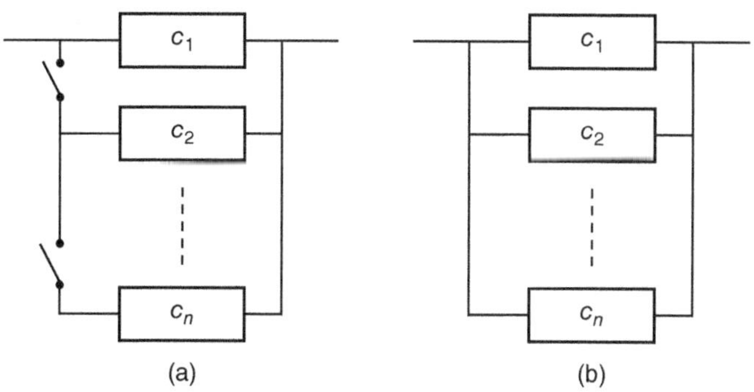

Figure 3.4 (a) Passive (standby) redundancy; (b) active redundancy.

The probability $R(t)$ that the standby system from Figure 3.4a will survive time t is equal to the probability that in the operational time interval $(0,t)$ any of the following events will be in place: A_0: no failure occurs; A_1: exactly one failure occurs; A_2: exactly two failures occur,..., A_{n-1}: exactly $n-1$ failures occur. These events are mutually exclusive and the probability of their union is equal to the sum of the probabilities of the separate events:

$$R(t) = P(A_0 \cup A_1 \cup \ldots \cup A_{n-1}) = P(A_0) + P(A_1) + \ldots + P(A_{n-1}) \tag{3.4}$$

If, for the sake of simplicity, all components are characterised by the same constant hazard rate λ, the lengths of the time intervals between failures follow the negative exponential distribution and the number of failures on the time interval $(0,t)$ follows the Poisson distribution. The probabilities of events A_0, A_1,...,A_{n-1} are therefore given by the Poisson distribution:

$$P(A_0) = \frac{(\lambda t)^0}{0!}\exp(-\lambda t), P(A_1) = \frac{(\lambda t)^1}{1!}\exp(-\lambda t), \ldots, P(A_{n-1}) = \frac{(\lambda t)^{n-1}}{(n-1)!}\exp(-\lambda t).$$

Substituting these expressions in Eq. (3.4) gives

$$R(t) = \exp(-\lambda t)\left[1 + \frac{(\lambda t)^1}{1!} + \frac{(\lambda t)^2}{2!} + \ldots + \frac{(\lambda t)^{n-1}}{(n-1)!}\right] \tag{3.5}$$

for the probability that the standby system will survive the time interval $(0,t)$.

Theoretically, by providing a sufficiently large number of standby components, the reliability of a standby system with perfect switching can be made arbitrarily close to 1. Indeed, as can be verified from Eq. (3.5), with increasing the number of components n, $\lim_{n\to\infty}\left[1+\frac{(\lambda t)^1}{1!}+\frac{(\lambda t)^2}{2!}+\ldots\right]=\exp(\lambda t)$ and, as a result, $\lim_{n\to\infty}[R(t)]=1$. The number of standby components is limited by constraints such as size, weight, and cost. Standby units may not necessarily be identical. An electrical device, for example, can have a hydraulic backup device.

Partial active redundancy (*k-out-of-n redundancy*) is present if the system works *if* and only if at least k out of the n components work. Partial active redundancy provides a very good reliability increase in safety-critical applications, which can be seen in the next case study.

3.2.1.1 Case Study: Improving Reliability by k-out-of-n redundancy

Suppose that within 168 hours, a chemical reactor requires at least 30 litres $\times$ s^{-1} cooling fluid from three pumps, each of which is capable of delivering cooling fluid with flow rate capacity of 10 litres $\times$ s^{-1}. Suppose that the reliability of each pump associated with the 168 hours is 0.9. The probability that the chemical reactor will remain without sufficient cooling in the 168 hour period is significant – equal to the probability that at least one of the pumps will fail: $1-0.9^3=0.271$. Suppose now that five pumps are working in parallel, each of which has a flow rate capacity of 10 litres x s^{-1}. Then, the probability P_{op} that the system will be in working state at the end of the 168 hour period is equal to the sum of the probabilities $P_{op}=P(A_0)+P(A_1)+P(A_2)$ of the following three mutually exclusive events: event A_0: zero failed pumps; event A_1: exactly one failed pump; and event A_2: exactly two failed pumps. Since $P(A_0)=(1-0.1)^5$, $P(A_1)=5\times 0.1^1\times(1-0.1)^4$ $P(A_2)=[5\times(5-1)/2]\times 0.1^2\times(1-0.1)^3$, $P_{op}=0.9914$. The probability that the chemical reactor will remain without sufficient cooling in the 168 hour period is equal to $1-P_{op}=0.0086$. Including three-out-of-five active redundancy increased dramatically the reliability of cooling.

Improving reliability by including redundancy has limitations. Replicating components and systems to carry the same function is expensive and increases the size and weight of the designs. Furthermore, redundancy provides effective protection only if the redundant components work independently from one another. Only then, their simultaneous failure would be very unlikely. If a particular event (common cause) causes the redundant components to fail simultaneously, for example due to fire, explosion, excessive vibrations, dust, and cold, redundancy does not provide effective protection against system failure. Another example of a common cause is a design fault common to all redundant components which causes redundant components to fail quickly. This is illustrated by the next example.

In 1982, a Boeing 747 flying above the Indian Ocean lost power from all of its four engines. The reason for the quadruple engine failure was a common cause: the airplane entered a cloud of volcanic ash which simultaneously affected all four engines. What was deemed to be nearly impossible – a simultaneous failure of all four redundant engines – occurred due to a common cause. Redundancy did not provide protection against a common cause which simultaneously affected all engines.

3.2.2 Fault-Tolerant Design

A fault may not precipitate failure if the component/system is *fault tolerant*. An example of a fault-tolerant component is a component made of composite material resistant to cracks, defects, and other imperfections. At the other extreme is a component made of material with low toughness (e.g. hardened high-strength steel) sensitive to different types of inclusions and micro-cracks. At a system level, a system with built-in redundancy is fault tolerant as opposed to a system with no redundancy. A digital circuit implementing *triple modular redundancy* is a typical example of a fault-tolerant design in digital electronics. Software including internal tests and exception handling routines which set up safe conditions in case of errors, is an example of fault-tolerant software.

Latent faults appear as a frequent *common cause*. Thus, faulty oil or a seal in several redundant units could cause failure in all of them. A latent fault in the software code recorded in a number of control devices could cause failure of all of the devices and a faulty design of a fuel tank could cause severe accidents with many cars of a particular make and model.

The *k-out-of-n* redundancy is an important type of redundancy because it makes the system fault tolerant. This was demonstrated with the example in section 3.2.1.1. Another example is a power supply based on six energy sources where only three working sources at any time are sufficient to guarantee adequate power supply. The system is resistant to failures associated with the separate power supply sources. Fault tolerance increases the capability of a system to resist overload and absorb shocks thereby increasing the resilience of the system. A k-out-of-n system is more resilient than a simple system performing the same function because $n - k$ components could fail and the k-out-of-n system will still be performing the required function.

A power distribution system with radial, tree-like structure is simple but also highly unreliable – a failure along any branch cuts off the power supply to the entire downstream part. An interconnected power grid with mesh-type topology is significantly more fault tolerant, because congestion and failure in one section do not have a critical impact on other sections of the grid. An interconnected power grid with mesh-type topology is more resilient compared with a power grid with tree topology. Even faults in several branches of the grid do not practically affect the capability of the grid to deliver electrical energy to consumers.

3.2.3 Fail-Safe Principle and Fail-Safe Design

The idea behind the fail-safe principle is to establish safe operating conditions for the system after the failure occurrence.

Hardware fail-safe devices are often an integral part of protection systems. A fail-safe gate valve, for example, uses the accumulated elastic energy of a spring to return the valve in safe 'closed' position should the hydraulic pressure keeping the valve open suddenly drop.

Lifts have breaks that are kept in a de-energised state by the tension of the cable holding the lift cabin. If the cable ruptures, the breaks are immediately activated and prevent the lift cabin from falling.

Failure of a pressure release valve to release pressure or a thermostat to switch off heating can, for example, be mitigated by ensuring that the fluid or heat supply will remain switched on for a limited period of time.

The fail-safe electrical contacts, part of machine guards, are designed to prevent a dangerous state of contacts sticking together. Without a fail-safe design, a failure resulting in electrical contacts sticking together permits the machine to be switched on in the absence of a machine guard.

The *leak-before-break concept* in designing pressure vessels is an application of the fail-safe principle – the pressure vessel must be able to tolerate without fast fracture a through crack of length twice the thickness of the shell of the pressure vessel. As a result, the leak from a crack penetrating the shell of the pressure vessel will be detected and the pressure will be safely reduced, thereby avoiding the damage caused by explosion of the pressure vessel.

Software fail-safe devices work by using internal program tests and exception handling routines which set up safe conditions in the case of errors. In another example, a control can be set up in a 'safe' position and an error indicated if an important component or a sensor has failed.

3.2.4 Reducing Risk by Eliminating Vulnerabilities

3.2.4.1 Eliminating Design Vulnerabilities

Design vulnerability is often present if a single failure of a component leads to a catastrophic failure or to damage escalation. Particularly sensitive to a single failure are systems concentrating a large amount of energy whose safe containment depends on the safe operation of one or several components or on a particular safe sequence of operations. Such systems can be compared with loaded springs accumulating a large amount of potential energy controlled by a single latch. Failure of the latch will release a large amount of stored energy with large destructive power.

Consider a dam containing a large amount of water, built from non-compacted material, whose strength critically depends on the reliable operation of the draining system. Such a dam is vulnerable because if the draining system fails (which can easily happen if the draining pipes are blocked by silt or debris), the strength of the dam will be eroded quickly which could trigger a catastrophic failure.

Another example of a vulnerable system is a system producing toxic substance whose safe containment depends on the reliable operation of a single control sensor. An accidental damage of the sensor results in a release of toxic substance with the associated grave consequences.

Yet another example of a vulnerable system is a system whose safe operation overly depends on the absence of human error. In this case, a human error during operation could trigger failure associated with grave consequences. One way of counteracting this type of vulnerability is to build in fail-safe devices or failure prevention systems that make the dangerous operation impossible. Possible solutions are the failure prevention interlocks that do not permit an operation until particular safety conditions are met.

In computer security, vulnerability is a weakness which allows an attacker to acquire access to a valuable service or data. Failure to include checks for a stack overflow or division by zero, for example, leaves vulnerabilities which could be exploited by attackers by executing malicious code.

Interconnected networks are particularly vulnerable to cascading failures that often bring down network services. Power, telecommunication and transportation networks are often interlinked. Globalisation processes entangle financial markets and create interlinked and interdependent financial networks. Interdependency means that nodes in one network depend on nodes of another network. As a result, failures of nodes in one of the networks could cause node failures in another network which in turn cause node failures in the first network. In a number of cases, this kind of interdependence was responsible for the collapse of large and complex systems.

Consider an interconnected power and telecommunication network. The telecommunication network is powered from the power network while the power distribution is controlled by the telecommunication network. Failure of a number of nodes in the power network due to overloading, causes failures of a number of dependent nodes in the interlinked telecommunication network. Failure of nodes in the telecommunication network cause, in turn, failures of dependent nodes in the power network and this cascading process continues until both networks collapse.

Traditionally, Monte Carlo simulation has been an important method for identifying and assessing the likelihood of vulnerabilities in interconnected networks and failures caused by an unfavourable stack up of values of risk-critical parameters. As it is shown in Chapter 9, this type of vulnerability is often driven by a positive feedback loop and can be eliminated by destroying the loop through decoupling.

Vulnerability of the equipment and systems is reduced by using protection barriers, housing, encapsulation, anti-corrosion and anti-erosion coatings, CCTV surveillance, exclusion zones, security systems for access, etc.

Vulnerability of humans is reduced by various barriers, guards, rails and by using personal protective equipment.

Vulnerability of data is reduced by using security systems, segmentation of the data storage, limiting access to personal records and confidential data, separating duties in handling data records, etc.

3.2.4.2 Reducing the Negative Impact of Weak Links

Consider a common example of a system with n components, logically arranged in series, with reliabilities $R_1,R_2,...,R_n$. The system contains a weak link with reliability r, logically arranged in series with the rest of the components. In other words, the inequalities $r < R_1$, $r < R_2,\dots,r < R_n$ are fulfilled and the reliability of the system $R_{sys} = R_1 \times R_2 \times \dots \times R_n \times r$ is smaller than the reliability of the weak link (Indeed, since $R' = R_1 \times R_2 \times \dots \times R_n < 1$, then $R_{sys} = R' \times r < r$). No matter how close to unity the reliabilities R_1, $R_2,\dots,R_n$ of the components are made, the reliability of the system is still limited by the reliability of the weak link. *The reliability of a system of components arranged in series cannot be improved before the reliability of its weakest link is improved.*

Interfaces often appear as weak links in the chain, thereby limiting the overall reliability of a system. Consider a common practical example related to two very reliable components with very high reliability R_c connected with an interface with a relatively low reliability $r << R_c$. The reliability of the system is smaller than the reliability r of the interface and in order to improve the reliability of the system, the reliability of the interface must be increased. One of the reasons why so many failures occur at interfaces, despite the high reliability of the interfaced components, is the circumstance that *often interfaces are not produced to match the reliability of the interfaced components.* Seals in mechanical assemblies and connectors in electrical assemblies, for example, commonly appear as weak links. An eliminated weak link cannot possibly fail. An excellent example, related to preventing the failure of a seal in installations for liquid helium, was given by Altshuller (1984). The seal needed lubrication in order to work properly but at the cryogenic temperatures, all lubricants freeze and the reliable work of the seal was impossible. The seal appeared as an insurmountable weak link. The solution found by P.L. Kapitsa is an excellent example of eliminating a weak link. Instead of looking for a suitable lubricant and seal, Kapitsa simply eliminated the seal. He deliberately increased the gap between the piston and the cylinder so that helium could freely leak through. During the leakage, the gas expands so quickly that a significant back pressure is created which prevents the leakage of new portions of gas.

Strengthening weak links to avoid failures and improve performance is a truly universal concept which transcends engineering applications. A weak link in a particular area can negate a big effort in other areas. The likelihood of failure-free performance of a group of individuals is often determined by the product of the probabilities of failure-free performance of all members of the group. A very weak performance in one particular area (sub-system) is effectively compromising significant effort in other areas. The human body can be viewed as a chain of individual elements. A weak link in an athlete, for example, might appear as faulty biomechanics or lack of joint mobility which could result in injury or prevents the athlete from achieving peak performance. To prevent injuries and maximise performance, weak links must be identified by qualified professionals and strengthened by specially prescribed corrective exercises. Fixing the weak performance in the faltering area before improving performance in the much stronger areas is the key to increasing the likelihood of peak performance.

The weak link should be strengthened *sufficiently, but not excessively*. Strengthening a weak link excessively, more than other areas, would add cost without increasing the overall reliability of the system.

3.2.4.3 Reducing the Likelihood of Unfavourable Combinations of Risk-Critical Random Factors

Reliable and smooth operation and consequences of failure often depend on not having a particular critical combination of values of the risk-critical random factors which control the performance of components and systems.

Monte Carlo simulation is a very good technique for revealing the probability of critical combinations of risk-critical random factors. This process is known as *virtual testing*. Some of the applications of the Monte Carlo simulation for virtual testing have been demonstrated elsewhere (Todinov 2016a). Consider, for example, a press-fit assembly. If a low value of the diameter of the shaft, a high value for the diameter of the hub and a low value for the coefficient of friction are combined, the press-fit assembly could lose its capability to transmit torque. A design should be made to minimise the probability of undesirable combinations of risk-critical random factors.

In some applications, several redundant components often wear out and fail almost simultaneously due to a common cause (e.g. intensive vibrations) which causes a catastrophic failure. This problem can be avoided by designing the redundant components in such a way that a simultaneous failure of all of them is excluded.

Suppose that failures of a pump delivering cooling liquid for a chemical process follow a homogeneous Poisson process with an expected number of 10 failures a year. (Any of the failures is equally likely to occur on any day of the year.) A single spare pump is kept as a standby redundant pump. In the case of failure of the working pump, it is replaced by the spare pump, a new spare pump is ordered immediately and the production continues without shutting down the chemical production. Suppose that the time for delivering the spare pump is one week. The probability that the chemical reactor will have to be shut down and production lost because no spare pump will be available, is surprisingly large: 77.5% (Todinov 2007). This counter-intuitive result shows how easy it is, without a proper calculation, to underestimate the probability of clustering of events which can result in poor risk management decisions. Methods for reducing the risk of clustering of risk-critical events are considered in Chapter 6.

3.2.4.4 Reducing the Vulnerability of Computational Models

Often, the computational models used for making safety-critical decisions are vulnerable to a particular combination of the input values. As a result, the models are unable to provide a correct value when it is needed, which can have serious safety implications. In the next case study, a little-known vulnerability of the popular *Newton–Raphson numerical method* for solving non-linear equations will be revealed.

Case Study: A little-known vulnerability of the Newton–Raphson Method for Solving Non-Linear Equations Consider the continuous function $f(x)$, accepting values with different signs at the ends of the interval $[a,b]$ $(f(a)\times f(b)<0)$. Suppose that the first derivative $f'(x)$ is a continuous function. According to the Bolzano theorem, a value $x^* \in [a, b]$ always exists such that $f(x^*)=0$. To solve the non-linear equation $f(x)=0$ on the interval $[a,b]$ by using the popular Newton-Raphson method, an initial approximation x_0 is selected from the interval $[a,b]$. The first approximation x_1 is obtained from $x_1 = x_0 - \frac{f(x_0)}{f'(x_0)}$, and the nth approximation is obtained from $x_n = x_{n-1} - \frac{f(x_{n-1})}{f'(x_{n-1})}$. Increasing the number of iterations n yields closer approximations x_n to the unknown root x^*. Because x^* is unknown, the iterations are stopped when two successive approximations are within the required precision ε of each other, i.e. when $|x_n - x_{n-1}| < \varepsilon$, which is equivalent to $\left|\frac{f(x_{n-1})}{f'(x_{n-1})}\right| < \varepsilon$.

In what follows, a little-known critical vulnerability of the Newton–Raphson method will be exposed. This vulnerability shows that the Newton–Raphson method is not a reliable method for solving non-linear equations and the iterations do not necessarily converge to the root.

If, for the function $f(x)$ in Figure 3.5 the initial approximation is selected to be $x=x_0$, the Newton–Raphson method will be trapped around the local maximum and the iterations will never reach any closer to the root x^*.

A simple counter-example which exposes this vulnerability of the Newton–Raphson method is the simple, continuously differentiable function $y=x^3-x^2-6x-9$, with initial value $x_0=0.5$.

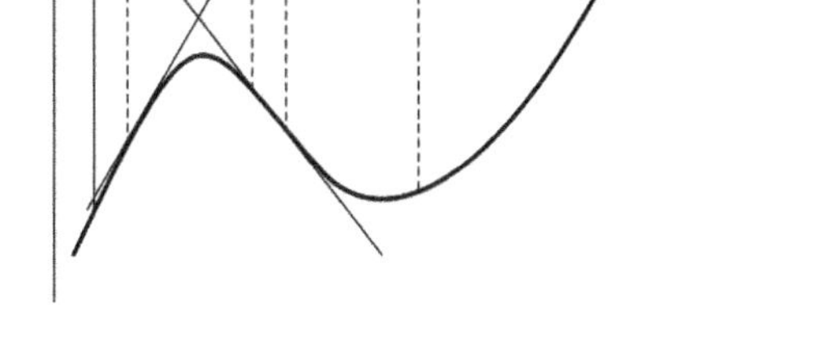

Figure 3.5 The Newton–Raphson iterations could be easily trapped around a local maximum.

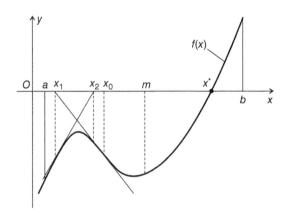

The value $x_0 = 0.5$ traps the Newton–Raphson method so it is never able to find the root of $y(x) = 0$. This vulnerability is present not only for the special selected initial value equal to 0.5 but for a whole range of initial values. For example, initial value x_0 selected from the interval [0, 0.9] also traps the Newton–Raphson iterations. However, if the initial approximation $x_0 = 3$ is specified, the solution x^* can be approximated to three decimal places within three iterations ($x_3 = 3.473$).

In order to avoid the Newton–Raphson method being trapped, the initial approximation x_0 needs to be specified so that convergence is guaranteed. Unfortunately, this possibility is not normally present in engineering models and applications. As a result, an initial approximation could be selected from an interval all of whose values trap the iterations. The inability to always find a solution by this method *carries risks*, for example in cases where the solution is used for making a decision or for control of equipment.

The exposed vulnerability can be eliminated if the repeated bisection method is selected to solve the non-linear equation $f(x) = 0$. It is always guaranteed to work because the function $f(x)$ is continuous and has opposite signs at the ends of the interval $[a,b]$. The Bolzano theorem guarantees the existence of a solution for $f(x) = 0$, where $f(x)$ is a continuous function in the interval $[a,b]$, with different signs at the ends of the interval $[a,b]$: $f(a) \times f(b) < 0$.

Suppose that $f(a) < 0$ and $f(b) > 0$ (Figure 3.5). The middle of the initial interval $[a,b]$ is found from: $m = \frac{a+b}{2}$ (Figure 3.5). The function $f(x)$ is calculated at m. If $f(m) < 0$, the search for the solution continues in the smaller interval $[m,b]$. The interval is halved and the procedure is repeated. If $f(m) > 0$, the search for the solution continues in the halved interval $[a,m]$. In each case, the procedure is repeated with a halved interval and this process continues until the solution is bracketed with the required precision.

Unlike the Newton–Raphson method, the repeated bisection method can never be trapped, for any continuous function $f(x)$. Unlike the Newton–Raphson method, the solution is always obtained with the desired precision. Unlike the Newton-Raphson method, the repeated bisection method always works, even if the derivative of the function does not exist or is zero at one or more points of the interval $[a,b]$.

Because of these advantages, despite its slower convergence compared with the Newton–Raphson method, the repeated bisection method is more reliable and therefore should be preferred to the Newton–Raphson method in safety-critical applications.

3.3 Improving Reliability and Reducing Risk by Protecting Against Common Cause

A common cause reduces the reliability of a number of components simultaneously. The affected components are then more likely to fail, which reduces the overall system reliability.

Typical conditions promoting common-cause failures are the common faults (e.g. common design, manufacturing, installation, assembly and software faults). Thus, two programmable devices produced by different manufacturers, assembled and installed by different people, can suffer common failures if the same faulty piece of software code has been recorded in the devices.

Shared acceleration stresses appear frequently as conditions promoting common-cause failures. *Acceleration stresses* lead to accumulation of damage and faster wearout and are typical examples of common causes. Examples of acceleration stresses are the temperature, humidity, cycling, vibration, speed, pressure, voltage, current, and concentration of particular ions (Nelson 2004). This list is only a small sample of possible acceleration stresses. Because acceleration stresses lead to a faster wearout, they entail higher propensity to premature failure for groups of components which reduces the overall system reliability.

A typical example of this type of common-cause failures is the high temperature which increases the susceptibility to deterioration of a group of electronic components. By simultaneously increasing the hazard rates of the affected components, the probability of system failure is increased. Humidity, corrosion, or vibrations also increase the joint probability of failure of the affected components and shorten the system's life. Even in blocks with a high level of built-in redundancy, in the case of a common cause, all redundant components in the block may fail within a short period of time and the advantage from the built-in redundancy is lost. Failure to account for the acceleration stresses acting as common causes usually leads to optimistic reliability predictions – the actual reliability is smaller than the predicted.

Common-cause failures may also be due to common-cause events: power supply cuts, fire, flood, explosion, failure causing other failures, incorrect maintenance, incorrect calibration, etc. Theft, sabotage bad weather, disease, political crisis, social unrest, and financial crisis appear often as common-cause events. Wrong maintenance and operating actions common to different components are common-cause events which appear as a source of common-cause failures.

Designing against common causes of failure, in order to reduce the risk of failure, can be done by (i) identifying and eliminating sources of common faults, (ii) decreasing the likelihood of the common-cause events, and (iii) reducing the exposure to acceleration stresses.

Common-cause failures are difficult to identify and are frequently overlooked if little attention is paid to the working environment and the possibility of latent faults. Latent faults appear as frequent common causes. Designing out common causes is not always possible, but it should be done if the opportunity arises. Simultaneous corrosion of various components in a cooling circuit or a hydraulic circuit, for example, can be eliminated by selecting non-corrosive working fluids. Erosion of several components caused by a production fluid can be reduced if a filter for catching the abrasive particles is installed. Destruction of all communication lines due to fire, accident, or vandalism can be avoided by avoiding placing all of the communication lines in a common conduit.

A typical example of reducing the impact of a common cause is the use of corrosion inhibitors which, when mixed with the cooling agent, reduce significantly the corrosion rate. The impact of a common cause can also be reduced by strengthening the components against environmental stresses. Such is, for example, the intention behind all corrosion protection coatings. Other examples are the water-tight connectors in underwater installations.

Common case failures can also be reduced by decreasing the likelihood of occurrence of common cause events, for example, fire, flood, and faulty maintenance.

Frequent design reviews and strict control of the manufacturing and assembly reduce the likelihood of latent faults which could be a common cause for expensive failures. A strict control of the maintenance operations reduces the maintenance faults which often initiate common failures. Furthermore, providing maintenance of components by different operators also reduces the likelihood of a common-cause failure due to faulty maintenance.

Providing diversity in design is a very efficient way of protecting against common cause and reducing common-cause failures. A common-cause failure due to a software bug, for example, can be avoided if an alternative algorithm and implementation is provided for the same task or if a different team is involved in developing the same piece of software, independently. If two cooling pumps (a main pump and an emergency pump) provide the cooling of a chemical reactor, failure of both pumps creates an emergency situation. If the two cooling devices are from different manufactures or operate on different principles, a number of common-cause faults will be blocked out. For redundant cooling devices, if one of them is powered by electricity and the other uses natural gravitation to operate, the common-cause event 'absence of power supply' will be eliminated. If, in addition, the two cooling devices are serviced/maintained by different operators, the common cause 'faulty maintenance' will also be blocked out. Similarly, a common cause due to an incorrect calibration of measuring instruments due to a human error can be avoided if the calibration is done by different operators. Finally, if the cooling devices are separated into different rooms, the common failure of the devices due to fire will also be blocked out.

Separating the components at distances greater than the radius of influence of the common cause is an efficient way of protecting against common-cause events bringing cascading failures. Thus, separating large fuel containers at safe distances from one another limits the extent of damage from accidental fire. Separating two or more communication centres at distances greater than the radius of destruction of a missile, increases the probability of survival of at least one of the centres.

Multiple backups of the same vital piece of information kept in different places protect against the loss of information in the case of fire, theft, or sabotage.

Another implementation of this principle is the separation of vital control components from a component whose failure could inflict damage. A typical example is separating the control lines at safe distances from aeroplane jet engines. In the case of engine explosion, flight controls will not be lost.

Insulating components from contact with environment characterised by excessive dust, humidity, heat, or vibrations is a well-known method of protecting against common-cause failures.

Investing in many unrelated sectors protects against a common cause event that reduces simultaneously the return from a number of sectors (e.g. agricultural sectors simultaneously affected by bad weather or disease, consumer sectors simultaneously affected by a health scare, investments in a country affected by a political crisis, financial crisis or social unrest, etc.).

Sundararajan (1991) suggests preliminary common-cause analysis which consists of identifying all possible common causes to which the system is exposed and their potential effects. The purpose is to alert design engineers to potential problems at the early stages of the designs.

3.4 Improving Reliability and Reducing Risk by Simplifying at a System and Component Level

Simplifying systems and components is an efficient way of improving reliability and can be done in various ways: reducing the number of components, simplifying their shape, simplifying their function, reducing the number of functions, etc.

Reducing the number of components increases the reliability of a system. The larger the number of components in a system, the more possibilities for failure exist, the smaller is the reliability of the system. Indeed, the reliability R of a system composed of n components logically arranged in series, with reliabilities R_i, $i = 1, ..., n$, is $R = R_1 \times R_2 \times ... \times R_n$. If a number of components are removed, the reliability of the initial system can be presented as $R = R' \times R_r$ where R' is the reliability of the simplified system and R_r is the product of reliabilities of the removed components. Since the product of the reliabilities of removed components is a number smaller than unity, for the reliability R' of the simplified system, $R' = R/R_r > R$ holds. In words, the simpler system has a larger reliability.

Often, the functions of the eliminated components can be transferred (integrated) into other components without compromising their reliability and the reliability and safety of the system. Furthermore, available resources in the environment or in the system can often be used to provide the functions of the removed components.

Removing components provides the added benefits of (i) reducing the weight of the system, (ii) reducing the cost of the system, and (iii) removing failure modes.

Complex mechanical designs are often associated with difficult maintenance and low reliability, due to the large number of interactions between components which are a constant source of faults and failures.

A typical example of simplifying at a component level is the simplification of a component's geometry which results in a larger load-carrying capacity and enhanced reliability. Simplifying the shape of components and interfaces aides manufacturing, creates fewer possibilities for manufacturing faults, reduces the number of regions with stress intensification, and improves the load-carrying capacity of components by a better distribution of the loading stresses.

Often, physical phenomena can be used to eliminate the introduction of complex control systems. If ferritic material is designed with a Curie temperature of 0 °C, it will be magnetic below the water freezing temperature and diamagnetic above this. This property can be used for heating electrical distribution lines to prevent failures caused by the formation of ice at sub-zero temperatures (Altshuller 1984). As a result, reliance on the Curie transition temperature eliminates the need of introducing a complex control system, simplifies the design of the heating system, and improves reliability.

An alternative approach to simplifying at a component level is to simplify the functions performed by the components, which also improves their reliability. Reducing the number of functions reduces the number of possible failure modes. Therefore, the effect from reducing the number of functions is similar to the effect from reducing the number of components in a system.

An essential part of simplifying at a system level is the reduction of the number of moving parts. Moving parts exhibit more failures compared with stationary parts. This is usually due to the increased kinetic energy, wear, fatigue, vibration, heat generation, and erosion associated with moving parts. The increased kinetic energy of moving parts (e.g. impellers, fans, turbines, etc.) makes them prone to overstress failures if their motion is suddenly restricted due to lodged foreign objects. Moving parts are also associated with large inertia forces which result in pulsating loading and increased fatigue. If out-of-balance forces are present in the rotating parts and excitation frequencies are reached, the associated large resonance amplitudes are a frequent cause of failure (Collins 2003).

Vibration is always associated with moving parts and promotes loss of contact retention, fast wearout and fretting fatigue. Moving parts are sensitive to tolerance faults because they require more precise alignment. The friction and heat generated by moving parts requires lubrication and cooling which makes moving parts also sensitive to failures of the lubrication or cooling system. As a result, reducing the number of moving parts leads to improving the reliability of a system.

3.5 Improving Reliability and Reducing Risk by Reducing the Variability of Risk-Critical Parameters

Reducing the variability of risk-critical parameters prevents them from reaching dangerous levels. Reliability and risk critical parameters vary and this variability can be broadly divided into the following categories: (i) variability associated with material and physical properties, manufacturing, and assembly; (ii) variability caused by the product deterioration; (iii) variability associated with the loads the product experiences in service; and (iv) variability associated with the operating environment.

Strength variability caused by production variability and variability of properties is one of the major reasons for an increased interference of the strength distribution and the load distribution (Figure 3.6) which results in overstress failures. Fracture stress of brittle materials, for example, is characterised by a very large variation (Bergman 1985).

A heavy lower tail of the distribution of properties usually yields a heavy lower tail of the strength distribution, thereby promoting early-life failures. Low values of the material properties exert stronger influence on reliability than do high or intermediate values.

Variability of critical design parameters (e.g. material properties and dimensions) caused by processing, manufacturing, and assembly is an important factor promoting early-life failures.

Reducing variability can be achieved by a statistical process control based on monitoring the variations of the output parameters (Montgomery et al. 2001). Another powerful tool for reducing variability is the six-sigma quality philosophy (Harry and Lawson 1992) based on production of a very small number of defective items. A comprehensive discussion related to the effect of variability on the reliability of products can be found in the literature (Haugen 1980; Fowlkes and Creveling 1995; Carter 1986, 1997; Booker et al. 2001). Due to the inherent variability of the manufacturing processes, the same component, of the same material, manufactured to the same specification is usually characterised by different properties. Because of the natural variation of critical design parameters, early-life failures are commonly caused by unfavourable combinations of parameter values (e.g. worst-case tolerance stacks).

Due to the inherent variability of the manufacturing processes, even items produced by the same manufacturer can be characterised by different properties. Production variability

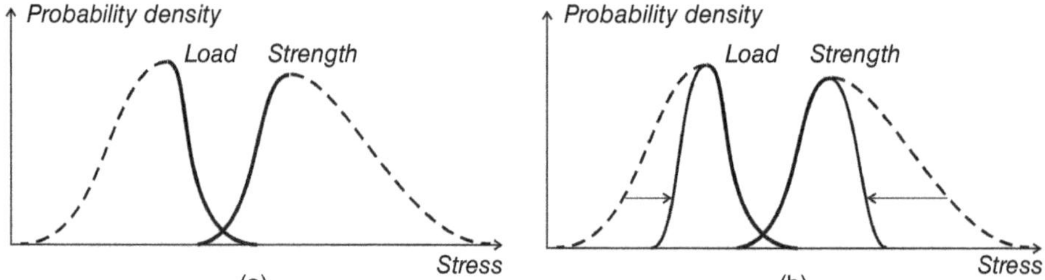

Figure 3.6 (a) Reliability is dependent on the degree of interaction of the upper tail of the load distribution and the lower tail of the strength distribution; (b) reducing variability does not always result in a reliability increase.

during manufacturing, not guaranteeing the specified tolerances or introducing flaws in the manufactured product, leads to a significant number of failures. Depending on the supplier, the same component, of the same material, manufactured to the same specification is usually characterised by different properties. Between-suppliers variation exists even if the variation of the property values characterising the individual suppliers are small (see Chapter 12). The variability associated with the lower tail of the strength distribution controls the load-stress interference, not the variability associated with the high or central values. Stress screening which eliminates substandard items is an efficient way of reducing the variability in the region of the lower tail of the strength distribution and increasing reliability.

Low reliability is often due to excessive variability of the load. If the load variability is large (rough loading), the probability of an overstress failure is significant. Altering the upper tail of the load distribution is often done by using stress limiters. Typical examples of stress limiters are the safety pressure valves, fuses and switches, activated when pressure or current reaches critical values. A common example of a stress limiter preventing surges in voltage from reaching dangerous levels are the anti-surge protectors used in the power supply of electronic equipment.

A large variation of the internal stresses in components and structures caused by variation of the loading forces is also a frequent cause of fatigue failures. Fatigue life is very sensitive to the amplitude of the internal stresses. As a consequence, in regions with large internal stress ranges, the fatigue crack growth rate is large and the life of the affected components is short. Design solutions restricting the variations of the internal stresses include but are not limited to stress limiters, appropriate modifications of the component geometry, appropriate modifications in the loading geometry, appropriate stress relieve notches, stress relieve grooves, etc.

Tolerances in geometric reliability-critical parameters must be controlled during manufacturing. Such control translates into fewer problems and failures during assembly, less possibility for a loss of precision, jamming, seizure, poor lubrication, and fast wearout. Often variability of geometric parameters causes *fit failures* resulting from interference of solid parts which makes the assembly impossible.

Material quality is positively correlated with the reliability of components. This correlation is particularly strong for highly stressed components. Sources of material supply must be controlled strictly, without relying on vendor trade names or past performance. Changes in the processing and manufacturing procedures often result in materials with poor quality.

The loading stress is a risk-critical parameter. In many cases, the reliability on demand is determined by the probability of a relative configuration of the load and strength, in which load is smaller than strength. Reliability on demand is controlled by the two risk-critical parameters 'load' and 'strength', characterised by distinct distributions.

Load and strength are much broader concepts than their direct mechanical interpretation. Any two interacting random factors can be interpreted as 'load' and 'strength'. Load and strength, for example, could stand for 'demand' and 'supply', 'rate of damage' and 'rate of recovery', 'corrosion rate' and 'corrosion resistance', 'stress intensity factor' and 'fracture toughness', etc.

Some authors equated improving reliability with reducing variability. Consequently, they view reliability improvement largely as an exercise in reducing variability or making the design insensitive to variation of design parameters.

Despite that variability of design parameters is a major source of failures, in some cases, reducing variability does not necessarily translate into improved reliability. Such a counter-example has already been given in Chapter 2 and in what follows, other counter-examples will be given.

Furthermore, reducing variability is certainly not the only method of improving reliability. The method of segmentation is an example where reliability is improved without decreasing variability or making the design insensitive to variability. Improving reliability by reducing the space and time exposure is another example where reliability is improved without reducing variability or making the design insensitive to variability.

3.5.1 Case Study: Interaction Between the Upper Tail of the Load Distribution and the Lower Tail of the Strength Distribution

The most important aspect of the load–strength interaction is the interaction of the upper tail of the load distribution and the lower tail of the strength distribution (Figure 3.6). Consequently, only information related to the lower tail of the strength distribution and the upper tail of the load distribution is necessary. The values from the lower tail of the strength distribution and the upper tail of the load distribution control reliability, not the values corresponding to the other parts of the distributions (Figure 3.6a).

To quantify the separation of the load and strength, data covering the lower tail of the load distribution and the upper tail of the strength distribution are not necessary. The variability of the load or the strength can be decreased by shortening the lower tail of the load distribution without altering its upper tail and by reducing the upper tail of the strength distribution without altering its lower tail (Figure 3.6b). As can be verified from Figure 3.6b, reducing the variability (standard deviation) of the load distribution and the strength distribution in the described fashion *will not alter the interference of the upper tail of the load distribution and the lower tail of the strength distribution and will not reduce the risk of failure*. As a result, not all variability reduction translated into increased reliability and reduced risk of failure.

The variability is of crucial importance only when it is associated with the lower tail of the strength distribution or the upper tail of the load distribution. Figure 3.7a depicts increased variability associated with the lower tail of the strength distribution. This variation is often caused by the deterioration of the material due to ageing processes, corrosion, fatigue, stress relaxation, wear, and other processes associated with damage accumulation.

Figure 3.7b depicts increased variability associated with the upper tail of the load distribution. This variation marks a transition from smooth to rough loading which is often caused

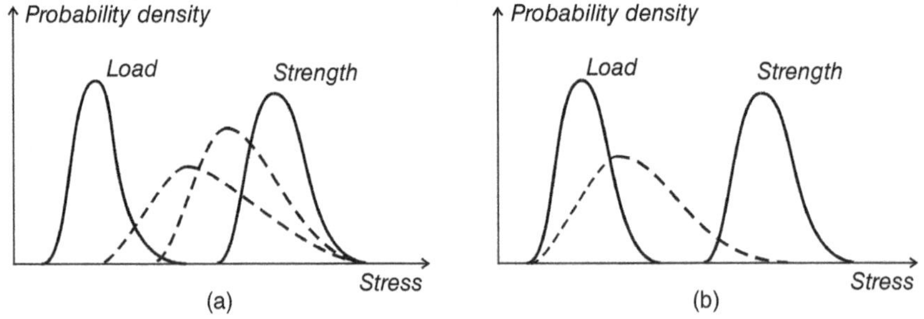

Figure 3.7 Variability is of critical importance when it is associated with (a) the lower tail of the strength distribution or (b) the upper tail of the load distribution.

by changes in the environment, the operating conditions, deviations from the prescribed area of application, lack of maintenance, etc. Consequently, an adequate model of the strength distribution should faithfully represent its lower tail and an adequate model of the load distribution should faithfully represent its upper tail.

In a load–strength interaction, risk is strongly dependent on the degree of relative separation of the upper tail of the load variation and the lower tail of the strength variation. The analytical treatment of the load–strength integral related to the interaction of the upper tail of the load distribution and the lower tail of the strength distribution can be found in (Todinov 2016a).

Separation of the load distribution and the strength distribution can, for example, be done by focusing on the strength distribution only. The well-known *burn-in* operation (O'Connor 2002) essentially increases the relative separation of the lower tail of the strength distribution and the upper tail of the load distribution thereby reducing the load–strength interference and reducing the risk of failure. The relative separation of the load distribution and the strength distribution can also be increased by introducing deliberate weak links or stress limiters.

Here, it needs to be pointed out that some measures of the relative separation of the load distribution and strength distribution, well-established in the reliability literature, are misleading for asymmetric distributions.

Consider a load distribution characterised by a mean μ_L and standard deviation σ_L and strength distribution characterised by a mean μ_S and standard deviation σ_S. A common measure quantifying the degree of relative separation of the load and strength is the *reliability index*, defined as

$$\beta = \frac{\mu_S - \mu_L}{\sqrt{\sigma_S^2 + \sigma_L^2}} \tag{3.6}$$

The reliability on demand R characterising the load–strength configuration is given by

$$R = \Phi(\beta) = \Phi\left(\frac{\mu_S - \mu_L}{\sqrt{\sigma_S^2 + \sigma_L^2}}\right) \tag{3.7}$$

where $\Phi(\bullet)$ is the cumulative distribution of the standard normal distribution (Carter 1986, 1997; O'Connor 2002).

From Eq. (3.7), it is easy to see that a larger difference $\mu_S - \mu_L$ between the means of the strength and load distribution and smaller variances σ_L^2 and σ_S^2 of the load and strength distributions, lead to a larger reliability index β and larger reliability on demand. The safety margin has been used as a measure for the relative separation of the load and strength distribution even for load and strength distributions which do not follow the Gaussian distribution (O'Connor 2002).

In what follows, it is demonstrated that for load and strength that do not follow the normal distribution, *the traditional reliability measure safety margin is misleading and cannot be used to measure the degree of relative separation between load and strength.*

Consider the load and strength distributions from Figure 3.8a. The figure shows a case where a low safety margin $\beta = (\mu_S - \mu_L)/\sqrt{\sigma_S^2 + \sigma_L^2}$ exists ($\mu_S - \mu_L$ is small and $\sigma_S^2 + \sigma_L^2$ is large) yet the reliability on demand is high. In Figure 3.8a, μ_S and μ_L are the mean values of

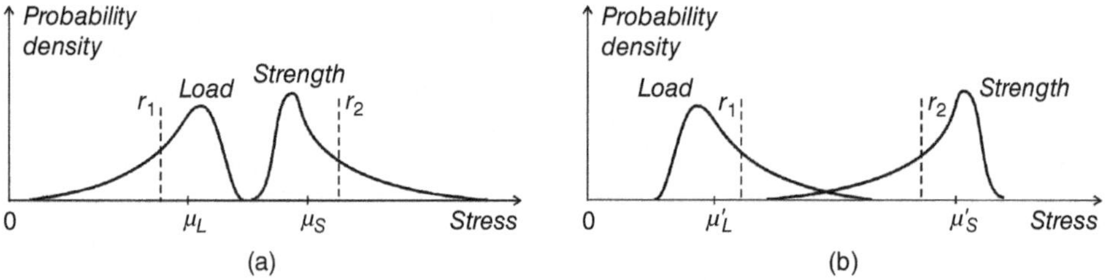

Figure 3.8 A counter-example showing that for skewed load and strength distribution, the traditional reliability measure 'reliability index' is very misleading.

the strength and load; σ_S and σ_L are the corresponding standard deviations. Now consider Figure 3.8b which has been obtained by reflecting symmetrically the distributions from Figure 3.8a with respect to axes r_1 and r_2, parallel to the probability density axis. Since the reflections do not change the variances of the distributions, the only difference is the larger difference of the means $\mu'_S - \mu'_L > \mu_S - \mu_L$ (Figure 3.8b). Despite the larger new safety margin

$$\beta' = \frac{\mu'_S - \mu'_L}{\sqrt{\sigma_S^2 + \sigma_L^2}} > \beta = \frac{\mu_S - \mu_L}{\sqrt{\sigma_S^2 + \sigma_L^2}}$$

the reliability on demand related to the load–strength configuration in Figure 3.8b is smaller than the reliability on demand related to the configuration in Figure 3.8a. Clearly, the safety margin concept applied without considering the shape of the interacting distribution tails can be very misleading.

The problems outlined do not exist if for load and strength which do not follow a normal distribution, a numerical integration is used to quantify the relative separation between load and strength.

3.6 Improving Reliability and Reducing Risk by Making the Design Robust

Robustness is a very important property of processes and products and can be defined as 'the capability of a process or a product to cope with variability of critical parameters with minimal loss of functionality'. Robust designs are insensitive to variations of the manufacturing quality, tolerances, drifts in parameter values, operating parameters, maintenance conditions, environmental loads, and deterioration with age (Lewis 1996). Often noise factors arise during production, use, and deterioration. The variability caused by the noise factors is often the primary reason for the unreliability. Consequently, the aim of the robust design is to create insensitivity to noise factors rather than eliminate or control them (Bergman et al. 2009). This is particularly important in cases where the likelihood of an adverse event or its consequences are associated with a great deal of uncertainty. In these cases, the robustness provides 'extra capacity' absorbing the variation of operating parameters and stresses from the adverse event. As a result, robustness enhances the system's resilience.

In many cases, the reliable work of components and systems occurs under too narrowly specified conditions. Slight variations in the material quality, the quality of manufacturing, the external load or the values of the design parameters are sufficient to induce failures or unacceptable deviations from the expected function/service.

Achieving high reliability levels by reducing the variation of parameters (e.g. material properties, geometric parameters) related to components building the systems, is not normally economically feasible. Thus, reducing the variation of material properties and geometric parameters through more expensive materials and more precise finishing operations (such as grinding, honing, and polishing) is not normally feasible because of escalating costs. Making the design insensitive to variations of material properties and geometric parameters or building in capacity for absorbing extreme variations of stresses, often achieves high reliability without excessive costs.

Suppose that an electronic device works only if the parameters of the electronic components used to build the device are precisely selected to be with strictly specified values. The design of such a device is not robust. The inevitable variation of the parameters of the electronic components building the device, will make a large number of manufactured devices faulty. A robust design is present when the electronic device is fully functional even in the presence of a relatively large variation of the parameters of the electronic components.

Similarly, if the reliable operation of a device is guaranteed only in a narrow temperature window, the inevitable variations of the temperature will make the device malfunction. A robust design has a built-in capacity that permits the device to operate reliably over a wide range of temperatures. An example of robust design with built-in capacity to absorb variations of critical parameters during operation are the *gear couplings* which can compensate simultaneously radial, angular, and axial misalignment.

To ensure robust design, capable of absorbing variations in geometric tolerances and deviations from correct alignment during operation, avoiding a double-fit is important. A double fit is present if a component is guided at the same time by two surfaces (machined separately). Because of the inevitable variation of tolerances, these assemblies are a source of problems during assembly. They are also a source of problems during operation, because of the assembly stresses and the uncertainty regarding the distribution and magnitude of the loading stresses during operation.

A typical example of a design, insensitive to variations of geometric parameters is the *involute gear system*, where the profiles of the teeth are involutes of a circle. The angular velocity ratio ω_1/ω_2 of two gears with involute gear profiles is insensitive to the inevitable variations of the distance between the gear axes. It remains constant. Another typical example is the design of self-adjusting bearing assemblies which can accommodate geometric imprecision and misalignment.

Design solutions requiring fewer parts with simple geometry reduce the susceptibility to manufacturing variability. Designs incorporating appropriate geometry, static determinacy, tolerances, and materials with high conductivity, reduce the susceptibility to large thermal stresses which are a common cause of failure. For example, making truss structures statically determinate makes them free of thermal stresses.

The statically indeterminate structures are subjected to thermal stresses or assembly-induced stresses if the components are characterised by a different coefficient of thermal expansion, or are kept at different temperatures. None of these problems exist for the statically determinate structures.

Further examples of robust designs are the sensors for measuring a particular property which are insensitive to variations of other parameters; for example, gas sensors insensitive to the variations of temperature and humidity.

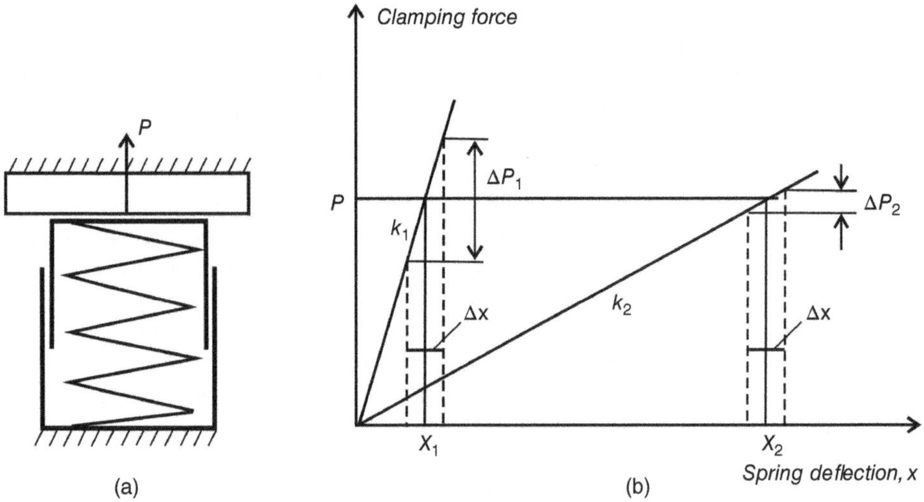

Figure 3.9 Clamping force variation for a stiff and a soft spring.

Often, the mean values of reliability-critical parameters are sought, that minimise the sensitivity of performance characteristic to variations of the input parameters.

This approach has been illustrated by the simple mechanical spring assembly in Figure 3.9a, required to provide a constant clamping force of specified magnitude P.

3.6.1 Case Study: Increasing the Robustness of a Spring Assembly with Constant Clamping Force

For the spring assembly in Figure 3.9a, a constant clamping force can be provided by a stiff spring, with a large spring constant k_1 and initial deflection $x_1 (P = k_1 x_1)$ or a softer spring, with a smaller spring constant $k_2 < k_1$, and larger initial deflection $x_2 > x_1$ (Figure 3.9b). The initial spring deflection is always associated with errors (errors in cutting the spring to exact length, imperfections associated with machining the ends of the spring coil, sagging of the spring with time due to stress relaxation, and variations in the length of the spring associated with the pre-setting operation).

As it can be verified from Figure 3.9b, for the softer spring (with a spring constant k_2), variations of magnitude Δx in the spring deflection cause much smaller variations ΔP_2 in the clamping force compared with the variations ΔP_1 in the clamping force of the stiffer spring, caused by variations of the same magnitude Δx in the spring deflection. Selecting a softer spring results in a robust design, for which the clamping force P is not very sensitive to variations in the spring deflection.

In the general case, determining the mean values of the control parameters that minimise the variation of the performance characteristic requires algorithms for constrained non-linear optimisation. For these optimal values of the design design parameters, the output performance characteristics will be least sensitive to variations of the design parameters. In many cases, the relationship between the performance characteristic and the values of the reliability-critical variables cannot be presented in a closed form or if it exists, it is too complex. Furthermore, the reliability-critical parameters may be interdependent, subject to complex constraints. In these cases, a simulation-based optimisation is a powerful alternative to other methods. It is universal, handles complex constraints and interdependencies between reliability-critical variables and does not require a closed form expression. Furthermore, the implementation of Monte Carlo simulation algorithms is relatively simple and straightforward.

3.7 Improving Reliability and Reducing Risk by Built-in Reinforcement

Reinforcement is added extra capacity to a component/system which makes it more difficult for the loading stresses to exceed the critical limit which precipitates failure. The additional capacity is constant and does not change with increasing the loading stress. This is an essential feature which distinguishes reinforcement from self-reinforcement. For self-reinforcement, increasing the loading stress creates a proportional response which decreases the resultant magnitude of the loading stress. As a result, the driving force towards failure precipitation is decreased. Self-reinforcement will be introduced in detail in Chapter 9.

3.7.1 Built-In Prevention Reinforcement

This principle in action can be found in the construction industry where reinforcing steel bars in concrete structures are placed in regions of high tensile stresses. Steel bars have a very good tensile strength and compensate for the lack of tensile strength in concrete. The result is a significant reduction in the likelihood of overstress failure. This principle is also used in the design of components from fibre-reinforced composite materials. The composite material is oriented so that the direction of its fibres is parallel to the direction of the expected high tensile stresses. Because the composite has a much higher tensile strength along its reinforcing fibres compared with the tensile strength in lateral direction, this layout increases the load carrying capacity of components and reduces the risk of overstress failure.

An example where the risk of failure is reduced by built-in prevention compensating a negative effect during service is with the allowances for lost wall thickness. The *corrosion, erosion*, and *wear allowances* added to the computed sections compensate for the loss of wall thickness and decrease the risk of failure.

Components working in close contact (e.g. piston–cylinder) and moving relative to each other generate heat which, if not dissipated, causes intensive wear, reduced strength, and plastic deformations. The risk of failure of such an assembly can be reduced significantly if one of the parts (e.g. the cylinder) is cooled to dissipate the released heat which reduces friction and wear.

Preloading with the same sign of the loading stresses is often used as a built-in prevention to counter effects increasing the risk of failure. It is frequently applied to bolted joints and flange-and-gasket assemblies. *Tensile preloading* increases the fatigue life of a part subjected to completely reversed zero-mean alternating stresses. The mean stress is increased but the completely reversed cyclic stress is reduced significantly and, as a result, the fatigue life is increased substantially (Collins 2003).

3.7.2 Built-In Protection Reinforcement

This method is often used for components and systems whose consequences of failure are significant.

This principle is the reason behind the safety practice of building residential areas beyond the radius of the harmful influence of toxic substances from chemical plants, compost production, fuel depots, etc.

Passive protective barriers physically separate the hazards (the energy sources) from targets. Physical barriers isolate and contain the consequences and prevent the escalation of damage. They provide passive protection against the spread of fire, radiation, toxic substances or dangerous operating conditions. A blast wall, for example, guards against the effects of a blast wave. Increasing the distance between sources of hazards and targets minimises damage in the case of an accident.

Examples of built-in protective barriers are: the safeguards protecting workers from flying fragments; the protective shields around nuclear reactors; the fireproof partitioning; the double-hulls in tankers limiting oil spillage, etc.

Active protective barriers. The consequences from an accident or failure can be mitigated significantly by activating built-in protective systems. Typical examples of active barriers designed to mitigate the consequences from accidents are: the safety devices activating sprinklers for limiting the spread of fire; extraction fans for reducing the concentration of toxic substances; surge barriers to limit the risk of floods; automatic brakes in the case of a critical failure; automatic circuit breakers in the case of a short cut, etc.

3.8 Improving Reliability and Reducing Risk by Condition Monitoring

Condition monitoring is used for detecting changes or trends in controlling parameters or in the normal operating conditions which indicate the onset of failure. By providing an early problem diagnosis, the idea is to organise in advance the replacement of components whose failure is imminent, thereby avoiding heavy consequences. Condition monitoring is particularly important in cases where the time for mobilisation of repair resources is significant. The early problem diagnosis helps to reduce significantly the downtimes due to unplanned intervention for repair. A planned or opportune intervention is considerably less expensive than unplanned intervention initiated after the failure occurrence.

Early identification and action upon detection of an incipient failure eliminates significantly the risk of environmental pollution, loss of life, loss of production assets, and losses caused by dependent failures. An important feature of condition monitoring is the capability to adequately read and interpret signals that are precursors of the adverse events. The better the interpretation, the earlier the warning, the larger the response time, the more efficient the loss prevention, and the more valuable the condition monitoring technique.

Because of the cost of monitoring devices, it is not feasible to deploy monitoring equipment to all components in a system. To make the monitoring process feasible, a limited number of optimal monitoring locations need to be identified in the system, within the allocated budget for monitoring equipment. This is a challenge because the number of possible distinct configurations of the monitoring equipment locations is huge. Consider a system consisting of M distinct components. In this system, n identical monitoring devices ($n < M$) can be placed on M available locations in the network in $M!/[(M-n)!n!]$ different ways, which is the number of combinations of n elements out of M. This can be a very large number. For $M = 100$ and $n = 5$, there are already more than one billion combinations.

To make the monitoring process feasible, a relatively small number of appropriate monitoring locations need to be identified in the network. Because of the early indication of a problem (e.g. lack of lubricant, excessive vibration, excessive temperature, excessive current, etc.) on the monitored component and the countermeasure taken to mitigate it, the presence of monitoring equipment increases the reliability of the monitored equipment. Early warning about excessive damage accumulation helps to choose an appropriate operational cycle, type, and frequency of maintenance so that failure is avoided.

Condition monitoring can be used to estimate the approximate time of failure occurrence which provides a basis for better planning of the necessary resources for repair. As a result, the number of required spare parts and the cost of their preservation and storage can also be reduced significantly. Orders for components whose failure is imminent can be placed for manufacture, thereby reducing the delays associated with their delivery. Because of the early indication of incipient failure, and the measures taken to order and replace the component, the presence of condition monitoring increases the availability of the monitored component by reducing the downtime for its replacement.

The cost of the monitoring equipment should be within the allocated budget and the locations of the monitoring devices should provide the maximum increase in the reliability/availability of the system. Brute-force testing related to the availability of the network at each of these possible locations, in order to select the location associated with the highest network availability, is not feasible considering the huge number of possible locations. Despite the intensive system reliability research over the last few decades, no viable solution has ever been presented for the long-standing problem of maximising the reliability or the availability of a system by determining the most appropriate places for network monitoring. For some systems (e.g. power networks), a reasonable solution can be found by a Monte Carlo simulation. This involves solving the electrical circuit many times, under different power demand profiles of the consumers, and determining the distributions of current/power in the separate components of the power network. Monitoring is then preferentially allocated to the components characterised by a large likelihood that the current/power will exceed a critical value. A key consideration related to condition monitoring is that the reliability of the monitoring system must exceed the reliability of the monitored equipment (Heron 1998).

A further discussion about the most appropriate monitored locations in a parallel–series system which maximise its reliability and availability is presented in Chapter 11.

Updating the probabilities of failure of components and assemblies on the basis of information provided by condition monitoring provides an excellent opportunity to reduce uncertainty about the time to failure of components and evaluate more precisely the probability of system failure. For example, knowledge of failed redundant components from status monitoring can be used to re-evaluate the probability of failure of assemblies. Such an example has already been discussed in Section 2.1. Knowledge of the current length of a fatigue crack from condition monitoring, can be used to re-evaluate the probability of failure of a part subjected to cyclic loading.

Condition monitoring is based on measuring values of specific parameters critical to the failure-free operation of the monitored equipment. Here are some examples of measured parameters:

- *Temperature measurements* to detect increased heat generation which is usually an indication of intensive wear, poor lubrication, failure of the cooling system, overloading or inappropriate tolerances.

- *Measurements of pressure and pressure differences* to detect leaks in a hydraulic control system, dangerous pressure levels, etc. Increased pressure difference before and after a filter, for example, is an indication of clogging or a blockage of the filter, while a pressure drop in a hydraulic line is an indication of leakage.
- *Measuring the displacement* of components and parts indicates excessive deformations, unstable fixtures, material degradation or design faults. Displacement transducers can be used for this purpose.
- *Vibration monitoring* detects incipient failures in bearings, increased wear, out-of-balance rotating components, etc. If a shaft rotating at a high speed goes out of balance, it is expected to produce a particular excitation frequency which will be identified from the monitored frequency spectrum, on the basis of signals obtained from accelerometers (Thompson 1999).
- *Corrosion monitoring* on test probes made of the same materials as the key components in the production system assesses the state corrosion damage affecting expensive components in the system (Thompson 1999).
- *Monitoring electrical parameters such as power, current, voltage, and resistance.*

Elevated power transmitted on a power line indicates elevated consumption of electrical energy, overloading, and increased likelihood of failure. A monitoring device provides an early indication of the problem which prompts the operator to take steps mitigating the problem. Decreasing the resistance of a high-voltage insulator indicates operational and safety issues. Increased current in electric motors, for example, indicates increased resistance from the powered equipment. This is an indication of jamming due to a build-up of debris or corrosion products, misalignment due to excessive wear, increased viscosity of the lubricant or lack of lubrication, damage of the bearings, clogged or blocked filters (in pumps), etc. Sticky valves that could seize can be detected by the increase of current required to operate their actuators (Thomson 1999).

Decreased conductivity of working fluids (hydraulic fluid) often indicates contamination which could induce poor heat dissipation, increased energy losses, accelerated erosion, and corrosion.

- *Monitoring the cleanliness* of lubricants and hydraulic fluids. Periodic sampling of lubricants and hydraulic fluids for debris is often used to determine the extent of wear undergone by the components. Another useful outcome of sampling hydraulic fluids for debris is reducing the possibility for jamming of control valves. Such failures could render inoperative large sections of a control system and could be very expensive if the cost of intervention is high (e.g. in subsea oil and gas production). This type of condition monitoring also reveals the degree of deterioration of the lubricants and hydraulic fluids because of poor cooling, or increased heat generation due to wear.
- *Ultrasonic inspection, radiographic examination, magnetic particles*, and *penetrating liquids* are used to detect various flaws (cracks, pores, voids, inclusions) in components, welded joints, and castings.
- *Measuring the degree of wear, erosion, and corrosion.* Since most of the mechanical systems undergo some form of degradation caused by wear, corrosion, and erosion, they certainly benefit from periodic monitoring of the extent of degradation. Deteriorated components can be replaced in time by opportunity maintenance.
- *Measuring the degree of fracture toughness deterioration* of materials due to ageing, corrosion, and irradiation. The fracture toughness of steels exposed to deterioration is

determined and the ductile-to-brittle transition curve is built. The degree of deterioration is indicated by the shift of the ductile-to-brittle transition region towards higher temperatures.

- *Monitoring manufacturing and process parameters.* Controlling the tolerances during manufacturing by monitoring the wear rate of cutting tools reduces the possibility of failures due to misfit during assembly, less possibility for jamming, poor lubrication, and accelerated wearout.

Condition monitoring is different from status monitoring which determines whether a component is in a working or failed state. Condition monitoring however, even used solely as a status-monitoring tool, has a high value and can improve the availability of the system immensely if combined with immediate intervention for repair/replacement of the failed components. This can be illustrated on a simple system with active redundancy which consists of two identical components logically arranged in parallel. Without status monitoring, repair is initiated only if the system stops production. The result from this breakdown-induced intervention is a sequence of uptime corresponding to the situation where at least one of the components works, followed by a downtime where both components have failed and mobilisation of resources for repair has been initiated. Status monitoring changes this availability pattern dramatically. If the status of both components is constantly monitored and repair is initiated whenever any of the component fails, the downtime associated with a critical (system) failure could be avoided almost completely.

Another example, related to dependent failures, is the common assembly fan-cooled device, where failure of the fan causes an overheating failure of the cooled device. Such a failure can be prevented if the status of the fan is monitored and in the case of fan failure, operation is discontinued until the fan is replaced.

It must be pointed out that while status monitoring benefits mainly systems with built-in redundancy, condition monitoring on the extent of degradation of components and structures also benefits systems without any built-in redundancy.

Finally, condition monitoring provides the basis for improved designs by feeding back information related to the actual times to failure, vulnerable components, root causes of failures, rate of material degradation and the impact of operating conditions.

A major limitation of the condition monitoring is its significant cost. Deploying various sensors and devices for recoding and storing collected data is associated with costs. Furthermore, condition monitoring requires trained personnel to analyse and interpret the recorded data, and trained personnel for preventing failures from occurring. Training and maintaining such personnel is associated with significant costs and the cost of the condition monitoring should be considerably lower than the cost of prevented failures. The infrastructure making it possible to react upon the data stream delivered from condition monitoring devices must be in place if the condition monitoring technique is to be of any use. There is little use for a condition monitoring system giving an early warning of an incipient failure if the infrastructure for interpreting the condition monitoring data and for preventive action is missing.

Condition monitoring can be used to limit the consequences from failure by transforming existing components, associated with large consequences from failure, into deliberate weaknesses cutting off the load so that the consequences from failure are limited.

As an example, consider a system transporting toxic fluid including a pump connected to a pipe through a flange connection. The component whose failure is associated with serious consequences is the seal of the flange. If a sensor is placed in the vicinity of the seal, a leak of toxic gas from the seal will be detected by the sensor and through a control circuit, a signal will be sent to a relay disconnecting the power supply to the pump. Once the power supply to the pump has been interrupted, the pressure in the pipe connection is reduced and the leak of toxic gas stopped. The component with severe consequences from failure has been transformed into a deliberate weakness whose failure causes immediate system failure limiting the consequences.

3.9 Reducing the Risk of Failure by Improving Maintainability

Many systems are subject to maintenance of some type. Maintenance actions can be divided broadly in two classes (Smith 2001): (i) corrective (unscheduled) maintenance; and (ii) preventive (scheduled) maintenance.

Corrective (unscheduled) maintenance is initiated only if a critical failure occurs (one or more production units stop production). Failed redundant components are not replaced or repaired until the system fails.

Preventive (scheduled) maintenance is performed at planned intervals with the purpose of keeping the built-in levels of reliability and safety and prevent the system failure rates from exceeding tolerable levels. This type of maintenance achieves its purpose by the following actions (Smith 2001):

- Regular service of operating components and subsystem: (e.g. lubrication, cleaning, adjustment).
- Replacement of failed redundant components.
- Replacement of components with excessive wear out.

Frequent inspection and replacement of failed redundant components enhances availability by improving the reliability of the system. Design to improve maintainability is vital to reducing the losses from failures. Improved maintainability reduces significantly the downtimes, the amount of lost production, and the cost of intervention and repair. Availability can be improved significantly by improving maintainability through improved accessibility and *modular design*. Sufficient number of spares should also be readily available. Using standard items with proven and tested properties also increases reliability.

A general principle in maintainability is the easy access to failed parts. Maintenance of a failed sub-system should not require the removal of another sub-system. The design must permit access to important components (Thompson 1999). This design aspect is particularly important in the case of maintenance and repair by remotely operated vehicles (ROVs) conducted underwater.

Modular design reduces the losses from failures by improving maintainability and reducing the cost of intervention and repair and the downtime. Consider an installation for subsea oil and gas production. In the case of failure of a subsea control module for example, only the failed module needs to be retrieved and repaired instead of retrieving the whole production tree. Compared with retrieving the production tree, replacing a failed control module usually requires less costly intervention, usually a ROV. Retrieving a production tree to the surface requires mobilisation of an oil rig which is a significantly more expensive operation compared with deploying an ROV. Moreover, while a repair by mobilising an oil rig may require months, repairs by deploying ROVs do not normally require more than several days.

Modules should be connected to other modules as simply as possible to reduce the downtime and cost associated with the repair/replacement. In this respect, the use of quick-release devices is beneficial (Thompson 1999).

A proper management of the maintenance resources for repair reduces significantly the downtimes associated with repair. Optimising the number of spares for critical components or components with high failure frequency is also important. The optimal number of spares should be estimated using modelling. Condition monitoring can significantly reduce the number of necessary spare components, which reduces the costs and the required storage space.

3.10 Reducing Risk by Eliminating Factors Promoting Human Errors

Human errors account for a significant number of technical failures. They are an inevitable part of each stage of the product development and operation: design, manufacturing, installation, and operation (Turner 1978; Reason 1997). Following Dhillon and Singh (1981), human errors can be categorised as: (i) errors in design; (ii) operator errors (failure to follow the correct procedures); (iii) errors during manufacturing; (iv) errors during maintenance; (v) errors during inspection; and (vi) errors during assembly and handling. A thorough analysis of the root causes, conditions, and factors promoting human errors is an important step towards reducing them. Some common factors promoting human errors are listed below:

- Time pressure and stress
- Overload and fatigue
- Distractions and high noise levels
- Poor work skills and lack of experience
- Unfamiliarity with the necessary procedures and equipment
- Inadequate professional knowledge or supplied information about the system operation and the environment
- Poor health
- Poor organisation, time management, and discipline
- Inattention and lack of concentration
- Making unwarranted assumptions and building a false picture of the reality
- Negative emotional states and disempowering beliefs
- Low confidence
- Poor motivation
- Poor communication
- Poor relationships with the other members of the team

Instructions and procedures must be clearly written, easy to follow and well justified. The procedures must also reflect and incorporate the input from people who are expected to follow them. It must always be borne in mind that human beings are prone to forgetting, misjudgement, lack of attention, creating false pictures of the real situation, etc. – conditions that are all difficult to manage. Hardware systems and procedures are much easier to manage and change than human behaviour. Therefore, the efforts should always concentrate on adapting the hardware to humans rather than adapting the humans to hardware.

Learning from past failures and making available the information about past human errors that have caused failures is a powerful preventive tool. In this respect, compiling formal databases containing descriptions of failures and lessons learnt, and making them available to designers, manufacturers, and operators are activities of significant value.

Frequent reviews, checks, and tests of designs, software codes, calculations, written documents, operations or other products heavily involving people are important tools for preventing human errors. In this respect, *double checking* of the validity of calculations, derivations or a software code are invaluable to preventing human errors. To eliminate common-cause errors associated with models and problem solutions, *double checking based on two conceptually distinct approaches* is particularly helpful. Such is, for example, the case in deriving an analytical solution to a probabilistic problem. Obtaining a very close result by a Monte Carlo simulation, of the same problem, provides very strong support that the analytical solution is correct.

A number of human errors arise in situations where a successful operation or assembly is overly dependent on human judgement. Human errors of this type can be avoided by using tools/devices which rely less on a correct human judgement. Poka Yoke design features, and special recording and marking techniques could be used to prevent assembling parts incorrectly. Blocking against common cause maintenance errors could be achieved by avoiding situations where a single person is responsible for all pieces of equipment.

A thorough task analysis reveals weaknesses in the timing of operations and is a key factor for improving their reliability. Additional training has a great impact on the probability of successfully accomplishing a task.

3.11 Reducing Risk by Reducing the Hazard Potential

The purpose is to limit the power possessed by hazards which limits their potential to cause damage. Thus, preventing the formation of large build-ups of snow reduces both the likelihood of an avalanche and its destructive power should it occur.

Processing small volumes of toxic substances at a time, as opposed to a single large volume, reduces the risk of poisoning in the case of accidental spillage.

Instead of investing in safety devices and passive barriers, often, it is much more cost efficient to passivate hazardous wastes or spilled hazardous substances. This eliminates or reduces significantly their hazard potential and with it, the associated risk. There are various methods by which this could be achieved:

- Treatment with chemicals which reduce the chemical activity and toxicity of the hazardous substances.
- Reducing the inherent tendency to ignite or burn (e.g. chemicals which cover spilled fuel and prevent it from catching fire).
- Reducing the capability to evaporate.
- Reducing the possibility of auto-ignition (e.g. by avoiding piles of flammable materials).
- Changing the aggregate state. Solidifying liquid toxic waste, for example, reduces significantly its potential to penetrate through the soil and contaminate underground water.
- Reducing the hazard potential by dilution.

3.12 Reducing Risk by using Protective Barriers

An example of a general risk-reduction approach can be given with the concept 'barrier' (Svenson 1991; Leveson 2011; Hollangel 2016). Barriers have been used as accident prevention tools and protection measures mitigating the consequences from an accident. A classification of barriers has been proposed in the literature (Eder and Hosnedl 2008).

Implementing appropriate protective barriers can prevent damage from escalating. Protective barriers control an accident by limiting its extent and duration. They can also arrest the evolution of the accident so that subsequent events in the chain never occur. Protective barriers can also prevent particular event sequences and processes which extend damage, by blocking the pathways through which damage propagates. The defence against a release of toxic substance, for example, combines:

- *Passive physical barriers* (machine guards, fences, protection equipment, clothing, gloves, respiratory masks).
- *Active physical barriers* (ventilation triggered by a detector).
- *Immaterial barriers* (handling rules minimising the released quantity in the case of an accident, e.g. handling a small amount of toxic material at a time).
- *Human actions barriers* and *organisational barriers* (e.g. evacuation).
- *Recovery barriers* (first aid, medical treatment).

A common way of preventing damage from escalating is by using *damage arrestors*. A typical example of a damage arrestor is the protection against buckling of a pipeline subjected to a high external hydrostatic pressure, discussed in Chapter 7.

The damage escalation can be avoided by avoiding concentration of hazards in close proximity. An example of this principle is avoiding building large containers for fuel storage in close proximity and leaving safe exclusion zones separating the containers. This measure makes the storage containers invulnerable to domino-type failures and damage escalation. It prevents an accidental explosion of a storage container initiating other explosions.

An efficient method of limiting the consequences from an accident or failure is *blocking the pathways through which the damage escalates*. This is done by studying the pathways through which the damage propagates and, where possible, automatically sealing them off in the case of an accident. A good example is the urgent quarantine measure of tracking and isolating infected individuals or animals to prevent the spread of infectious disease. *Protection systems*, designed to limit the consequences by blocking automatically the pathways through which the consequences propagate, are also used. Examples are the shut-down systems and fail-safe devices which automatically close key valves in the case of a critical failure, thereby isolating toxic or flammable production fluids and reducing the consequences from failure. Various stop buttons halting the production cycle in the case of failure along a production line are also part of protection systems. Other examples of protection devices are the cut-off switches or fuses which disconnect a circuit if the current exceeds a maximum acceptable value.

An example of barriers *delaying the rate of deterioration in the case of accident* are the fireproof coatings of steel supporting structures, limiting the consequences should fire break out. Without the fireproof protection, in the case of fire, the steel quickly loses its strength and causes the entire structure to yield and collapse.

3.13 Reducing Risk by Efficient Troubleshooting Procedures and Systems

Computer-based expert troubleshooting systems are a powerful tool for reducing the downtimes in the case of failure. Expert systems capture and distribute human expertise related to solving common problems and the correct operating procedures in particular situations. Compared with people, these systems retain all the time knowledge about a vast number of situations, problems and operating procedures. Furthermore, the troubleshooting prescriptions are objective and not coloured by emotions. Troubleshooting systems can help in training staff to handle various problems or accidents. They also help counteract the constant loss of expertise as specialists leave or retire (Sutton 1992).

3.14 Risk Planning and Training

The purpose of risk planning is to specify the most appropriate response should a failure scenario occur. Risk planning guarantees that the optimal course of action will be taken for dealing with the consequences from failure. Usually, in the absence of planning, the quickest and the most obvious actions are taken which are rarely the optimal ones.

Risk planning prepares for the unexpected thereby increasing the speed of recovery and the resilience of systems and infrastructures. It yields contingency plans for the course of action in the case of failure or accident. Planning guarantees proactive rather than reactive attitude to risk. It is closely linked with the research preparation involving a careful study of the system or process, identifying possible sources of risk and training for providing an emergency response. The time invested in risk planning and training pays off because the response time, the chances of taking the wrong course of action, and the chances of providing inadequate response to the materialised risks are reduced significantly. Risk planning and training help to avoid panic and hasty actions which could otherwise promote errors aggravating the consequences from failure. This increases the capacity of the system to better absorb the impact from adverse events, thereby increasing its resilience.

Planning also provides an answer to the important question of how much resource to allocate now given the possibility of failure scenarios in the future, in order to remove the maximum amount of risk within the specified budget (Todinov and Weli, 2013). In this sense, quantifying the risks associated with the different scenarios and deciding upon a mitigating strategy is at the heart of risk planning.

Good accident response management, based on well-established rules and training, is a major factor mitigating the consequences from materialised risks. Evacuation procedures, fast rescue operations, fast response to an extreme emergency, and fast response to crime, reduce significantly the consequences from accidents. Various types of emergency training help reduce casualties and losses should accidents occur. Adequate first-aid training, security training, and crime combat training are important factors mitigating the risks.

4

Improving Reliability and Reducing Risk by Separation

4.1 The Method of Separation

This chapter introduces an important domain-independent reliability improvement and risk reduction method referred to as 'the method of separation'.

Harmful interaction of factors critical to reliability and risk is a major source of failures. Separating risk-critical factors to reduce this harmful interaction is therefore a major avenue for improving reliability and reducing risk. Surprisingly, the method of separation has not yet been discussed as a risk-reduction tool. Although a number of risk-reduction techniques used in engineering are clearly instances of the method of separation, they have never been linked with this method.

The *method of division of tasks* featured in Pahl et al. (2007) is effectively an application of the method of separation despite the fact that it has not been linked with this method.

Another example can be given with the concept 'barrier' (Svenson 1991; Eder and Hosnedl 2008; Leveson 2011; Hollangel 2016). Even though barriers are also instances of separation, no link has ever been made with the method of separation. Barriers distancing triggers from hazards reduce the likelihood of an accident, while barriers distancing hazards from targets reduce the consequences given that an accident has occurred.

Separation has been applied in the TRIZ methodology for inventive problem solving (Altshuller 1984, 1996, 1999) for resolving physical contradictions in engineering of the type: 'the object must have attribute *A* during one mode of use or during one stage of a particular process and the opposite attribute (not *A*) during another mode of use or another process stage'. However, the separation principle in TRIZ is not oriented towards reliability improvement and risk reduction. The separation on time, space and condition in TRIZ is oriented towards resolving physical contradictions appearing in inventive problem solving. No treatment is provided in the TRIZ methodology related to the mechanisms through which the separation works in increasing reliability and reducing risk, which is central to the systematic application of this method for improving the reliability of engineering designs. No treatment of the mechanisms through which the method of separation could increase reliability has also been presented in more recent literature related to TRIZ (Terninko et al. 1998; Savransky 2000; Orloff 2006, 2012; Rantanen and Domb 2008; Gadd 2011).

TRIZ never considered *stochastic separation*, for which the separation is guaranteed only with certain probability. TRIZ also never considered logical separation, where no time, space, or separation on a condition is present yet the dangerous interaction of critical actions is prevented.

Methods for Reliability Improvement and Risk Reduction, First Edition. Michael Todinov.

Here, it needs to be pointed out that the method of separation *is not equivalent* to the traditional method of barriers separating hazards from targets. While introducing barriers between hazards and targets is a special case of the method of separation and has been widely used before, *a number of mechanisms and techniques of the method of separation are fundamentally different and virtually unknown to design engineers and reliability practitioners*. Here are some examples:

- The method of 'stochastic separation' which is a special case of the method of separation.
- *Separation based on the cost of failure* and its application aspects is still not well understood considering the large number of high-consequence failures, for which the reliability built-in a critical component did not correspond to its high cost of failure.
- Separation to counter poor performance caused by homogeneity is rarely used in design.

A substantial gap in the existing reliability and risk literature is that no coverage of the mechanisms through which the method of separation achieves reliability improvement and risk reduction has ever been presented. The lack of knowledge of the mechanisms through which the method of separation works does not permit its systematic implementation for improving reliability and reducing risk.

4.2 Separation of Risk-Critical Factors

Full separation of risk-critical factors is important for scheduling critical random events for which any overlapping has grave consequences and, because of this, a simultaneous presence of events must be excluded completely.

Familiar examples of a full separation of risk-critical factors are: (i) the time separation provided by traffic lights, preventing collision between intersecting flows of traffic and flows of pedestrians and (ii) the space separation provided by isolating intersecting flows of traffic and pedestrians at different levels in order to eliminate the risk of accidents.

4.2.1 Time Separation by Scheduling

Full separation in time is required from risk-critical factors or events whose simultaneous presence must be excluded completely.

The time separation by scheduling enforces consistent time spacing between hazardous events. Full time separation is used in air traffic control, where it enforces consistent time spacing between arriving aircraft, on the basis of real-time information about the weather, headwinds, altitude, and speed. In this respect, the air-traffic control clearance is an important instrument providing the necessary separation in order to avoid collisions within controlled airspace. The separation provided by the air-traffic control clearance also guarantees a sufficient runway approach capacity, which keeps the risk of accidents low.

Suppose that n random events, each with duration d, need to be scheduled over the time interval 0, L such that the possibility of overlapping of random events is excluded completely and, at the same time, the start times s_i of the events are truly random along the time interval (0, L).

4.2.1.1 Case Study: Full Time Separation with Random Starts of the Events

Consider the configuration for which the random start times $s'_1, s'_2, \ldots, s'_n$ are generated along the smaller length $L-(n-1)d$, by a random generator of uniformly distributed

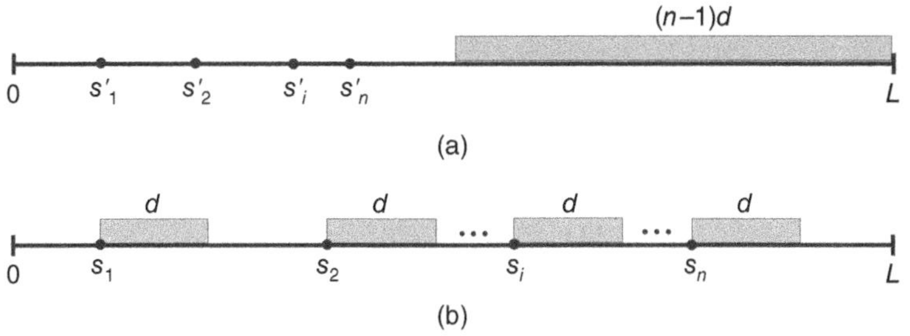

Figure 4.1 Full separation with random start times of the events.

random numbers (Figure 4.1a). From this configuration, the configuration in Figure 4.1b can be obtained by placing after each randomly generated s'_i time, a duration interval with length d (Figure 4.1b). The start times in the obtained X-configuration in Figure 4.1b are $s_1 = s'_1, s_2 = s'_2 + d, s_3 = s'_3 + 2d, \ldots, s_n = s'_n + (n-1)d$.

As a result, the random events are scheduled over the time interval 0, L in such a way that no overlapping of random events is present and, at the same time, the start times s_i of the events are truly random along the time interval (0, L).

4.2.2 Time and Space Separation by Using Interlocks

Preventing the simultaneous occurrence of two events can be done by implementing a sentinel whose state is returned as 'busy' if an event is still active. The state of the sentinel is checked, and a new event can be launched only if the current state of the sentinel is 'free'. Upon termination of the event, the state of the sentinel is changed to 'free'. The sentinel can also be built in an interlock which prevents any overlapping of events.

4.2.2.1 Case Study: A Time Separation by Using an Interlock

A realisation of this mechanism for full time separation has been given in Figure 4.2 with the three buttons $B1$, $B2$, and $B3$ which activate corresponding electromotors $e1$, $e2$, and $e3$, each of which is responsible for a motion in a particular direction. Once a button is pushed, the circuit of the corresponding electromotor must stay latched until the stop button S is pressed, and the circuit of the electromotor is opened. While the circuit of any particular electromotor remains energised, pushing any other button must not have any effect.

Consider the circuit in Figure 4.2a. Pushing button $B1$, closes the normally open contact $k1$ and the circuit of electromotor $e1$ is latched into an energised state. At the same time, the normally closed sentinel contact F is opened, which prevents energising the circuit of any other electromotor ($e2$ or $e3$) (Figure 4.2b). Electromotor $e1$ will be running until the stop button S is pressed. Pressing the stop button S de-energises the circuit of electromotor $e1$ and returns the open contact F into its normally closed state (Figure 4.2c). This is effectively a process of moving the sentinel into a 'free' state. Only now will the circuit react to pressing any other button. Pressing button $B2$, for example, closes the normally open contact $k2$ and the circuit of electromotor $e2$ is latched into an energised state (Figure 4.2d). At the same time, the normally closed sentinel contact F is opened, and the process is repeated.

The dangerous simultaneous energising of two electromotors has been excluded by the time interlock implemented through a sentinel.

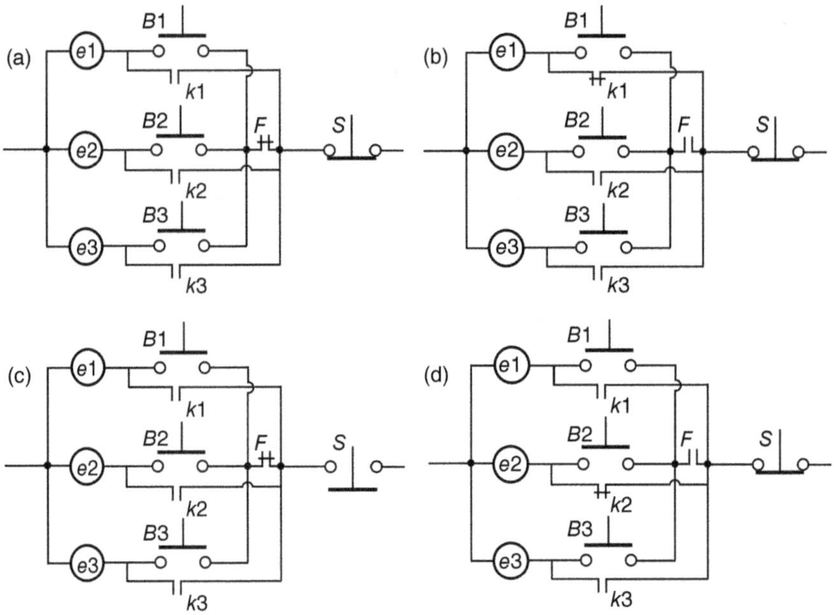

Figure 4.2 Full time separation by using an interlock.

A full space separation can also be based on physical interlocks. A mechanism by which space separation can be achieved is through opening a power circuit or control circuit.

A common implementation of this separation mechanism is the interruption of the power supply by removing the protective shield of rotating machinery. As a result, it is impossible to operate the machine without positioning first the protective shield in place, which reduces the likelihood of injuries.

A more sophisticated example of a space separation interlock is the presence-sensing safeguard interlocks which stop the operation of hazard equipment if a person is detected in a location where injury can occur. The presence-sensing system could be based on laser beams, light, or infra-red beams. Beams of light forming a curtain are generated and if any of the beams is blocked by a person moving towards the hazard equipment, a control circuit is opened and the power to the hazard equipment is switched off.

Another mechanism of space separation is the trapped-key interlock. In one of the possible implementations, the access to repair hazard equipment is through a door operated by a key which is held trapped on the door until the door is firmly closed again by operating the key. While opening the door, the key also operates a switch interrupting the power supply. The hazard equipment cannot be re-energised until the door is closed, and the key released.

4.2.3 Time Separation in Distributed Systems by Using Logical Clocks

The time separation in a distributed computer system with time delays can be a problem because of the delay associated with the transmission of messages. Suppose that a process *A* issues first a request for a service from device *D* through process *B* and later the process *A* issues another request for a service from the same device *D* through another process *C*.

The two requests need to be fulfilled in the order in which they have been issued by process *A*. Because of time delays in the distributed system, the request for the service through process *C* may arrive at device *D* earlier than the request through process *B*. To resolve granting the resource in the order the requests have been issued, the Lamport algorithm based on logical clocks can be used (Lamport 1978). Before sending a message, process *A* increments a logical clock (counter) time_A, whose initial value is zero to time_A = time_A + 1 and sends the message together with this time stamp to process *B*. Next, process *A* increments its logical clock (counter) time_A (time_A = time_A + 1) and sends another message with this timestamp to process *C*, this time with the time stamp time_A = 2.

Upon receiving the message, process *B* compares the value of its logical clock (time_B) which is zero and the time stamp of the received message, selects the larger time stamp and increments by one its counter. As a result, the logical clock of process *B* is updated to 2.

Upon receiving the message, process *C* compares the value of its logical clock (time_C) which is zero and the time stamp of the received message (which is 2), selects the larger time stamp and increments by one its counter. As a result, the logical clock of process *C* is updated to 3.

Consequently, upon receiving messages from processes *C* and *B*, device *D* can easily determine which event occurred earlier.

4.2.4 Space Separation of Information

This separation mechanism reduces the risk of overloading with too much information, which is a commons source of costly human errors. To mitigate the risk of human errors induced by information complexity and overload, information present in machine–human interfaces is separated into layers and only the layer of immediate relevance to the task at hand is displayed.

A common error in design of machine–human interfaces is to present to the user all available options which leads to cognitive overload and human errors. With the use of the mechanism of information separation, the interface is designed in such a way that users who do not need particular options never see them.

4.2.5 Separation of Duties to Reduce the Risk of Compromised Safety, Errors, and Fraud

The essence of this separation mechanism is the division of a particular key task, subject to abuse into sub-tasks assigned to different individuals.

Separation of duties among different individuals in an organisation enforces accountability, eliminating conflicts of interest and errors. The result is a significantly reduced risk. Not following properly, the separation of duties often entails serious consequences.

Thus, a unit engaged in dangerous operations should not be solely responsible for the safety control associated with the operations. The significant efforts and costs invested in improving safety often result in confirmation bias and such units tend to exaggerate the safety of their operations and play down the risks. Separating the safety control functions by introducing an independent unit which monitors and reports directly to the top management on the safety of operations performed by the operational unit increases significantly the level of safety.

In another example, a person with a duty of providing a security system cannot design the security system and also test it. The separation of duties in this case, avoids the conflict of interest where a person has to report on the weaknesses of their own creation. This principle is also followed widely in the delivery of software which combines designing the software and testing it. To avoid costly bugs, the task of designing the software must be separated from the task of testing the software.

Separation of duties in handling financial transactions (e.g. authorising the transaction, receiving the funds, depositing the funds, recording the transaction, reconciling the bank statement) reduces the possibility of fraud. No single individual has the ability to both perpetrate and conceal fraud because, due to the separation of duties, committing fraud would require collusion between several people.

Sales transactions include conducting a sale and managing the customer delivery. Separating the duty related to conducting the sale from the duty of managing the customer delivery, prevents the risk of falsifying sales records to non-existent customers and stealing funds from the company.

4.2.6 Logical Separation by Using a Shared Unique Key

A logical separation is present when it is *logically impossible* for a dangerous operation to occur at a given point in time or at a given space location. A logical separation is also in place if it is logically impossible for two or more objects to be in a dangerous proximity at a given location or at a given time.

In logical separation, no barriers of any kind are set between the different parts of the system, yet separation is still present. The dangerous contact between hazards and triggers and hazards and targets is made to be logically impossible.

Consider the safety problem related to preventing the hand of an operator from being in the cutting area of a guillotine. If the cutting action is activated only by a simultaneous pressure on two separate knobs/handles which engage both hands of the operator, it is logically impossible for the operator's hand to accidentally reside in the cutting area, at any time. The operator's hands have been separated from the cutting area through the logic of guillotine activation. This is an example of a full logical separation reducing the risk of an accident.

Logical separation avoiding dangerous simultaneous occurrence of processes can be implemented relatively easily by using the mechanism of the *shared unique key*. The same unique key is required for activating each process in accomplishing a particular task. As a result, a dangerous overlapping of processes cannot occur because the unique key cannot be simultaneously available to activate more than one process.

4.2.6.1 Case Study: Logical Separation of X-ray Equipment by a Shared Unique Key

The realisation of this idea can be seen in Figure 4.3. Suppose that a particular task requires an adjustment of a specimen before switching on X-ray beam. A dangerous simultaneous occurrence of actions is possible if a person appears in Room *A* and switches on the X-ray beam while the operator is still adjusting the specimen under the X-ray head in room *B* (Figure 4.3).

The dangerous overlapping of actions can be prevented from occurring by a logical separation implementing a shared unique key. The design of the X-ray equipment must be such that the release of the X-ray shield in order to position the specimen and the switching of the X-ray beam in room *A*, can only be done by using the same shared unique key. Switching on the X-ray beam and adjusting the specimen would then require the same object (the

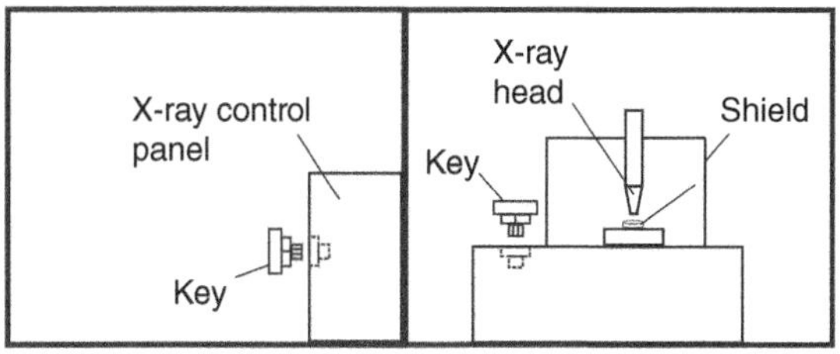

Figure 4.3 An example of logical separation by a shared unique key.

shared unique key) to be in two different places, at the same time, which is impossible. The safety risk has been eliminated by a logical separation based on a shared unique key.

The logical separation leads to low-cost yet very efficient designs eliminating safety risks. It is a simple yet underutilised generic tool for improving safety and reducing risk.

4.2.7 Separation by Providing Conditions for Independent Operation

The increased complexity of modern technology causes increased coupling and interdependence among components and building blocks. The increased coupling and interdependence causes failure of a component to affect negatively other components and cause their failure too. If failure of one component makes the failure of another component more likely, the reliability of the system can be improved by making the two components operate independently from one another. Providing conditions for independent operation of components is an instance of a full separation.

Consider the operation of two devices where the second device is powered from the first device. Failure of the first device will cause a loss of power for the second device. The reliability of the system can be improved if powering of the two devices is separated.

Often, to cut the cost, devices share control/communication channels. Consider a dual control system based on two control modules CM1 and CM2 controlling an electromechanical device M. To cut the cost, the two control modules share the same control channel (Figure 4.4).

The two control channels in Figure 4.4a are not independent because failure of the cable connecting the electromechanical device will cause both control channels to fail which entails a loss of control over the electromechanical device M. Separating (decoupling) the control channels (Figure 4.4b) ensures independent operation of the control channels and improves the reliability of the system. Failure of any of the connecting cables will not result in a loss of control over the electromechanical device M.

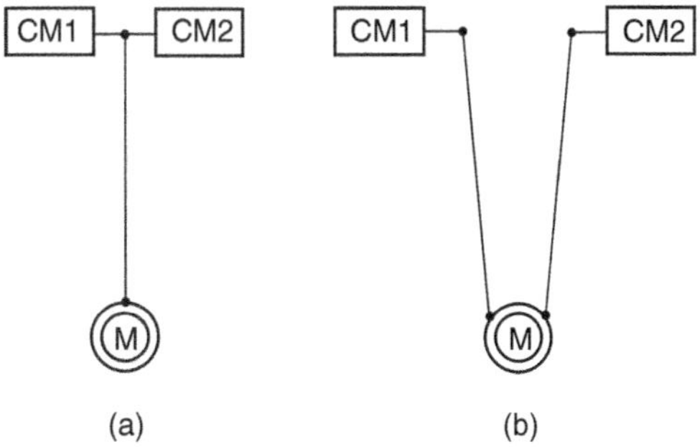

Figure 4.4 Improving the reliability of a control system by ensuring independent operation.

Decoupling of circuits is usually done by a decoupling capacitor and is often used in electronics when a portion of a circuit is prevented from being affected by fluctuations of the power supply due to switching occurring in another portion of the circuit.

Spreading vibrations through mechanical and electrical components is an important common cause which simultaneously affects many components and induces various failure modes. Decoupling (separating) from vibrations has always been an important aspect of improving the reliability of mechanical and electrical components.

Decoupling is an important mechanism for destroying positive feedback loops which cause a fast deterioration of many interconnected networks (see Chapter 9). Decoupling destroys the self-reinforcing loops between networks and helps bring the systems into a stable condition. Since the interdependency between systems is increasing, the importance of decoupling to eliminate positive feedback loops with negative impact is also increasing.

Separation to block a common cause is present when a component or a group of components are distanced (insulated) from the action of a common cause. A common cause reduces the reliability of a number of components simultaneously. The affected components are then more likely to fail, which reduces the overall system reliability. A detailed discussion of separation mechanisms for blocking a common cause has been presented in (Todinov 2015).

4.3 Separation of Functions, Properties, or Behaviour

This is an important type of separation that can be implemented through the following basic mechanisms and techniques.

4.3.1 Separation of Functions

Separation of functions consists of assigning different functions to different parts of a component or system. Separating critical functions among different parts/components improves reliability and reduces risk through several distinct mechanisms.

4.3.1.1 Separation of Functions to Optimise for Maximum Reliability

The reliability of a single component is difficult to optimise for each function. Consider a seal that carries two basic functions: (i) sealing – isolating particular fluids; and (ii) load-carrying – resisting tensile loads or bending moments. In designing a joint, the load-carrying function is often assigned to the part carrying the sealing function. Despite the simplified design, the reliability of such joints is often insufficient. This is a common design error that has caused a number of high-impact failures. Achieving high reliability requires separating the sealing function and the load-carrying function to distinct parts. Each of the distinct parts can then be optimised for the respective function and the result is a highly reliable joint assembly.

The separation of functions will be illustrated with the design of a pipe or container containing corrosive fluid (Figure 4.5a). It is difficult to optimise the material of the pipe/container so that high levels of strength and corrosion resistance are attained simultaneously. Selecting a material combining high levels of strength and high corrosion resistance is very expensive.

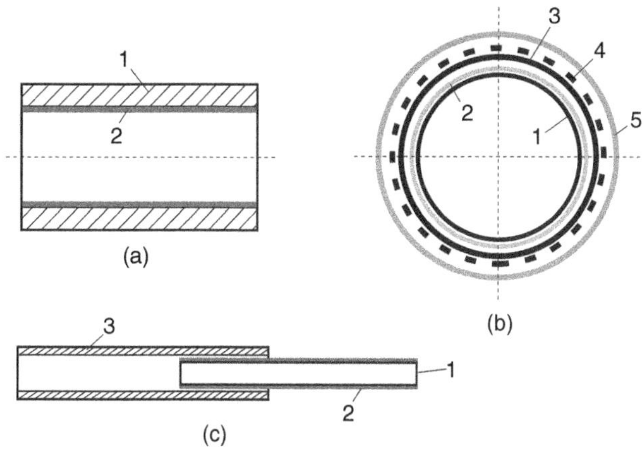

Figure 4.5 Separation of functions in: (a) a pipe carrying corrosive fluid; (b) a flexible pipe transporting hydrocarbons under water; and (c) a component experiencing wearout from friction.

Low-carbon steels, for example, have high strength and low cost but their corrosion resistance in contact with corrosive fluids is low. Conversely, plastics have a high corrosion resistance in contact with corrosive fluids but possess low strength. What is difficult to achieve by a single component, can be achieved by separating the functions strength and corrosion resistance into two distinct components: a steel component (1), providing the required strength and a corrosion-resistant plastic lining (2), providing the corrosion resistance (Figure 4.5a). Plastic and resin coatings on wooden panels provide effective separation from the harmful influence of rain water and are commonly used to insulate and prevent wooden structures from water damage. Plastic sheets can also be used to provide effective separation of sink cabinets from water damage, etc.

By using separation, the reliability is improved at a reduced cost. This is a big advantage over conventional methods for reliability improvement (e.g. upgrading components) invariably associated with additional investment.

The illustration of this technique has been extended with the flexible pipe carrying hydrocarbons under water (Figure 4.5b).

The pipe is composed of a stainless steel *internal carcass* (1); an *internal sheath* (2) which is extruded polymer barrier; *a pressure armour* (3) – a carbon-steel interlocked circumferential layer; and *a tensile armour* (4) – helically wound carbon-steel layers for axial strength. Externally, the pipe is protected by *extruded sheath* (5). The internal carcass (1) prevents collapse of the internal sheath (2) due to the hydrostatic pressure of the water and also ensures mechanical protection. The internal sheath (2) ensures the integrity of the transported fluid while the function of the pressure armour and the tensile armour is to provide resistance against radial and tensile loads. The external sheath (5) is a mechanical barrier shielding the pipe's internal structural elements from the marine environment.

As a result, the different parts of the flexible pipe carry different functions: to protect against external corrosion, to resist tensile loads, to resist radial loads resulting from internal pressure, to make the pipe leak-proof, and to prevent collapse due to external pressure. It is difficult to optimise a homogeneous pipe with respect to each of these functions. The separation of functions to different parts permits the optimisation of each part with respect to the single function it carries. The result is increased overall reliability of the pipe.

Finally, the illustration of this technique has been extended with component 1 moving into another component 3 and experiencing friction (Figure 4.5c). The intensity of the wearout caused by friction, can be reduced significantly by a separation of friction properties. This can be done by applying coating (2) made by a chemical vapour deposition (CVD) or by other means. As a result, the high coefficient of friction between the surface of component 1 and the inside of component 3, is replaced by the low coefficient of friction coefficient between the coating (2) and the inside of component 3. The result is a reduced likelihood of jamming and reduced rate of wear.

Separation of functions can be readily applied to the design of electrical connectors. An electrical connector must combine a low electrical resistance and a minimum contact pressure which maintains the electrical contact. If these two functions are assigned to a single component made of copper alloy, the contact will deteriorate substantially with time because of the low fatigue resistance, resistance to stress relaxation and resistance to vibrations. What is difficult to achieve by a single material can be achieved if the functions of maintaining the electrical contact and conducting electricity are separated. The copper alloy elements are used only to provide the electrical contact while the contact pressure is maintained by using elastic elements (springs) or by using kinematic coupling. Separating functions eliminates important failure modes of electrical connectors: loss of contact due to wear, fatigue failure, loss of contact due to stress relaxation and loss of contact due to vibrations.

4.3.1.2 Separation of Functions to Reduce Load Magnitudes

This separation mechanism works by separating the load into several components instead of load being carried by a single component. The strength of a component performing several functions can be easily exceeded if combined multiple loads are present. The increased resultant load increases the rate of degradation of the component. By separating the load into several components, the load on any single component is reduced.

For example, a single bearing carrying both high-magnitude radial force F_r and high-magnitude axial force F_a is often overloaded (Figure 4.6a). Because of the high-magnitude resultant load F (Figure 4.6b), the rate of degradation of the bearing is elevated and its reliability is reduced.

By separating the functions of carrying radial and axial loads into two types of bearings '1' and '2', the load is effectively split between roller bearings (1), resisting only radial loads and no axial loads, and an axial ball bearing (2), resisting only axial loads and no radial load. The load on each of the bearings is reduced, the degradation rate is reduced, and reliability is increased.

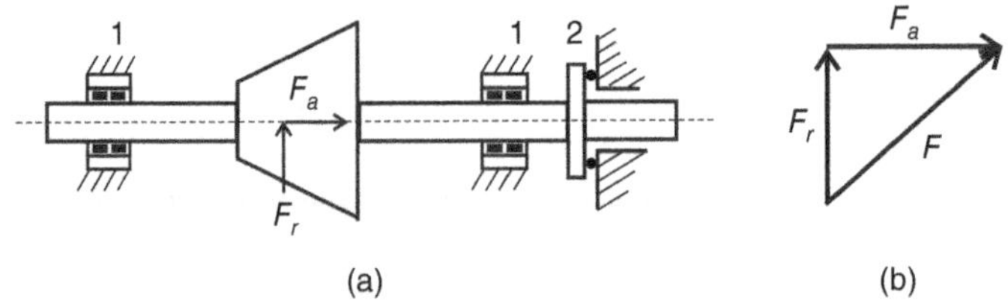

Figure 4.6 Separating the functions carried out by the bearings (carrying only radial and only axial load) reduces the load magnitudes and improves reliability.

4.3.1.3 Separation of a Single Function into Multiple Components to Reduce Vulnerability to a Single Failure

This separation mechanism is particularly relevant to the same function assigned to several identical function carriers. Separation of the function decreases the vulnerability to a single failure. If the function is assigned to a single component, failure of the component will cause a loss of the function. If the function is assigned to several identical components such that each component performs the same function, a failure of the component will not entail a loss of the function.

The traditional method of improving reliability by implementing redundancy *is effectively a special case of the method of separation where the same function is carried out by identical components working in parallel.*

Separation of a function carried out by a single component, to be carried by multiple smaller parts into which the component has been divided, is a process of *segmentation*, discussed in Chapter 7. Segmentation increases the tolerance of components to flaws causing local damage, reduces the rate of damage accumulation and damage escalation and reduces the hazard potential. Segmentation essentially replaces a sudden failure on a macro-level with gradual deterioration of the system on a micro-level through non-critical failures. As a result, the method of segmentation can be thought of as a special case of the method of separation.

4.3.1.4 Separation of Functions to Compensate Deficiencies

The separation of functions can be used to compensate deficiencies associated with the different components building a system. A typical example is the hybrid joint, combining an adhesive joint (1) and mechanical fixing (2) (Figure 4.7). There is a clear separation of functions: the adhesive part reduces the stress concentration along the joint while the mechanical fixing increases the peel resistance of the adhesive joint and its stiffness.

Such a separation is often present in the design of complex alloys where some of the microstructural constituents provide wear resistance, while other constituents provide toughness (resistance to crack propagation).

4.3.1.5 Separation of Functions to Prevent Unwanted Interactions

This separation mechanism (also known as 'separation of concerns') is well-known in the design of computer programs (Reade 1989). A concern is a relatively simple, self-contained task, addressed by a program section. Separation of functions in programming is achieved by encapsulating data and statements inside a section of code that has a well-defined interface. This results into a modular program, consisting of procedures and functions. The encapsulation means that the variables defined into the encapsulated module (procedure or a function) remain only visible within the module and can be altered only within the module. Encapsulation avoids unwanted interactions between different pieces of code in

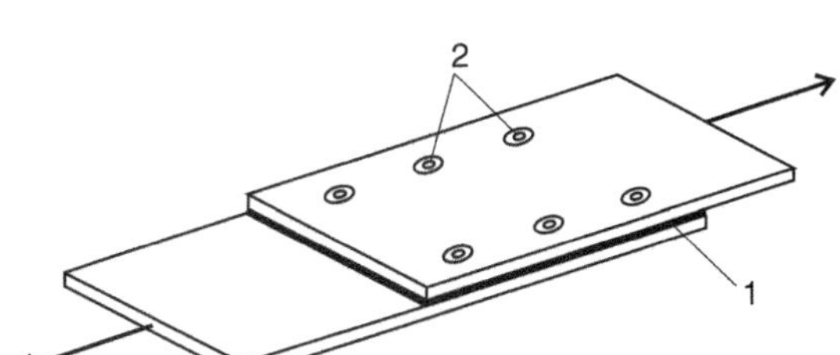

Figure 4.7 Separation of functions to compensate mutual deficiencies.

the same program. Avoiding unwanted interactions reduces the possibility of side effects and difficult to rectify bugs, if a variable from one particular section of code is altered from another section of code. Furthermore, the encapsulated sections of code can be updated and tested independently, without having to alter code in the rest of the sections, which decreases significantly the possibility of introducing bugs. The encapsulated piece of code is essentially a black box with specified input and output, whose content can be independently developed and replaced without affecting the logic of the program.

4.3.1.6 Separation of Methods to Reduce the Risk Associated with Incorrect Mathematical Models

Separation is a powerful method that can be applied to enhance the reliability of a new methodology/model. Suppose that a particular value has been estimated from a complex theoretical model. The reliability of the theoretical model can be significantly increased, if the same value is also estimated by another, fundamentally different method. If the calculated value from the two separate methods agree, the reliability of the theoretical model is enhanced significantly.

An application example can be given with the equations regarding the probability of no overlapping of a fixed number of random events considered in Chapter 6. These equations have been derived by using a chain of probabilistic arguments. The probability of no overlapping has also been evaluated by a Monte Carlo simulation. Both methods yield the same result which significantly enhances the reliability of the theoretical model.

Hand calculations on a simple loaded component can be used to check the correctness of the boundary conditions in a software package performing finite element analysis. A computer algebra package can be used for checking the validity of algebraic manipulations, the results from integration, differentiation, finding maxima and minima of complex functions, etc.

4.4 Separation of Properties to Counter Poor Performance Caused by Inhomogeneity

4.4.1 Separation of Strength Across Components and Zones According to the Intensity of the Stresses from Loading

Separation countering the drawbacks of homogeneity is necessary in cases where the average property characterising a homogeneous state cannot provide the required reliability in terms of effective resistance against combined hazards or damage accumulation factors. This mechanism is needed in engineering design, for assuring that the different parts of a component have the appropriate properties needed to successfully resist the local loading. A homogeneous material with average value of the strength (resistance) is not optimised according to the local type of loading and cannot provide sufficient resistance in all parts of the component. If the type of loading at a particular location in the component is a cyclic loading with large amplitude, the required response from the material is high fatigue resistance. If the type of loading at a particular location is static and no corrosion is present, there is no need for a high fatigue resistance because the conditions for the development and propagation of fatigue cracks are missing. Since the loading is not homogeneous, the resistance cannot be homogeneous.

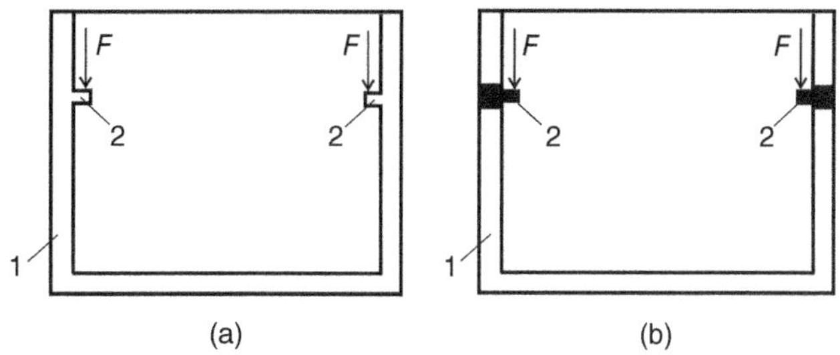

Figure 4.8 Improving the reliability of a container with a rotating lid.

The mechanism for implementing this type of separation requires uneven distribution of the resistance against a particular stress for different components or different zones of the same component, depending on the type and intensity of the local stress the component/zone experiences.

The application of this separation mechanism is illustrated with the application in Figure 4.8. The container 1 and the cylindrical hinges 2 in Figure 4.8a are made of the same material, with relatively low fatigue resistance. An asymmetric lid (not shown) is rotating around the hinges 2 and loads the hinges with the pulsating force F. Because of the pulsating force and the small fatigue resistance of the material of the container, cracks appear at the base of the hinges and soon one of the hinges fails. According to the method of separation of strength according to the experienced local stress, the material of the hinges must have a significantly larger fatigue resistance compared with the rest of the container. If the hinges are made of material with high fatigue resistance (inserted as shown in Figure 4.8b), the reliability of the container will be increased significantly.

The example in Figure 4.8 can be extended to a case of overloading from a concentrated force F applied directly through on a component 2 inserted into a low-strength material 1 (e.g. a bolt inserted into a ceiling; Figure 4.9a). If the pulling force F is large, the excited shear stresses $\tau_1 = F/(2\pi rt)$ along the surface of the cylinder (marked by dashed lines) could easily overcome the resistance of the material and cause shear failure. Inserting a plate 3 of material with superior strength provides a separation of properties (Figure 4.9b). Because of the superior strength of plate 3, shear failure through the plate is excluded. If a pulling force F is present, the excited shear stress $\tau_2 = F/(2\pi Rt)$ is along the surface of a cylinder (marked by dashed lines) with much larger radius ($R > r$) and therefore the shear stress $\tau_2 < \tau_1$ is smaller compared with the shear stress characterising the design from Figure 4.9a. The reliability of the connection has been improved by separation of properties, at a low cost.

Separation of strength according to the experienced stress can be illustrated with objects likely to impact other objects. The likelihood that a prismatic object will enter into contact with its corner is significant (Figure 4.10a) and, as a result, the object can be damaged. If a separation of properties is applied, the corners will be made more resistant to damage.

Figure 4.9 Improving the reliability of a connection inserted into a low-strength component.

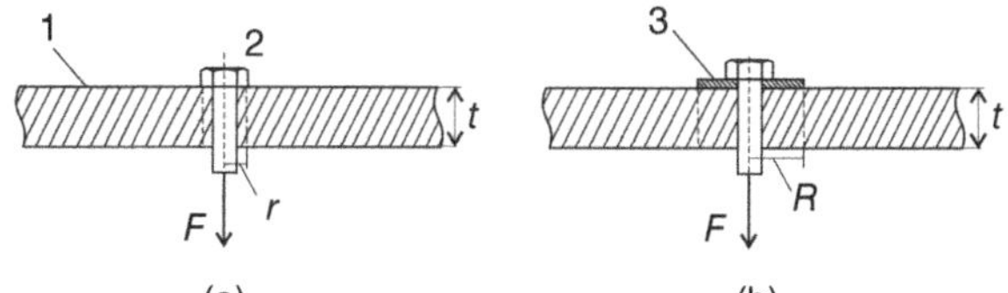

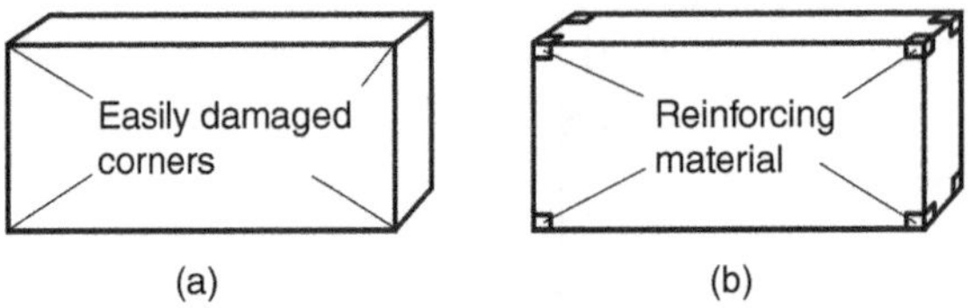

Figure 4.10 Improving the reliability of a prismatic object by separation of strength. The corners have been strengthened by a material resistant to contact damage.

In Figure 4.10b, separation of properties has been applied by reinforcing the corners with material resistant to contact damage.

4.4.2 Separation of Properties to Satisfy Conflicting Requirements

A gear must be hard, to endure large contact stresses and intensive wear and soft, to endure impacts. These conflicting requirements require conflicting material properties: the surface of the gear must be hard while the core of the gear must be soft. In a compromise, with no separation of properties, a homogeneous material will be selected which has satisfactory hardness to resist wear and satisfactory toughness to resist shock loading. The result is an inferior solution which is neither optimised against wear nor against shock loading. These contradicting requirements can be simultaneously guaranteed if a separation of properties is implemented. These conflicting properties can be guaranteed by the separation of properties achieved through *case hardening* (Kalpakjian and Schmid 2013). This consists of local induction heating of the surface layers of the gear teeth followed by quenching. Case hardening improves the resistance of the surface to large contact stresses and wear, while leaving the core tough which makes it resist impact loads and improves resistance to fracture.

The demand of materials which combine conflicting properties is increasing in modern technology. Functionally graded materials, for example, combine such conflicting properties (Mahmood et al. 2012). These are materials where combining conflicting properties has been done by a separation of properties through a deliberately created gradient of chemical composition microstructure and phase distribution. The gradient in the composition and structure results in a corresponding gradient in the properties of the material. As a result, functionally graded materials can withstand without failure very high *thermal gradients*, giving rise to high thermal stresses, which makes them suitable for aerospace applications. Their ability to inhibit crack propagation improves the reliability of components. The functionally graded materials are useful in applications where the traditional homogeneous materials are powerless. *Ceramics-metal* functionally graded materials combine irreconcilable properties such as high *hardness* and *wear resistance* at the surface where the temperature is high and toughness beneath the surface to withstand the large *thermal stresses* (Figure 4.11).

The gradient in their properties makes part of their structure biocompatible while the rest retains significant mechanical strength. This makes them suitable for dental and orthopaedic applications. Furthermore, functionally graded coatings are characterised by better reliability compared with conventional coating – they provide a stronger bond, reduced thermal stresses, and residual stresses.

The *vapour deposition technique*, *powder metallurgy*, the *centrifugal method*, and the *solid freeform fabrication method* are the most important methods for creating thin and bulk functionally graded materials with the required gradient of composition and microstructure (Mahmood et al. 2012).

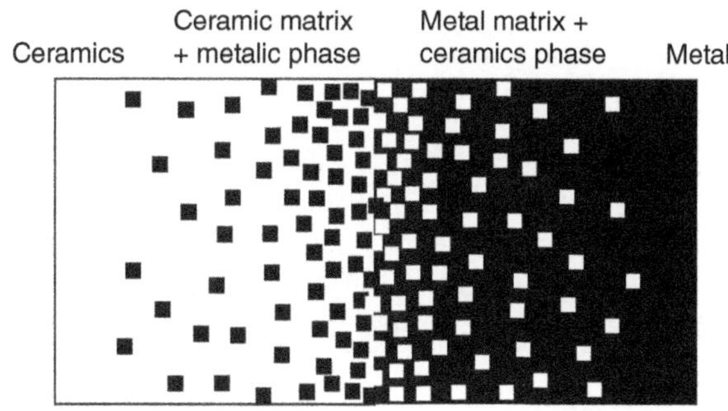

Figure 4.11 Ceramic alloy functionally graded material.

4.4.3 Separation in Geometry

Separation in geometry is present when different parts of an object or assembly have different geometry to provide optimal conditions maximising reliability and minimising the risk of failure. An example of separation in geometry is shown in Figure 4.12.

4.4.3.1 Case Study: Separation in Geometry for a Cantilever Beam

Consider the cantilever beam with length L and uniform rectangular cross section with thickness t and width b (Figure 4.12a) which has been overloaded by a force with magnitude P. Suppose that the tensile strength of the material is σ_s. According to the theory of elasticity, the maximum tensile stress σ_t acting on the beam, is at the cantilever support and is given by

$$\sigma_t = \frac{6PL}{bt^2} \tag{4.1}$$

Consequently, the maximum magnitude of the overloading force which the beam in Figure 4.12a can sustain is

$$P_{\max,a} < \frac{\sigma_s bt^2}{6L} \tag{4.2}$$

Suppose that the beam has been tapered in the way shown in Figure 4.12b, such that the right end of the beam has thickness $0.5t$ and the left end has thickness $1.5t$. If the width b and the length L of the initial beam remain unchanged, the volume of material $V_b = \frac{(0.5t+1.5t)}{2}L \times b = tLb$ used for the beam in Figure 4.12b is equal to the volume of material $V_a = tLb$ used for the beam in Figure 4.12a. However, the maximum overload stress which the beam in Figure 4.12b can sustain is now 2.25 times bigger than the

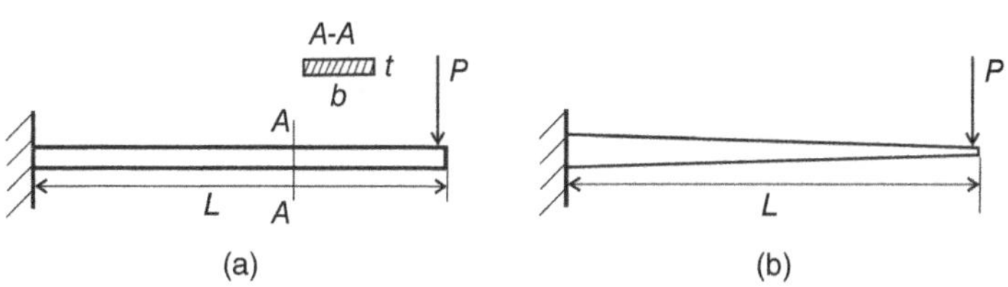

Figure 4.12 Separation in geometry to improve the reliability of a cantilever beam.

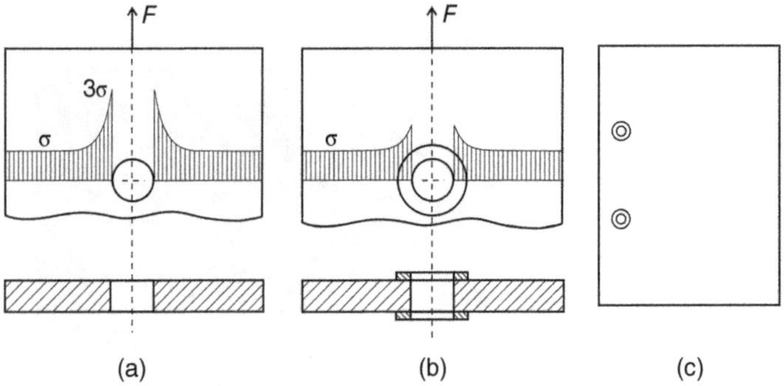

Figure 4.13 Improving reliability by a separation in geometry resulting in reducing the maximum stress in the vicinity of a hole.

maximum overload stress characterising the beam in Figure 4.12a:

$$P_{\max,a} < \frac{\sigma_s b(1.5t)^2}{6L} = 2.25\frac{\sigma_s bt^2}{6L} \tag{4.3}$$

The load-carrying capacity of the beam has been increased and the risk of failure has been reduced by a separation in geometry.

By tapering the beam, its load resistance has been improved for the same volume of material. The beam is made thicker towards the cantilevered end where the stresses are the largest and thinner towards the free end where the stresses are the smallest. By using separation on geometry, the reliability of the cantilever beam has been improved at no extra cost.

With the advancement of technology, the importance of separation in geometry is increasing. By using 3D-printing technique, for example, components with complex shapes can now be produced using less material than traditional methods.

Figure 4.13 demonstrates increasing the resistance to fracture by a separation. Extra material is added around the hole, in the vicinity of which the magnitude of the loading stress is approximately three times larger than the magnitude of the loading stress farther from the edge of the hole (Figure 4.13a). The result is significantly reduced stresses at the edge of the hole and improved fracture resistance (Figure 4.13b). This method has application for reinforcement of punched sheets of paper (Figure 4.13c).

4.5 Separation on a Parameter, Conditions, or Scale

Reducing risk by a separation, assuring distinct behaviour at different values of a risk-critical parameter, is present when different characteristics at different values of a risk-critical parameter are ensured.

4.5.1 Separation at Distinct Values of a Risk-Critical Parameter Through Deliberate Weaknesses and Stress Limiters

Introducing deliberate weak links and stress limiters is an important mechanism for assuring separation from dangerous levels of risk-critical parameters. Because this is a very important method of full separation, the next chapter (Chapter 5) is covering this technique in detail.

4.5.2 Separation by Using Phase Changes

Separation by using phase changes is an important mechanism to provide a separation on 'temperature' which is a common risk-critical parameter. A very reliable separation on operating temperature can be made by using the changes in the aggregate state of the material with temperature. Thus, freezing volatile and flammable substances during transportation eliminates the risk of spillage and explosion during an accident. The low melting point of special alloys can be used for highly reliable triggers for sprinkler systems in the case of fire. Often, the reliability of complex devices based on thermostats, sensors, and electronic circuits, etc., is not sufficient.

Phase changes of the magnetic state of materials can also be used for separation. During induction heating, for example, to prevent overheating of an object beyond a critical temperature T_{cr}, the alloy could be selected to be with Curie temperature equal to T_{cr}. At a temperature T_{cr}, from ferromagnetic (temperatures $T < T_{cr}$), the material becomes diamagnetic (temperatures $T > T_{cr}$).

Because the magnetic properties of the material change at $T = T_{cr}$, induction heating beyond $T = T_{cr}$ is no longer possible, therefore overheating is not possible.

Separation based on evaporation has been used to protect the control equipment in rockets from overheating. Rockets are placed in a foam shell which evaporates after the rocket has been launched.

4.5.3 Separation of Reliability Across Components and Assemblies According to Their Cost of Failure

A very important separation mechanism for reducing risk is that the *allocation of reliability across components should correspond to their cost of failure*. Components whose failure is associated with a large cost of failure should have a proportionally large built-in reliability. This is the underlying principle of the cost-of-failure-based design (Todinov 2007). Components, processes, and operations used in safety-critical applications should have higher reliability compared with analogous components used in non-critical applications.

This is an important domain-independent risk reduction principle and failure to understand it has caused catastrophic oil spills and serious industrial accidents. *The common mistake underlying these accidents was insufficient reliability built-in products, not corresponding to their high cost of failure.*

Failure of the cement used for sealing an oil production subsea well, for example, causes a catastrophic pollution of the environment. Consequently, the cement seal should have a proportionately high built-in reliability. Because of the extremely high cost of failure, the reliability of the cement seal in a production oil well *cannot be similar to the reliability of a cement seal in a non-critical application, not associated with such a high cost of failure.* The reliability of the cement seal in a subsea oil production well must be very high, so that the probability of failure is significantly reduced which leads to a proportional reduction of the potential loss. *Separation of reliability based on the cost of failure is therefore a very important mechanism of reducing risk.*

The argument that the failure of the cement seal is a low-probability high-impact event and, consequently, is something difficult to predict is fundamentally flawed and has caused a lot of damage to many industries. The purpose of the risk management of components whose cost of failure is high is to reduce the probability of failure of such components to an extent proportional to the cost of failure of the component so that the potential loss is acceptable. This is done by implementing appropriate risk control measures followed by an evaluation of the probability of failure in order to prove that the potential loss is low. This is rarely done for components with high cost of failure.

The separation of reliability according to the cost of failure has also a direct application in distributing a fixed budget for improving the reliability of components in order to achieve a maximum overall risk reduction. Distributing the investment uniformly (homogeneously) across all components is, as a rule, an inferior strategy. To achieve a maximum risk reduction, the available risk reduction budget should be preferentially distributed to improve the reliability of components whose failure is associated with the largest consequences and for which a unit investment yields the maximum risk reduction. A special dynamic programming algorithm has been developed for this purpose (Todinov 2014).

4.5.3.1 Case Study: Separation of the Reliability of Components Based on the Cost of Failure

This principle can be illustrated with the simple control system shown in Figure 4.14.

In the control system in Figure 4.14, there are six sources of production (generators, compressors, pumps, etc.) s1–s9, supplying commodity (electricity, gas, fluid). The system consists of two identical control modules CM1 and CM2 and nine identical blocks K1–K9. The amount of commodity supplied per day from each production source is M units.

Suppose that the failure rate of control module CM1 is λ_{CM1} year^{-1} and the failure rate of control module CM2 is λ_{CM2} year^{-1}. Suppose that the two modules have the same downtime for repair of d days. For a period of operation of a years, the lost production due to failure of control module CM1 is $L_{CM1} = 8\lambda_{CM1}\, a\, d\, M$ and the lost production due to failure of control module CM2 is $L_{CM2} = \lambda_{CM2}\, a\, d\, M$. A single failure of control module CM1 causes eight times greater loss of production compared with a single failure of control module CM2. The ratio of the lost production from failures over the operational time interval (0,a) is

$$L_{CM1}/L_{CM2} = 8\lambda_{CM1}/\lambda_{CM2}.$$

To make the losses from failures of the two control modules equal: $L_{CM1} = L_{CM2}$, the failure rate λ_{CM1} of the first control module must be eight times smaller than the failure rate λ_{CM2} of the second control module: $\lambda_{CM1} = (1/8)\lambda_{CM2}$. This means that the reliability level

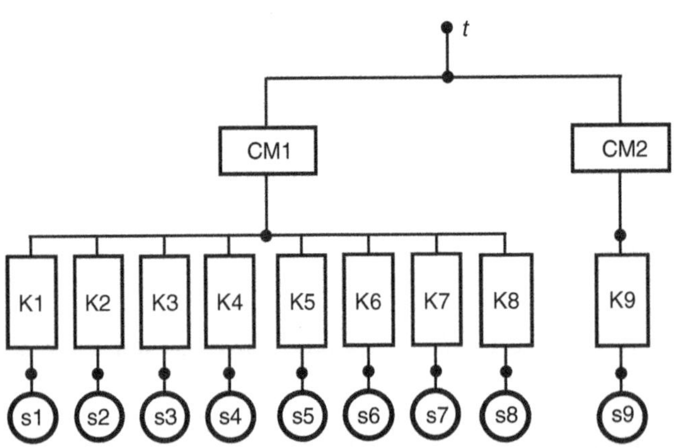

Figure 4.14 A production system with hierarchy based on nine production sources.

of control module CM1 should be significantly higher than the reliability level of control module CM2. The principle of separation based on the cost of failure requires measures that reduce the failure rate of control module CM1 eight times.

In developing safety-critical software, separation according to the cost of failure should be implemented by allocating the amount of testing of a software module to be proportional to the cost of failure of the module.

The classification of the different separation mechanisms and techniques has been summarised in Figure 4.15. Stochastic separation will be discussed in Chapter 6.

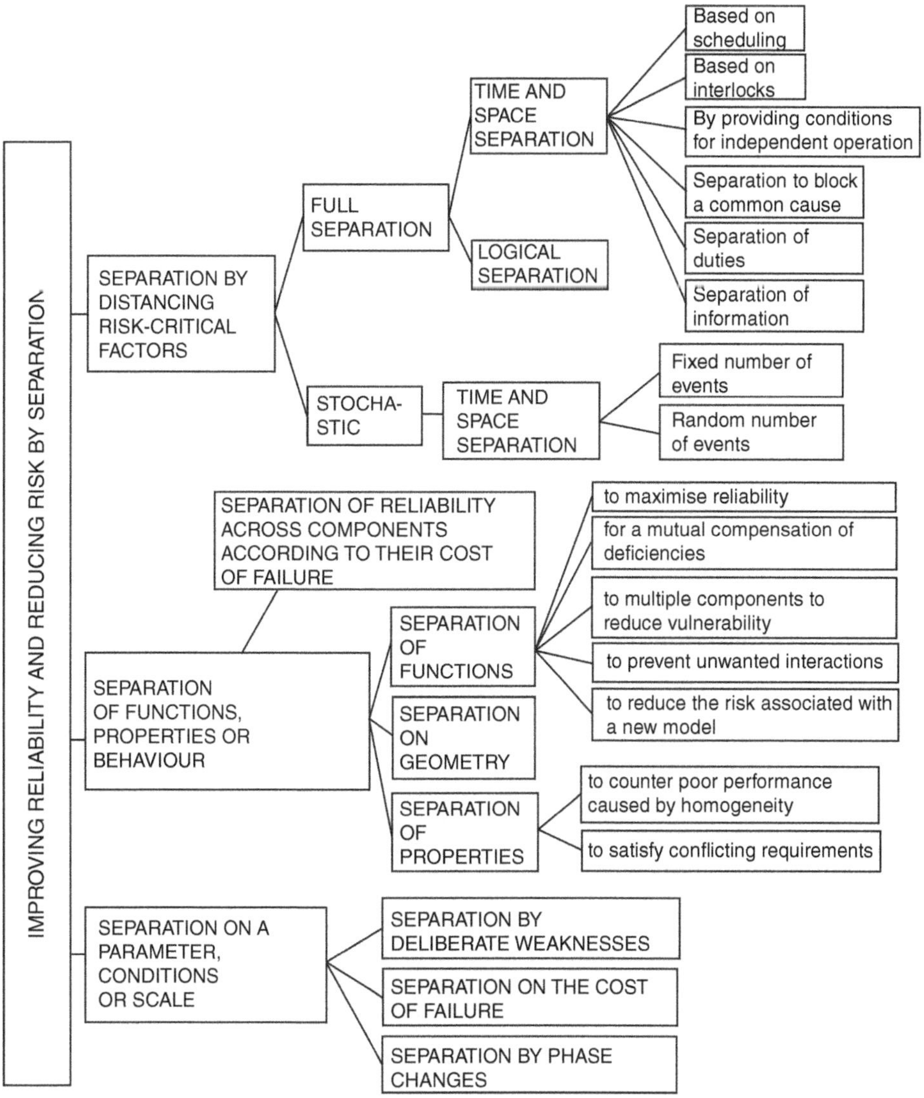

Figure 4.15 Classification of different separation mechanisms and techniques.

5

Reducing Risk by Deliberate Weaknesses

5.1 Reducing the Consequences from Failure Through Deliberate Weaknesses

On many occasions, separation through *deliberate weaknesses* can be used successfully to limit the consequences from failure. Deliberate weaknesses can be induced purposefully to induce failure automatically, whenever stress, damage, or other dangerous factors exceed a predetermined level. The purpose of the deliberate weakness is to separate a valuable entity from damage, to stop the escalation of damage, or to reduce the rate of damage accumulation.

Separation by introducing deliberate weak links and *stress limiters* has already been used for preventing stresses from reaching dangerous levels (Eder and Hosnedl 2008). The deliberate weak links are deliberately created points of weakness towards which a potential failure is channelled. By channelling failure into weak spots, designed to fail in a predictable way, the consequences from failure are decreased. Should the unfavourable conditions occur, the weak link is the one to fail and protect the expensive component or system. In this way, the conditional losses are limited. Deliberate weak links possess a maintenance-free constant readiness to operate.

Although the deliberate weak links technique is effectively an instance of separation on a risk-critical parameter, it has never been linked with the method of separation.

The discussion regarding reducing risk by deliberate weaknesses in the literature is very limited, restricted around a few well-known applications: *electrical fuses and circuit breakers, crumple zones* in road cars, *crash cones* in racing cars, *shear pins, sacrificial anodes, rupture discs, and blowout panels.*

More importantly, the mechanisms through which the separation through deliberate weaknesses reduces risk have never been discussed in the literature. This is probably the primary reason why this risk-reducing technique is rarely used in designs. The discussion in the literature is centred around reducing risk by the removal of weak links, not by introducing deliberate weak links.

However, the importance of the method of deliberate weaknesses is increasing. The modern tendency towards lightweight designs means reducing to a minimum the cross sections and the safety factors of components. As a result, components are fully loaded and utilised which results in working stresses that are close to the critical stresses triggering failure. Compared with the old and heavy designs with large safety factors, the lightweight modern designs have small or non-existent safety factors. As a result, the critical stresses in modern lightweight designs can be exceeded even by a moderate overload. Furthermore, the

Methods for Reliability Improvement and Risk Reduction, First Edition. Michael Todinov.

globalisation processes in the modern world increase the links between entities and create strong interdependency in various types of networks: financial networks, energy distribution networks, communication networks, etc. As a result, damage in one part of the network quickly escalates in other parts of the network.

Introducing separation through deliberate weaknesses and failures is an important barrier assuring the protection of valuable entities from damage due to excessive stresses.

In the case of M mutually exclusive failure modes, the expected conditional loss $\overline{C}_f$ (given that failure has occurred) is given by

$$\overline{C}_f = p_{1|f}\,\overline{C}_{1|f} + p_{2|f}\overline{C}_{2|f} + \ldots + p_{M|f}\overline{C}_{M|f} \tag{5.1}$$

where $\overline{C}_{k|f}$ is the expected conditional loss associated with the kth failure mode ($k = 1, 2, \ldots, M$) and $p_{k|f}$ is the conditional probability that given failure, it is the kth failure mode that has initiated it ($\sum_{k=1}^{M} p_{k|f} = 1$). Indeed, the loss from failure $\overline{C}_f$ can take its values in M distinct, mutually exclusive ways: if the first failure mode materialises and the loss is equal to the loss $\overline{C}_{1|f}$ associated with the first failure mode; if the second failure mode materialises and the loss is equal to the loss $\overline{C}_{2|f}$ associated with the second failure mode;…;and finally, if the Mth failure mode materialises and the loss is equal to the loss $\overline{C}_{M|f}$ associated with the Mth failure mode. Considering the conditional probabilities $p_{k|f}$ ($k = 1, 2, \ldots, M$) associated with the failure modes (given that failure has occurred) and applying the total probability theorem (DeGroot 1989) yields Eq. (5.1).

Without loss of generality, suppose that a deliberate weakness has been designed and $\overline{C}_{M+1|f}$ is the conditional loss associated with the failure of the deliberate weakness. The loss $\overline{C}_{M+1|f}$ is the smallest among all conditional losses $\overline{C}_{k|f}$, ($\overline{C}_{M+1|f} \ll \overline{C}_{k|f}$, $k = 1, 2, \ldots, M$). Suppose that the conditional probability $p_{M+1|f}$ of the deliberately built weakness has been made to be significantly larger than any other conditional probability ($p_{M+1|f} \gg p_{k|f}$, $k = 1, 2, \ldots, M$). Given failure, it is highly likely that the deliberate weakness has caused it (failure mode $M + 1$, associated with the smallest conditional losses $\overline{C}_{M+1|f}$). If a deliberate weakness is present, it is highly likely that the conditional loss $\overline{C}_f$ given failure will be equal to the conditional loss of the deliberate weakness and the consequences of failure will be limited.

5.2 Separation from Excessive Levels of Stress

5.2.1 Deliberate Weaknesses Disconnecting Excessive Load

Effectively, this type of deliberate weakness separates from excessive levels of a particular stress by disconnecting the loading stress. If the load (stress) increases to a particular value p^*, the deliberate weakness fails and prevents a further increase of the loading stress (Figure 5.1) on the component/system. As a result, components/systems are protected from overloading.

Usually, there is little control over the excessive levels of load which form the upper tail of the load distribution. If the uncertainty associated with the level of loading stress is high, introducing a deliberate weakness at an appropriate place curtails the loading stress distribution and ensures that the expensive parts of the system will not be affected.

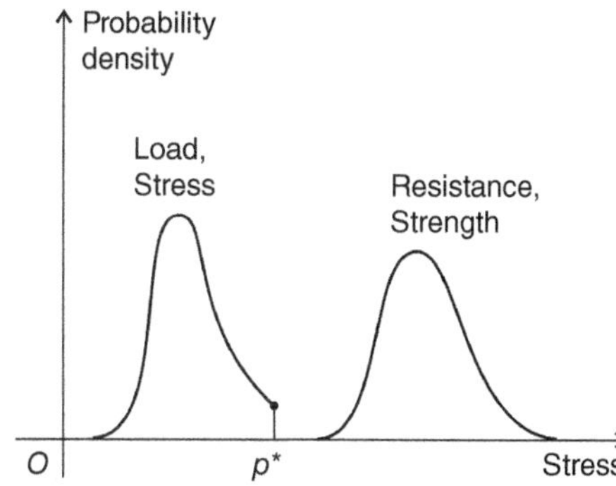

Figure 5.1 Separation of the loading stress from the resistance by including a deliberate weakness.

The stress–resistance interference can be (temperature) / (temperature resistance), (pressure) / (pressure resistance), (stress) / (strength), (deformation) / (deformation tolerance), (misalignment) / (misalignment tolerance), (corrosion) / (corrosion resistance), (stress amplitude) / (fatigue resistance), etc.

Depending on the load–resistance pair, an appropriate physical process can be used for inducing failure through a deliberate weakness and separating the load from resistance. Such are, for example, the following physical processes: current creating magnetic field; pressure activating a sensor; melting an alloy; stress reaching the yield stress limit; shear stress causing plastic deformation, etc. In addition, stress intensifiers such as notches and pre-cracks can also be used on mechanical components to provide stress intensification zones and guarantee that the strength will be exceeded first in the designated zone of the deliberate weakness.

The electrical fuses and circuit breakers are a well-known examples of deliberate weaknesses of this type. They protect electrical circuits against current/voltage exceeding critical tolerable levels. The fuse is normally activated when current exceeds a critical value at which the inserted wire connection, with deliberately reduced cross section, melts and protects the expensive circuit from dangerous levels of current.

Non-resettable thermal fuses fail and protect equipment from excessive temperature rise. Unlike electric fuses, they do not react directly to excessive current; they react to excessive temperature rise and interrupt electric current if heated beyond a specific temperature. One of the possible realisations of a thermal fuse is an alloy pellet containing, for example, specific mass fractions of copper, beryllium, and silver, which form an alloy composition that melts at a precise temperature. When the pellet melts, a holding spring separates contacts and interrupts the flow of current. Thermal fuses are used to protect heat-generating equipment from overheating and fire, in the case of a defective thermostat, for example.

Another common example of this type of deliberate weakness is the shear pin in a mechanical coupling, which transmits torque up to a specified level p^* (Figure 5.1), beyond which the shear pin fails and disconnects the driving shaft from the mechanical device. The shear pin is deliberately weakened by reducing its cross section at a selected location. This guarantees that during an overload, failure will be initiated on the cheap shear pin and not on the expensive motor, gearbox, or connected mechanical device. In the case of overload caused by jamming of the mechanical device, the shear pin shears, and disconnects the mechanical device from the motor. As a result, the mechanical device and the motor are separated from the overload. The consequences of failure are limited to replacing a cheap shear pin, as opposed to replacing an expensive mechanical device or gearbox.

Shear pins are often used in drive trains or on propellers attached to marine engines. They are also commonly used in snow blowers to connect the auger to the shaft. On single-stage snow blowers, the auger pulls snow into the machine and directs it through a discharge chute. If an object under the snow is pulled by the auger and jams the snow blower, the shear pin fails and disconnects the auger from the shaft. As a result, damage to the expensive auger and motor is prevented.

Shear pins are also used in torque wrenches in cases where it is critical to reach and not exceed a particular pre-set torque during a precision assembly of two components. Polyethylene shear pins with diameter calculated for the required assembly torque are also used. A cutting blade splits the shear pin when the specified torque is reached. As a result, the torque wrench does not need to be recalibrated and offers a high level of reliability.

Shear pins are often subjected to embrittlement tempering so that they break rather than bend at the predetermined level of load. Shear pins should be resistant to fatigue because shear pins prone to accumulating fatigue damage are gradually weakened and could break well below the specified threshold level of the load. Shear pins are maintenance-free and retain their reliability for long periods of time. This is why it is important not to use a sharp notch that creates a severe stress concentration zone and, as a result, promotes fatigue failure.

Rupture discs have also been used as deliberate weaknesses disconnecting overloading stress. They protect vessels from over-pressurisation by providing separation from excessive pressure. Compressed gas systems are ubiquitous in industry. However, operating them at even the slightest overpressure could result in surprisingly large amounts of stored potential energy, which is a significant hazard that could cause severe damage in the case of an explosion (Hedlund et al. 2016). The advantages of rupture discs to pressure-relief valves is that they are leak-tight and relatively inexpensive. Rupture discs are widely accepted in many industries and prescribed in the design codes of various pressure vessels.

Blowout panels are intentionally weakened components that fail in a predictable manner and are used in situations where sudden overpressure may occur. The pressure wave is channelled through the weakened area, thereby protecting the rest of the structure and the neighbouring structures from catastrophic damage. Examples of blowout panels are the deliberately weakened wall in a room used to store compressed gas cylinders and the deliberately weakened roof of a bunker used for ammunition storage.

Deliberate weaknesses can also be used to prevent the difference in the values of a parameter p in two parts of a system (part A and part B) exceeding a particular critical value Δp^* (Figure 5.2). The parameter p could be pressure, temperature, potential, force, etc.

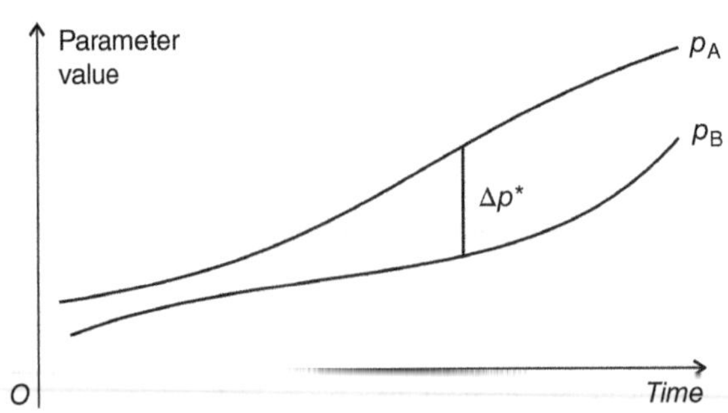

Figure 5.2 Separation by a deliberate weakness if a difference of the parameter values exceeds a critical value Δp^*.

Unlike many conventional methods for reducing risk based on upgrading the reliabilities of components, introducing redundancy or condition monitoring, introducing deliberate weaknesses leads to a risk-reducing solution at a considerably lower cost.

5.2.2 Energy-Absorbing Deliberate Weaknesses

This type of deliberate weakness maintains the rigidity of the structure from which they are a part, until an overloading shock arrives. In the case of shock overloading, a fracture or deformation is initiated at the deliberate weakness. In the case of fracture, energy is consumed to propagate the cracks from the deliberate weakness. Alternatively, if elastic, plastic, or elastic–plastic deformation mechanism is initiated, part of the energy of the shock is absorbed as deformation energy and heat.

Well-known examples of deliberate weaknesses absorbing the energy of the shock through plastic deformation are the crash cones in racing cars and the crumple zones in road cars. Effectively, the crash cones and the crumple zones are deliberate weaknesses designed to fail by absorbing large amounts of impact energy. Their deliberate failure separates from excessive deceleration during an impact by consuming a significant amount of the kinetics energy of the car in the form of deformation work. The result is a less severe impact and protection of the car occupants from fatal injury. In this respect, honeycomb sandwich panels with carbon fibre skins have found wide application in motor racing to minimise the consequences of an impact.

Similar deliberate weaknesses are the deliberately weakened beams used in earthquake engineering. The deliberately weakened section on the beam reduces the bending moment at which the beam yields. During an earthquake, the weakened sections create plastic zones whose failure (deformation) absorbs seismic energy and makes the structure more resilient to seismic waves.

Deliberately reduced cross sections of struts maintain the rigidity of the structure. In the case of impact, the supporting struts buckle, and the energy consumed during buckling reduces the severity of the impact to which the structure is subjected.

A windscreen including deliberate rupture zones, obtained by using a high-energy beam penetrating the glass, reduces the likelihood of severe injury during an impact with a pedestrian.

5.2.2.1 Case Study: Reducing the Maximum Stress from Dynamic Loading by Energy-Absorbing Elastic Components

The energy-absorbing deliberate weaknesses can be used to reduce the maximum stress in components subjected to impact or large inertia forces due to sudden stopping.

Consider an object with weight W (1) supported by a cable (3) with cross-sectional area A, wound on a rotating hoist drum (2) (Figure 5.3a). The weight is moving downward with a particular speed v when the rotation of the drum is suddenly stopped (for example, the drum is suddenly locked).

The sudden stopping of the drum creates stress σ_{max} in the cable (3) which is significantly larger than the static stress $\sigma_{st} = W/A$ created in the cable in case of a stationary drum.

Gere and Timoshenko (1999) derived the expression

$$\sigma_{max} = \sigma_{st}\left(1 + \sqrt{\frac{v^2 EA}{gWL}}\right) \tag{5.2}$$

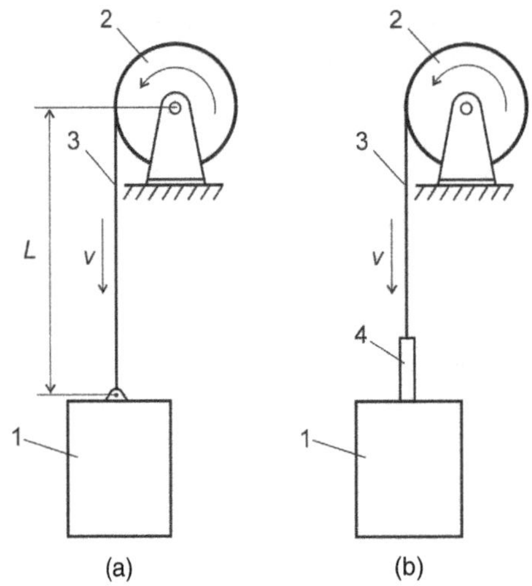

Figure 5.3 Hoist drum (a) without and (b) with a deliberate weak link.

where E is the elastic modulus of the material of the cable, L is the length of the cable, and g is the gravitational acceleration.

From Eq. (5.2) it can be seen that reducing the elastic modulus E of the material of the cable will result in reduced maximum stress σ_{max}. However, reducing the elastic modulus E of the material of the rope requires a selection of material whose yield strength may be low and therefore inappropriate for the application.

Suppose that a deliberate weak link (4) is introduced (Figure 5.3b) in the form of an additional component (effectively acting as a spring) whose elastic modulus E_w is significantly smaller than the elastic modulus E of the cable. The deliberate weak link (4) and the cable (3) are effectively springs arranged in series and their equivalent elastic modulus will be smaller than the elastic modulus E_w of the deliberate weak link (4). The smaller equivalent elastic modulus will reduce the maximum stress σ_{max} in the cable and the risk of rupture of the cable upon sudden locking of the drum.

5.2.3 Designing Frangible Objects or Weakly Fixed Objects

Frangible objects are objects of low strength, deliberately designed to break or yield in the case of collision with a more valuable object. As a result, the more valuable object is protected and the damage minimised. Frangible objects are effectively deliberate weaknesses because their design is contradictory to the normal specifications of high strength and stiffness required from objects with their function.

For example, in the design of lighting towers on airports, meteorological equipment, radio navigational equipment, etc., the structure should break, distort, or yield in the case of collision with an aircraft. The result is minimum damage to the aircraft and passengers.

The idea behind weakly fixed or weakly attached objects is similar. They are designed to detach quickly upon impact and reduce the extent of damage. Thus, the detachment of an aircraft engine upon an emergency landing minimises the consequences of fire initiated by the engine. The detachment of a weakly fixed object upon impact with a human minimises injury in the case of collision. Weakly fixed light poles fail during collision with a vehicle. This minimises the damage to the vehicle, to the driver, and to passengers.

5.3 Separation from Excessive Levels of Damage

Introducing deliberate weaknesses is an important method for assuring separation from dangerous levels of risk-critical parameters. There are several mechanisms by which deliberate weaknesses separate a valuable entity from damage.

5.3.1 Deliberate Weaknesses Decoupling Damaged Regions and Limiting the Spread of Damage

One of the most important mechanisms of limiting the spread of damage is *by decoupling the regions where damage occurred from the rest of the system*. Separation by using deliberate weaknesses limits the spread of damage, helps to isolate the source of damage, and brings the system into a stable condition.

The separation can be executed by using deliberate weaknesses or by a deliberate disconnection upon receiving a warning from condition monitoring devices.

The essence of decoupling valuable hubs in a network through deliberate weaknesses is to connect the hubs with deliberate weak links that are strong enough to propagate the benefit generated within the hubs but sufficiently weak to fail in the case of damage. In a network with hubs connected with strong links (Figure 5.4a), damage appearing in a hub (e.g. hub 1 in Figure 5.4a) easily spreads through the network and affects the other hubs. In a network with hubs connected with deliberate weak links (Figure 5.4b), damage appearing in a hub (e.g. hub 1 in Figure 5.4b) is prevented from spreading because of the failure of the deliberate weak links. The deliberate weak links connecting the hubs with the rest of the network burn ('fail') and contain the damage within a single hub thereby preventing the propagation of damage through the rest of the network. As a result, damage is contained within a small section of the network while the rest of the network is unaffected. This mechanism also operates in reverse. If a weak link is damaged, it fails and prevents damage from propagating and affecting the hubs.

The deliberate weaknesses effectively reduce the interdependency in the network. Failure of a single hub or even a group of hubs does not significantly affect the rest of the hubs. By introducing deliberate weaknesses, the capability of the network to absorb failures of its hubs is enhanced, and therefore its resilience is enhanced.

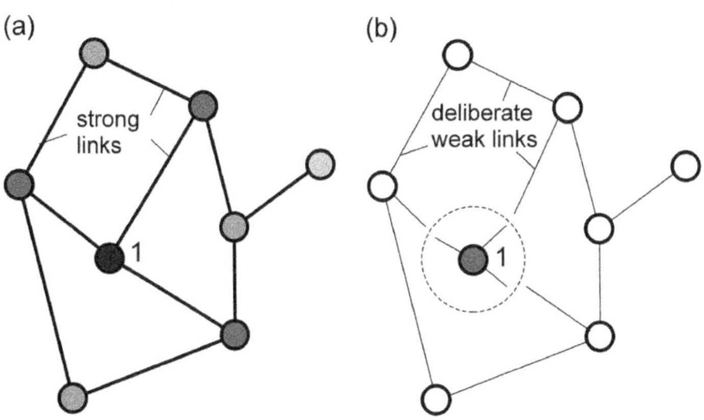

Figure 5.4 Networks built with (a) strong links and (b) deliberate weak links.

Separation by a deliberate disconnection can also be executed, upon receiving a warning from condition monitoring devices about imminent damage in some part of the network.

Power grids are often affected by random failures caused by lightning, fire, extreme weather conditions, failures in generators, malicious acts, etc. Furthermore, the power demand from consumers often varies drastically. Failure of a power line can cause an overloading of neighbouring power lines and further power line failures. The damage can spread quickly and the entire network may fail in a cascading mode. This is a type of failure that is particularly relevant to interconnected systems with positive feedback loops where damage can be reinforced by mutual dependencies and spread very quickly. Because the triggering events are unavoidable, optimal management of the power grids in real time is required to guarantee an immediate response and stable operation in the case of triggering events.

In this process, separation by disconnection or by failures of deliberate weaknesses is an important mechanism for limiting the spread of damage throughout the network and preventing rapid deterioration. It is also a low-cost solution compared with traditional methods based on improving reliability, introducing redundancy, or condition monitoring.

Deliberate weak links between control centres or censors collecting and transmitting information provides the basis for a decentralised network and guarantees its survival if a number of control centres/sensors are simultaneously eliminated.

Globalisation of trade and financial operations increases connectivity and interdependency between financial institutions, therefore, the importance of separation by using deliberate weaknesses is also increasing. Separating financial institutions from the rest of the financial network could limit the spread of defaults through the financial network and prevent a global crisis.

An example of a separation from exposure to excessive losses by disconnection is the *stop-loss order* placed with a broker. It gives an instruction to the broker to sell the security held by the investor when its price falls to a particular predetermined level.

5.3.2 Deliberate Weaknesses Providing Stress and Strain Relaxation

The next important mechanism through which deliberate weaknesses limit the spread of damage is by relaxation of accumulated stress.

Connecting modules by deliberate weak links provides an important safety mechanism in the case of increased tension in the system. Instead of triggering failure in the expensive modules, a deliberate weal link fails instead thereby dissipating energy and decreasing the tension in the system. This prevents tension from reaching critical levels in the expensive modules.

Consider an example where the risk of fast fracture of a loaded structural component is significant. In such applications, the reasoning given next justifies the selection of a low-strength steel, characterised by a relatively low yield strength, over a high-strength steel, characterised by a high yield strength.

Selecting a low-strength soft material surrounding a source of stress concentration (e.g. the tip of a crack) is effectively an act of introducing a deliberate weakness. Increasing the load, increases the stress around the stress concentrator. However, before the stress magnitude increases to a level capable of severing the interatomic bonds and propagating the crack through the material, the yield stress σ_{S1} of the material is reached (Figure 5.5a), the low-strength material deforms, and stress relaxation occurs. Fracture is prevented

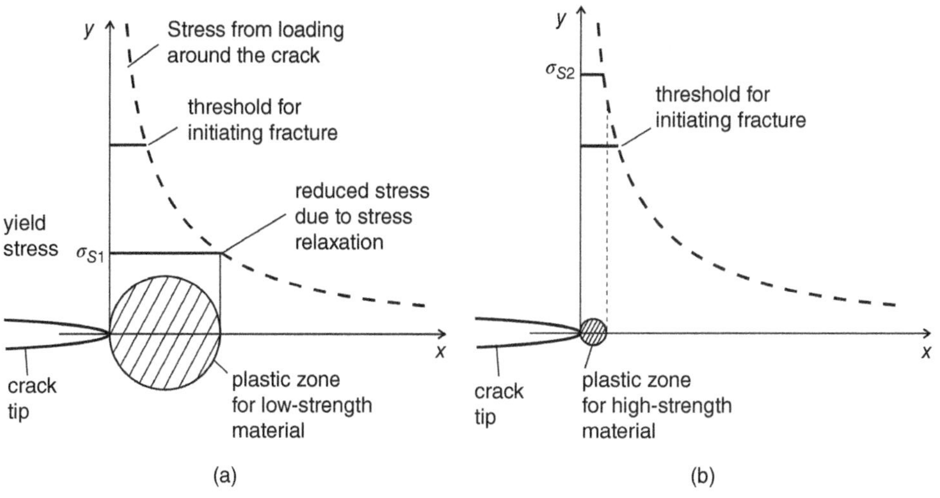

Figure 5.5 Stress relaxation provided by a ductile alloy acting as a deliberate weakness.

because the relatively small stress surrounding the crack tip is not sufficient to overcome the interatomic forces needed for propagating the crack (Figure 5.5a). A large plastic zone means that the stress in the vicinity of the crack tip is small, hence insufficient to overcome the inter-atomic bonds and cause fast fracture. Materials with large plastic zones will not be susceptible to fast fracture because they will yield first.

In contrast, high-strength materials are characterised by a large yield stress σ_{S2} (Figure 5.5b). This results in a small plastic zone at the crack tip, which often means that the stress in the vicinity of the crack tip is large enough to exceed the threshold for severing the interatomic bonds and initiating fracture (Figure 5.5b). High-strength materials have increased susceptibility to fast fracture. They fracture before they yield.

The ability of the material to deform is critical. In ductile materials, the zones near the tip of the crack also deform resulting in a blunt crack tip, which increases the resistance to fracture.

Expansion gaps, expansion offsets, expansion joints, and control joints are examples of deliberate weaknesses designed to absorb dilatation, thermal strains and axial, angular, and lateral strains. These deliberate weaknesses are commonly used in railways, buildings, and piping systems. Bridges, metal girders, and railways all have expansion gaps to accommodate thermal strains and prevent the formation of large stresses in hot and cold weather. The control joints are also examples of deliberate weaknesses permitting stress relaxation. There are cracks and gaps introduced at regular intervals between pavement slabs. In hot and cold weather, these deliberate weaknesses accommodate thermal strains and prevent damage.

Expansion offsets are a common generic solution to accommodate thermal expansion in pipes and prevent the formation of excessive thermal stresses. The shape and size of the offsets depends on the amount of thermal expansion that needs to be accommodated.

Expansion joints on pipes are often made of metal and rubber. They are strong enough to withstand internal pressure and flexible enough to provide relaxation of axial, lateral, or angular strains.

5.3.3 Deliberate Weaknesses Separating from Excessive Levels of Damage Accumulation

The essence of this mechanism consists of introducing deliberate weaknesses where damage accumulates. Instead of accumulating in the expensive component (with a large cost of failure or cost of replacement) the damage accumulates in a cheap component which is replaced when the accumulated damage reaches a critical level. The deliberate weaknesses must be designed to undergo a rate of damage accumulation higher than the rate of damage accumulation in the expensive components.

Good examples of this type of deliberate weakness are sacrificial anodes. Sacrificial anodes form a galvanic couple with the protected component and by corroding preferentially they protect the component from corrosion. Effectively, sacrificial anodes separate components (underground and underwater pipes, underwater installations, submerged structures, ship hulls, internal surfaces of storage tanks, etc.) from excessive corrosion. Most commonly, magnesium, zinc, and aluminium are used as sacrificial anodes. Sacrificial blocks of magnesium alloy, for example, are used to protect the steel legs of oil rigs from corrosion. As a result, large supporting metal structures are cheaply protected from corrosion in a highly corrosive sea-water environment.

Galvanisation (applying protective zinc coating to steel) is most commonly used on outdoor steel structures to prevent rusting. A metal anode can be used to protect another metal part as long as a sufficient difference in electropotential is present, and the metal anode has a more negative potential than the protected metal part. Thus, components made of copper can be protected by iron anodes because a sufficient difference in electropotential is present, and iron has a more negative potential than copper in the electropotential series.

Deliberate weak links of this type also protect against failure from excessive wear. For example, the inserts in a journal bearing take most of the wear. Excessive wear requires only the replacement of an insert rather than the replacement of the entire journal bearing.

Spray nozzles are often subjected to intensive accumulation of wear damage which requires a replacement of the nozzle. The damage accumulation in the expensive spray nozzle can be avoided by a replaceable nozzle insert. The nozzle insert takes all the wear damage and acts as a deliberate weakness. Instead of replacing the entire nozzle, only the nozzle insert is replaced.

Rubber segments bolted on top of a metal conveyor belt also act as deliberate weaknesses reducing the damage accumulation to the expensive conveyor belt. They take most of the wear and their failure requires the replacement of a cheap rubber segment rather than the replacement of the expensive metal parts of the conveyor belt.

For two components in contact, characterised by high contact stresses, one of which is cheap and the other is expensive, the less expensive component should be manufactured from softer material. As a result, the wear and deformation due to the high-magnitude contact stresses will be concentrated in the cheaper component. The consequences of failure will be significantly reduced because the cheap component will be replaced and not the expensive one.

In project management, this technique can be applied by dropping (sacrificing) tasks that are not essential to the main objectives of the project. This is particularly important when lots of 'nice to have' project tasks have been planned. Some of these can be sacrificed to leave more time for improving the quality of the core deliverables on the project and protect the project from the risk of inferior quality.

5.4 Deliberate Weaknesses Deflecting the Failure Location or Damage Propagation

5.4.1 Deflecting the Failure Location from Places Where the Cost of Failure is High

This mechanism deflects the failure location away from places where failure is expensive. Effectively, this mechanism separates the failure location from expensive parts of the system. An efficient implementation of this type of deliberate weakness is through appropriately placed stress raisers or stress-relief gaps around the protected valuable part of the system. Upon loading, the deliberately introduced stress raisers initiate cracks and channel the main crack away from the valuable part of the system. As a result, the main crack is deflected along the deliberate weakness and the damage bypasses the valuable part of the system.

In this case, a large stress concentration effect is beneficial. It guarantees that the main crack will develop at the raiser and not in the expensive part of the system.

An interesting solution of this type has been discussed in Altshuller (1999) for protecting underground electrical lines from damage by ground freezing. Narrow ditches made parallel to lines buried underground were proposed. These effectively act as deliberate weaknesses. The thermal stresses from freezing cause the cracks to form along the ditches. As a result, the formation of cracks cutting across the electrical lines, is prevented.

Consider another example related to a rigid floor tiling over a concrete slab. Concrete slabs shrink as they dry out. The shrinkage is often sufficient to form cracks which propagate through the slabs. The floor tiles laid over these cracks will also crack which results in costly damage. Deflections in suspended floors can also induce high compressive stresses in rigid floor tiling which cause delamination and cracks.

The costly damage could be avoided easily if deliberate weaknesses are created within the flow coverings in the form of gaps (filled with plastic) which allow stress relief and deflect potential cracks in the tiling induced by shrinkage.

Deliberate weaknesses can be used to deflect the damage accumulation from the expensive parts of the system. This can, for example, be done by selecting material that wears/deforms more easily than the material of the expensive component. Deflecting the failure location can also be done by selecting material that corrodes more easily compared with the material of the expensive component. (Some of these techniques have already been covered in Section 5.3).

Deflecting failure location outside expensive parts of the system, by deliberate weaknesses, transcends the mechanical engineering domain. In business, carrying out a small-scale trial of a new business idea, with a small amount of initial funding is effectively an implementation of a deliberate weakness. If the business idea is inappropriate and fails to deliver, there will be a loss of a relatively small investment instead of a large loss. Lessons useful for the future development of the business are then learned at a small price.

5.4.2 Deflecting the Failure Location from Places Where the Cost of Intervention for Repair is High

This mechanism forces the failure location away from places where the cost of intervention for repair is high. Thus, for a system combining components located deep under water and components located above water, a location of the deliberate weakness above water reduces the cost of intervention because the cost of intervention for repair above water is significantly smaller than the cost of intervention for repair underwater.

Consider a device composed of two blocks *A* and *B*. Block *A* can be accessed easily while block *B* is difficult to access. Accordingly, retrieving block *A* requires a much smaller number of hours and less expensive equipment compared with retrieving block *B*. To reduce the cost of repair, the deliberate weakness should be located in block *A* rather than in block *B*.

5.4.3 Deliberate Weaknesses Deflecting the Propagation of Damage

Deflecting damage from a particular direction diminishes its rate of propagation, its magnitude, and the extent of impact on the target. Deflecting cracks along weak grain boundaries or along deliberately created weak interfaces increases the resistance to crack propagation and increases the toughness of materials.

One of the reasons why cutting tool inserts fail are the microcracks that start at the surface of the inserts and progress towards the substrate. The layered structure achieved by introducing relatively weak interfaces improves the durability of the inserts by introducing microcrack deflection barriers preventing excessive crack growth. Such interfaces can be manufactured by a series of coatings, for example by a chemical vapour deposition (CVD) process.

The technique of deliberate weaknesses deflecting the propagation of damage is not only reserved for fatigue cracks. Thus, in the case of failure of a dam, deflecting the path of a powerful water stream by deliberate weaknesses originating multiple branching streams, dissipates the stored potential energy of the water and reduces the impact on the built-in area in the vicinity of the dam.

5.5 Deliberate Weaknesses Designed to Provide Warning

Effectively, this type of deliberate weakness separates from failure by providing a warning.

It is a well-known fact that components can fail in a 'ductile' or 'brittle' manner. Many fractures contain characteristics of both types.

Ductile fracture is accompanied with a considerable amount of plastic deformation. The crack will not normally extend unless increased stress is applied. Ductile fracture is associated with a substantial amount of absorbed energy. This is indicated by the large area beneath the *load–displacement curve* which is numerically equal to the work done to break the component.

Brittle failure is accompanied with little or no plastic deformation. Once initiated, the crack extends at a high speed without the need for increased stress. The component fails quickly, without any warning. Brittle fracture is associated with a small amount of energy to break the component, which is indicated by the small area beneath the load displacement curve.

In engineering applications, where safety concerns are involved, materials with ductile behaviour are the obvious choice for the following reasons:

- Ductile fracture is preceded by plastic deformation, redistribution, and relaxation of the high stresses in the zones of stress concentration.
- The component deforms before fracture which gives early warning and sufficient time for intervention and repair.
- Ductile fracture requires more strain energy in order to develop.
- In the process of plastic deformation, the strength of the material is enhanced through strain hardening.

The deformation preceding ductile fracture can be used for designing deliberate weaknesses issuing warning upon overloading thereby separating from overloading and fracture. A mechanism through which this type of separation can be implemented is to ensure that before the main failure, a secondary failure in a ductile manner occurs that alerts the operator to decrease the load and avoid catastrophic failure.

An implementation of this separation mechanism is the retrofitable cable mechanical fuse shown in Figure 5.6 (Tunno and Larsen 2011a).

A cable (1) forms a slack inside the device and is attached securely to the device by clamps (2). In normal operation conditions, the load is carried only by the weakened part (3) of the device. When the load exceeds a critical level, the weakened part bends, then breaks and finally, the load is transferred to the cable (1). This provides ample warning to the operator that the permitted operating load has been exceeded.

The importance of deliberate weak links is increasing in design. While the design is normally based on static loading, altering the operating conditions may give rise to impact loading or sudden applications of the load. This produces a significant increase in the actual loading stress. Operating loads are easily exceeded if, for example, the load is applied suddenly. The magnitudes of the stresses resulting from suddenly applied loads are twice those resulting from statically applied loads (Hearn 1985).

Security tapes with layers which easily delaminate or fracture to show entry is another example of a deliberate weakness designed to provide warning.

Early detection of erosion in pipelines due to abrasive particles in the stream is crucial in preventing damage to equipment resulting in dangerous leaks of toxic substances to the environment. An example of a deliberate weakness providing an early warning about critical metal loss due to erosion is the *sand probe* which is a corrosion-resistant sacrificial tube inserted in the pipeline. Should excessive amounts of abrasive particles appear in the fluid, the sand probe fails due to excessive wear. The pressure inside the sand probe then becomes equal to the working pressure of the pipeline. This is registered by a pressure gauge assembly which sends a warning signal to the operator about the presence of abrasive particles (e.g. sand) in the flow. As a result of the early warning sent by the deliberate weakness, the flow in the pipeline can be stopped, which prevents the expensive pipeline and valves from fast wear out and early-life failure.

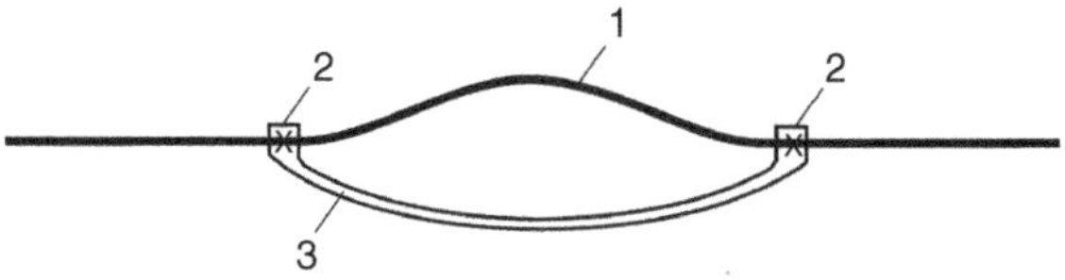

Figure 5.6 Retrofitable cable mechanical fuse.

5.6 Deliberate Weaknesses Designed to Provide Quick Access or Activate Protection

There are situations where a specific resource is required in the case of an emergency: access to first aid, fire extinguishing fluid, alarm button, ammunition, hydraulic dampening in the case of an earthquake, etc.

For some of these applications, the seal providing access to the emergency resource is deliberately weakened to speed up access in the case of an emergency. The deliberately weakened seal does not prevent normal operation. For example, the air-tightness and the possibility for carrying emergency supplies remain unaffected. The alarm button, behind the weakened cover, is protected from accidental activation. In normal operating conditions, a structure incorporating a hydraulic dampener is secured by shear pins and is rigid. In the case of large earthquake forces, the pins shear and activate the hydraulic dampening system which protects from excessive earthquake damage.

A deliberately weakened plug on a reservoir with fire-extinguishing fluid under pressure (a sprinkler system) made of a mixture of Wood's metal (eutectic alloy including bismuth, lead, tin, and cadmium with melting point of approximately 70 °C) seals the fire-extinguishing fluid at normal operating temperature. In the case of a fire, the plug melts due to the temperature rise and opens the sprinkler with the fire extinguishing fluid. An alternative solution is a small glass bulb full of glycerine-based liquid, which expands if the temperature rises, shatters the bulb, and opens the sprinkler. Medical gas cylinders could also be protected by Wood's metal seals. In the case of a fire, the seals melt and release the gas thereby reducing the risk of explosion. In all cases, the deliberate weakness fails and activates a protection process which reduces the consequences from failure.

5.7 Deliberate Weaknesses and Stress Limiters

Similar to the deliberate weaknesses, stress limiters also separate from excessive levels of stress. A common example of a stress limiter is an anti-surge protector preventing voltage from reaching dangerous levels that could damage electrical equipment. A safety pressure valve, activated when pressure reaches a critical level, is another common example of a stress limiter separating the loading stress from the strength of the material. Friction

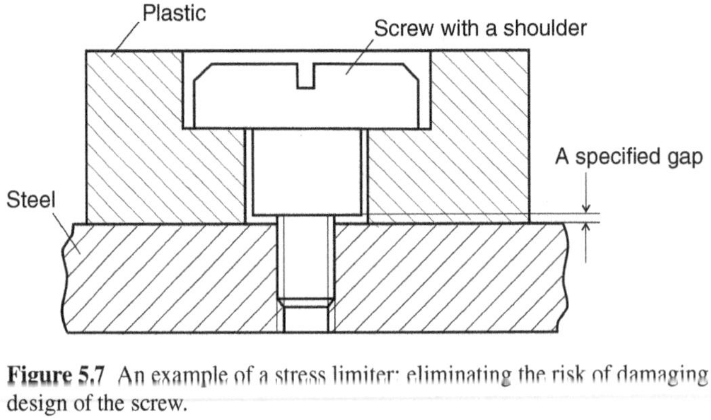

Figure 5.7 An example of a stress limiter: eliminating the risk of damaging the plastic part by a special design of the screw.

clutches are another example of a stress limiter that has been specifically designed to slip during a torque overload. While deliberate weaknesses are designed to fail and separate from excessive loading stress, stress limiters separate from excessive loading stress without necessarily suffering failure.

It is tempting to consider stress limiters as instances of deliberate weaknesses because some of them also channel failure towards a deliberate weakness in order to separate from excessive levels of loading stress. However, there are essential differences. Not all stress limiters are deliberate weaknesses, just as not all deliberate weaknesses are stress limiters.

The retrofitable cable mechanical fuse shown in Figure 5.6, for example, is a deliberate weakness without being a stress limiter. The device only provides warning and does not limit the load.

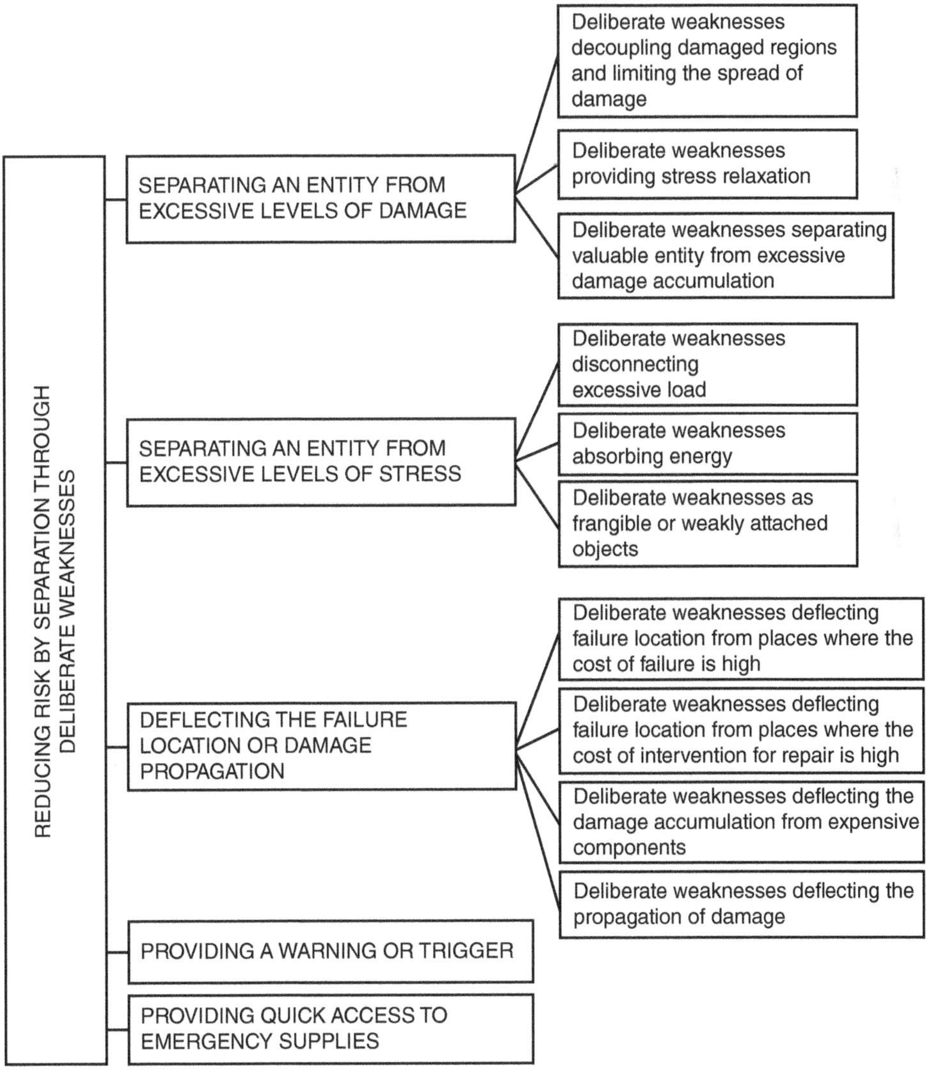

Figure 5.8 A classification of mechanisms for reducing risk by deliberate weaknesses.

Alternatively, the specially designed shoulder on the screw in Figure 5.7 (Erhard 2006), which prevents over-tightening and damaging the plastic components, is a stress limiter without being a deliberate weakness. The magnitude of the loading stress on the clamped part is limited without introducing a deliberate weakness.

In summary, the classified mechanisms of reducing risk by introducing deliberate weaknesses are presented in Figure 5.8.

6

Improving Reliability and Reducing Risk by Stochastic Separation

6.1 Stochastic Separation of Risk-Critical Factors

6.1.1 Real-Life Applications that Require Stochastic Separation

Risk is often the result of the simultaneous presence of risk-critical events. Reducing risk then depends on reducing the degree of overlapping of the risk-critical events. Here are a number of real-world examples illustrating this point:

- A source servicing a number of randomly arriving requests (demands), where the source can only service a single request at a time. Unsatisfied demand occurs if two or more demands cluster within a critical distance s (Figure 6.1a).
  The competition of random demands for a particular resource/service on a finite time interval is a common example of risk and reliability controlled by the simultaneous presence of critical events. The appearance of a critical event engages the servicing resource and if a new critical event occurs during the service time of the first critical event, no servicing resource will be available for the second event.
  Suppose that only a single repair unit is available for servicing failures on a power line. In the case of a power line failure, the repair resource will be engaged and if another failure occurs during the repair time s associated with the first failure (Figure 6.1c), no free repair resource will be available for recovering from the first failure. The delay in the second repair could lead to overloading of the power distribution system thereby inducing further failures.
  There are cases where the probability of simultaneous presence (overlapping) of risk-critical events must be low. A low probability of a simultaneous presence of random demands is required, for example, in situations where people in critical condition demand a particular piece of life-saving equipment for a time s (Figure 6.1b). If only a single piece of life-saving equipment is available, simultaneous demands cannot be satisfied and the consequences could be fatal.
- Stored spare equipment servicing the needs of customers arriving randomly during a specified time interval. After a demand from a customer, the warehouse needs a minimum time to restore/return the dispatched equipment before the next demand can be serviced. In this case, the probability of unsatisfied demand equals the probability of clustering of two or more customer arrivals within the critical period needed for making the equipment available for the next customer (Figure 6.1a).
- Supply systems which accumulate the supplied resource before it is dispatched for consumption (compressed gaseous substances for example). Suppose that, after a demand

Methods for Reliability Improvement and Risk Reduction, First Edition. Michael Todinov.

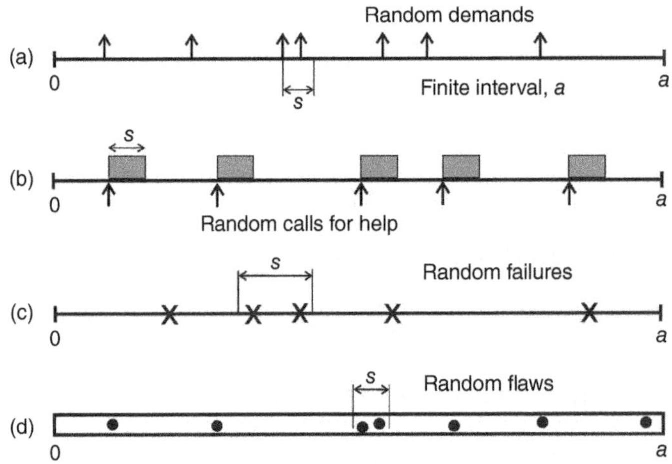

Figure 6.1 Common examples where risk depends on the existence of separation between risk-critical factors/events.

for the resource, the system needs a minimum period of specified length s to restore the amount of supplied resource to the level existing before the demand. In this case, the probability of unsatisfied demand equals the probability of clustering of two or more random demands within the critical recovery period s (Figure 6.1a).

- A related risk, controlled by overlapping (simultaneous presence) of critical events, is present when the appearance of one critical event (e.g. a shock) requires a particular time which the system needs in order to recover. If another critical event appears before the system has recovered, the system's strength/capacity is exceeded which results in a system failure. Consider, for example, shocks caused by failures associated with pollution to the environment (e.g. a leakage of chemicals) (Figure 6.1c). Failure, followed by another failure associated with leakage of chemicals before a critical recovery time interval has elapsed, could result in irreparable damage to the environment. For example, clustering of failures associated with a release of chemicals in sea water could result in a dangerously high acidity which will destroy marine life.
- Forces acting on a loaded component which fails if two or more forces cluster within a critical time interval s.
- Clustering of two or more random flaws within a small critical distance s (Figure 6.1d) dangerously decreases the load-carrying capacity of thin fibres and wires. As a result, a configuration where two or more flaws are closer than a critical distance cannot be tolerated during loading. Reliability in this case depends on the probability of clustering of the random flaws.

In all of the listed examples, the overlapping of the risk-critical random events cannot be avoided therefore, risk is directly related to the probability of overlapping of the risk-critical events. Reducing risk is achieved by reducing the probability of overlapping of the risk-critical events. In all these cases, it is important to guarantee with a specified probability, a separation of minimum length between the risk-critical events.

The probability with which the risk-critical random events should be separated varies significantly. It is directly related to the magnitude of the consequences resulting from the overlapping of two or more risk-critical events. The level of the probability of separation must be set individually by risk experts in the specific application domain.

Guaranteeing with a specified probability the existence of separation between risk-critical factors/events will be referred to as *stochastic separation* on the probability of overlapping.

The problem related to estimating the risk of simultaneously present critical events is essentially a problem from geometric probability, where a large segment of specified length L is covered by randomly positioned small segments with different lengths. The segment with length L represents the operational time interval while the smaller segments represent the durations of the critical random events. Simultaneously present critical events exist if and only if overlapping among randomly located small segments is present.

There have been a number of publications related to covering the circumference of a circle with segments or a linear segment with segments (Stevens 1939; Shepp 1972; Solomon 1978; Huffer and Shepp 1987; Justicz et al. 1990; Coffman et al. 1994, 1998; Calin and Udriste 2014). However, none of the existing publications treats the key question related to the dependence of the probability of overlapping of random events on a given time interval on the durations of the random events.

In queuing theory, the Poisson process has been traditionally used as a statistical model for random events occurring in a time interval. Despite the abundance of publications on queuing systems, none of the published classical studies (Khintchine 1969; Cooper 1981; Gross and Harris 1985; Thompson 1988; Gnedenko and Kovalenko 1989), nor more recent texts on different problems in queuing theory (Allen 1990; Giambene 2005; Bhat 2008; Kalashnikov 2010), nor recent texts on probability (Rosenthal 2006; Weiss 2006; Stroock 2011; Venkatesh 2013) treat the key question related to the dependence of the probability of overlapping of random events on a time interval on the distributions of their durations. Coverage of the theory of stochastic separation of random events on a time interval has been recently presented in Todinov (2017d).

Furthermore, in queuing theory, a central assumption in deriving the expected lengths of queues is the negative exponential distribution of the waiting (servicing) times. As it will be demonstrated in Chapter 12, contrary to the conventional assumptions, *the distribution of the servicing times for a system which includes different types of components is not the negative exponential distribution, the normal or log-normal distribution. The distribution of the servicing times of such a system is a distribution mixture.*

Accordingly, the present chapter provides a comprehensive treatment of the key questions related to: (i) the probability of overlapping of risk-critical random events on a time interval; (ii) the expected time fraction of overlapping of a particular order for risk-critical random events on a time interval; and (iii) the methods for achieving stochastic separation of risk-critical random events at a minimum cost.

6.1.2 Stochastic Separation of a Fixed Number of Random Events with Different Duration Times

Consider a common case where demands for a single resource arrive from a number of consumers, at random times, during a time period. The demand times are with different durations for each consumer. Because of the single piece of resource, a simultaneous demand from more than a single consumer cannot be satisfied.

This statement of the problem is formally presented as n random events, with durations $d_1, \ldots, d_n$, appearing randomly during a time interval $0,L$ $(d_1 + d_2 + \ldots + d_n < L)$. The

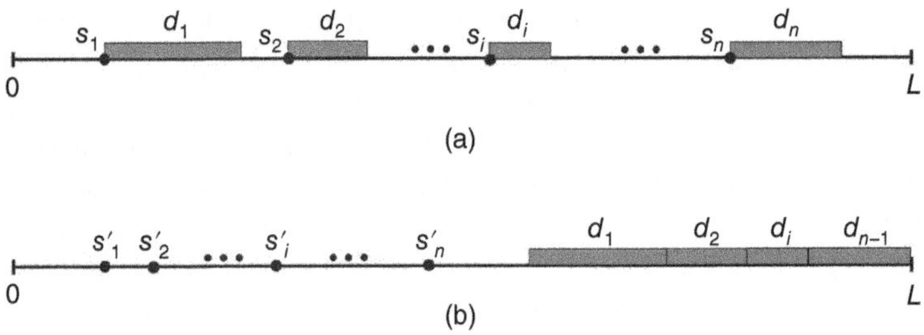

Figure 6.2 (a) X-configuration: fully separated random events in a finite time interval (0,L); (b) Y-configuration: random points generated along the interval (0,L) all falling in the interval $0,L-(d_1+d_2+\ldots+d_{n-1})$.

objective is evaluating the probability that no overlapping of random events will be present (Figure 6.2a). The event start times $s_1, s_2, \ldots, s_n$ are uniformly distributed along the interval (0,L). The configuration in Figure 6.2a will be referred to as the X-configuration if no overlapping of random events is present. Note that the duration of the last event cannot possibly contribute to overlapping in the time interval (0,L) and therefore can be ignored. The first $n-1$ duration intervals $d_1, \ldots, d_{n-1}$ can then be 'cut out' of the time interval (0,L) and the remaining parts of the time interval can be 'brought together' to form a shorter length $L-(d_1+d_2+\ldots+d_{n-1})$ (Figure 6.2b). As a result of this operation, the points $s_1, s_2, \ldots, s_n$ marking the start of the random events for the X-configuration, transform into a unique Y-configuration where the points $s'_1, s'_2, \ldots, s'_n$ are randomly distributed along the shorter length $L-(d_1+d_2+\ldots+d_{n-1})$. Consequently, a single Y-configuration corresponds to each X-configuration. Thus, for the set of all possible X-configurations and the set of all possible Y-configurations, the relationship $X \subseteq Y$ holds.

Now suppose that a set of n random points $s'_1, s'_2, \ldots, s'_n$ are randomly generated (uniformly distributed) along the length L. If all n randomly generated points fall within the length $L-(d_1+d_2+\ldots+d_{n-1})$, a Y-configuration is present. By inserting the corresponding duration intervals d_i ($i = 1, \ldots, n-1$) after each start time s_i, an X-configuration will be obtained with random non-overlapping events along the interval 0,L (Figure 6.2a).

Thus, from each Y-configuration, a unique X-configuration can be obtained by placing after the randomly generated points s'_i duration intervals with lengths $d_1,\ldots, d_{n-1}$, correspondingly. The start times in the X-configuration (Figure 6.2a) are therefore: $s_1 = s'_1$, $s_2 = s'_2+d_1$, $s_3 = s'_3+d_1+d_2, \ldots,$ $s_n = s'_n+d_1+d_2+\ldots+d_{n-1}$. In the obtained X-configuration, all random events are non-overlapping. Consequently, each Y-configuration corresponds to a single X-configuration. Thus, for the set of all possible Y-configurations and the set of all possible X-configurations, the relationship $Y \subseteq X$ holds.

Since for the sets of the X-configurations and Y-configurations $Y \subseteq X$ and $X \subseteq Y$ hold simultaneously, there exists a one-to-one correspondence between the X-configurations, characterised by randomly located non-overlapping random events and the Y-configurations, characterised by randomly located random points on the smaller time interval $L-(d_1+d_2+\ldots+d_{n-1})$. Because of the one-to-one correspondence between X-configurations and Y-configurations, the probability of an X-configuration can be estimated by the probability of a Y-configuration.

Suppose that the start times of the events are uniformly distributed along the length of the time interval (0,L). Let $A_1, A_2, \ldots, A_n$ denote the events 'the last event has a duration d_1,

$d_2, \ldots, d_n$, correspondingly. The probability of the event B, that there will be no overlapping of random events, can be determined by the following probabilistic argument.

Initially, the conditional probability $P(B|A_n)$ is determined – the probability that there will be no overlapping of random events, given that the last event has a duration d_n. Because each random event has an equal chance to be the last event, the probabilities $p(A_i)$ of the events A_i are all equal to $1/n$ ($p(A_i) = 1/n$, $i = 1, \ldots, n$).

The probability of a Y-configuration that n uniformly distributed random points $s'_1, s'_2, \ldots, s'_n$ along the interval $(0,L)$ will all fall in the shorter interval $L - (d_1 + d_2 + \ldots + d_{n-1})$ is given by $\left(\frac{L-(d_1+\ldots+d_{n-1})}{L}\right)^n$. This is also the probability of an X-configuration (no overlapping of random events) given that the last random event has a duration d_n.

$$P(B|A_n) = \left(1 - \frac{(d_1 + \ldots + d_{n-1})}{L}\right)^n \tag{6.1}$$

The absence of overlapping (event B) however, can occur in n different ways. The absence of overlapping can occur given that the last event has a duration d_n, given that the last event has a duration $d_{n-1}, \ldots$, and so on. The probabilities $P(B|A_i)$, $i = 1, \ldots, n-1$, are determined in a similar fashion. According to the total probability theorem,

$$P(B) = P(B|A_1)P(A_1) + \ldots + P\left(B|A_n\right)P(A_n) \tag{6.2}$$

As a result, the expression

$$P(B) = \frac{1}{n}\left[\left(1 - \frac{d_2 + d_3 + \ldots + d_n}{L}\right)^n + \left(1 - \frac{d_1 + d_3 + \ldots + d_n}{L}\right)^n + \ldots + \left(1 - \frac{d_1 + d_2 + \ldots + d_{n-1}}{L}\right)^n\right] \tag{6.3}$$

is obtained for the probability of a full separation (no overlapping) of random events with durations $d_1, \ldots, d_n$. Eq. (6.3) can also be presented as

$$P(B) = \frac{1}{n}\left[\left(1 - \frac{D - d_1}{L}\right)^n + \left(1 - \frac{D - d_2}{L}\right)^n + \ldots + \left(1 - \frac{D - d_n}{L}\right)^n\right] \tag{6.4}$$

where $D = d_1 + d_2 + \ldots + d_n$.

Equation (6.4) has been confirmed by the results from computer simulations. Thus, for four consumers demanding a particular resource for $d_1 = 5$ minutes, $d_2 = 10$ minutes, $d_3 = 20$ minutes, and $d_4 = 35$ minutes, respectively, during a time interval of 10 hours, the probability of no overlapping calculated from Eq. (6.4) is 0.7. This probability has been confirmed by the probability of 0.7 estimated from the simulation. (The details of the simulation algorithm have been omitted.) For a given set of events, with duration times $d_1, \ldots, d_n$, three principal mechanisms of stochastic separation can be implemented:

- Providing stochastic separation with a specified probability $P(B)$, by decreasing the durations of the events.
- Providing stochastic separation with a specified probability $P(B)$, by increasing the length L of the time interval.
- Providing stochastic separation with a specified probability $P(B)$, by decreasing the number of events.

All of these mechanisms can be implemented by solving the non-linear Eq. (6.4) with respect to L, or with respect to n, or with respect to the durations of the demand times.

6.1.2.1 Case Study: Stochastic Separation of Consumers by Proportionally Reducing Their Demand Times

Suppose that three consumers are demanding a particular resource for $d_1 = 35$ minutes, $d_2 = 45$ minutes and $d_3 = 60$ minutes, during an interval of $L = 500$ minutes. The factor $0 < k < 1$ by which the demand times d_1, d_2, and d_3 need to be proportionally reduced is sought so that the probability of full separation (no overlapping) is equal to a specified level of 0.7.

To determine the factor k by which the demand times need to be reduced, the non-linear equation

$$f(k) \equiv \frac{1}{3}\left[\left(1 - \frac{k(D - d_1)}{L}\right)^3 + \left(1 - \frac{k(D - d_2)}{L}\right)^3 + \left(1 - \frac{k(D - d_3)}{L}\right)^3\right] - 0.7 = 0 \tag{6.5}$$

can be solved with respect to k by a repeated bisection in the interval $k_{min} = 0.01 \leq k \leq k_{max} = 1$, because at the ends of this interval $f(k)$ has different signs ($f(k_{min}) > 0$ and $f(k_{max}) < 0$). The solution of Eq. (6.5) with respect to the parameter k, obtained by using a standard repeated bisection algorithm, is $k = 0.6$. Thus, in order to provide the required separation probability of 0.7, the durations of the demand times need to be reduced to $d'_1 = kd_1 = 0.6 \times 35 = 21$; $d'_2 = kd_2 = 0.6 \times 45 = 27$; and $d'_3 = kd_3 = 0.6 \times 60 = 36$. Considering that $D' = d'_1 + d'_2 + d'_3 = 84$, substitution in Eq. (6.4) where $n = 3$ gives:

$$\frac{1}{3}\left[\left(1 - \frac{84 - 21}{500}\right)^3 + \left(1 - \frac{84 - 27}{500}\right)^3 + \left(1 - \frac{84 - 36}{500}\right)^3\right] = 0.7$$

Suppose that the durations $x_1, x_2, \ldots, x_{n-1}, x_n$ of the first, second, …, nth random event are realisations of a random variable X following a statistical distribution with mean μ and standard deviation σ. According to Eq. (6.4), the probability that the random events will be separated (will not overlap) is given by

$$P(B) = \frac{1}{n}\left[\left(1 - \frac{D - x_1}{L}\right)^n + \left(1 - \frac{D - x_2}{L}\right)^n + \ldots + \left(1 - \frac{D - x_n}{L}\right)^n\right] \tag{6.6}$$

where $D = \sum_{i=1}^{n} x_i$.

The mathematical model related to stochastic separation of a fixed number of risk-critical events provides the opportunity to gain insight into the fact that the increase of the demand times x_i by a particular factor $k > 1$ and the increase by the same factor $k > 1$ of the length of the operational interval L *will have no impact on the probability of overlapping*. Thus, an increase of the demand times by 10% can be compensated by a corresponding increase by 10% of the length of the operation interval L.

Even for a relatively small number of events, the sum of the event durations $\sum_{i=1}^{n} x_i - x_1$, $\sum_{i=1}^{n} x_i - x_2, \ldots, \sum_{i=1}^{n} x_i - x_n$ in Eq. (6.6) can be approximated reasonably well with $(n-1)\mu$, where μ is the mean of the duration times x_i. As a result, the probability $P(B)$ of a full separation (non-overlapping) of random events becomes

$$P(B) \approx \left(1 - \frac{(n-1)\mu}{L}\right)^n \tag{6.7}$$

This probability is practically insensitive to the variance of the random variable X standing for the durations x_i of the demand times. The probability $P(B)$ depends strongly on the expected value $\mu = E(X)$ of the demand times. The probability of event separation is practically insensitive to the variance (standard deviation σ) of the demand times X.

This conclusion has been verified by numerous computer simulations where the probability of a full separation has been plotted as a function of the variance of the duration times.

The simulations involved demand times following a log-normal distribution with mean 140 minutes, coming from 36 users, over a time interval with length 60 000 minutes (1000 hours). Each of the 36 users initiates exactly one demand, randomly located on the operational time interval (0,60 000 minutes). A single source for servicing the random demands is available, capable of servicing only a single random demand at a time.

The simulation results shown in Figure 6.3 have been obtained after incrementing the standard deviation of the demand times by a step of five minutes. The calculated probabilities of full separation were very close to 0.046 and correspond to a single source servicing the demands.

If a constant duration of the random demands is used, equal to the mean of 140 minutes of the log-normal distribution, the value 0.046 is obtained for the probability of no overlapping.

In the next simulation experiment, the log-normal distribution of the demand times was replaced by a normal distribution with the same mean (140 minutes). The number of users (36) and the length of the operation interval (1000 hours) were kept the same as in the previous simulation experiment. The standard deviation of the normal distribution was varied with a step of five minutes. The results were almost identical to the results in Figure 6.3.

The calculated probability of fully separated random demands was again very close to 0.046.

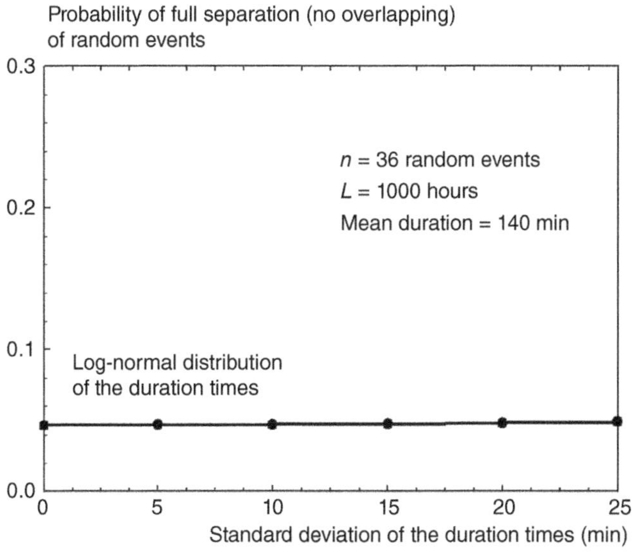

Figure 6.3 Probability of full separation of random events (no overlapping) as a function of the variance of the log-normal distribution modelling the duration times.

In the next simulation experiment, a uniform distribution for the duration of demand times has been selected, with a probability density function $f(t) = \frac{1}{2a}$ if $140 - a \leq t \leq 140 + a$ and $f(t) = 0$ if $t > 140 + a$ or $t < 140 + a$, where t is the time. The parameter a determines the spread of the uniform distribution. The uniform distribution and its parameter have been specified in such a way that its mean (140 minutes) coincides with the mean of the log-normal distribution and the normal distribution used in the previous simulations. Again, the random demands come from 36 users over a time interval with length 60 000 minutes (1000 hours). Each of the 36 users initiates exactly one demand, randomly located along the operational time interval of 1000 hours. The results for the probability of a full separation of the random demands, for a different spread a of the distribution, are almost identical to the ones shown in Figure 6.3.

In the next simulation experiment, a triangular distribution, with probability density function $f(t) = 0.004762 \times (1 - t/420)$ was used as a model of the demand times. The distribution function and the parameters of the triangular distribution have been specified in such a way that its mean (140 minutes) coincides with the means of the log-normal distribution, the normal distribution and the uniform distribution used in the previous simulations. Random sampling from the triangular distribution has been done by using Von Neumann's rejection method (Ross 1997).

The result from the simulation using demand times following a triangular distribution, were again very close to 0.046, for the probability of a full separation, if a single source is present.

The final simulation involved a single source and only $n = 2$ random demands during a time interval of 17 hours. The random demand times were sampled from a normal distribution with mean $\mu = 140$ minutes. The standard deviation was varied within the range (0–25 minutes).

As can be verified from the results in Figure 6.4, even for the smallest possible number of random events ($n = 2$) which could produce overlapping, the simulated probability of full

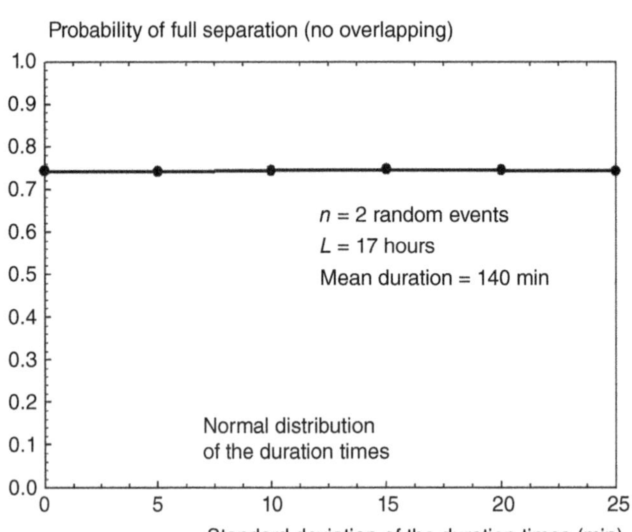

Figure 6.4 Probability of full separation of two random events (no overlapping) as a function of the variance of the normal distribution modelling the duration times.

separation (no overlapping) is still practically insensitive to the variance of the durations of the random demands.

Finally, the duration times were sampled from a discrete distribution with mean equal to 140 minutes. The discrete distribution was defined as follows: with probability 0.3 the event duration was 200.67 minutes and with probability 0.7, the event duration was 114 minutes. The mean duration time of the simulated events is therefore $\mu = 0.3 \times 200.67 + 0.7 \times 114 = 140$ minutes. The simulated probability of no overlapping (full separation) was 0.745 and was, once again, very close to the probability of no overlapping simulated with normal distribution of the event durations with mean 140 minutes.

For two random events and small ratios x_i / L $(i = 1, 2)$, the probability of full separation can be approximated very well by

$$P(B) \approx \left(1 - \frac{\bar{x}}{L}\right)^2 \tag{6.8}$$

The practical insensitivity of the probability of full event separation to the variance of the duration times, for a very small number of random events (two random events), has been analysed in detail in Todinov (2017d).

Equation (6.8) has also been verified by numerous computer simulations.

The simulation results and the theoretical analyses demonstrate that the probability of non-overlapping random events is practically insensitive to the type of distribution of the duration times, provided that the means of the distributions are the same.

The simulation results and the theoretical analyses also demonstrate that the probability of non-overlapping of random events is practically insensitive to the variance of the duration times.

The mathematical model of stochastic separation of a fixed number of risk-critical events also shows that the variance of the demand times X practically has no impact on the level of risk as long as the mean of the demand times remains the same.

These results provide *the valuable opportunity to work with random demand times characterised by their means only and not requiring information related to the variance of the demand times or their distribution.*

6.1.3 Stochastic Separation of Random Events Following a Homogeneous Poisson Process

Suppose that the times of the risk-critical events follow a homogeneous Poisson process in the time interval $(0,L)$ and each event has a duration equal to s. In other words, the number of events in the time interval is a random variable. According to an equation rigorously derived in Todinov (2004a), the probability p_0 that there will be no clustering of two or more random events within a critical distance s is

$$p_0 = \exp(-\lambda L)\left(1 + \lambda L + \frac{\lambda^2 (L-s)^2}{2!} + \ldots + \frac{\lambda^r [L-(r-1)s]^r}{r!}\right) \tag{6.9}$$

where r denotes the maximum number of time gaps of length s, which can be accommodated into the finite time interval with length L ($r = [L/s] + 1$, where $[L/s]$ is the greatest integer which does not exceed the ratio L/s).

6.1.3.1 Case Study: Stochastic Separation of Random Demands Following a Homogeneous Poisson Process

Consider now the real-life problem of requests for a single piece of control equipment during a 24-hour period. The requests arrive randomly (follow a homogeneous Poisson process) with density 0.2 hour^{-1} (on average 2 requests per 10 hours) and the control equipment can service only a single request at a time. Assume that initially $s = 1$ hour has been allocated for servicing each request. Without using a stochastic separation model, the actual probability of unsatisfied request for the control equipment could not be guessed correctly even by experts. The calculation using Eq. (6.9) yields the highly counter-intuitive result that there is approximately 52% probability that within 24 hours there will be a case for which no control equipment will be available on demand. This counter-intuitive result has actually been confirmed by Monte Carlo simulation which also yields 52% for the probability that the equipment will be unavailable on demand.

To reduce the probability that the control equipment will be unavailable on demand to 20%, the equation

$$0.2 = 1 - \exp(-\lambda L)\left(1 + \lambda L + \frac{\lambda^2(L-s)^2}{2!} + \ldots + \frac{\lambda^r[L-(r-1)s]^r}{r!}\right) \tag{6.10}$$

can be solved with respect to s by using the method of repeated bisection. The solution yields, $s \approx 0.25$ hours. In order to provide a stochastic separation that satisfies the tolerable probability of unsatisfied demand of 20%, the use of the control equipment must be restricted to 0.25 hours.

This case study demonstrates that by creating a mathematical model (or algorithm) a significant benefit can be extracted from the method of stochastic separation.

Running the model reveals an interesting behaviour. While for a fixed number of demands, increasing the length of the operational time interval reduces the probability of overlapping, with random demands following a Poisson process on a time interval, increasing the length of the operational time increases the probability of overlapping.

Equation (6.9) can also be used for setting reliability requirements to provide a stochastic time separation (avoiding clustering) of at least s time units, with high probability. For any specified duration of demand s and a minimum probability p_0 with which the separation intervals of length at least s must exist, solving the equation with respect to λ yields an upper bound λ^* (an envelope) for the number density of the random events. The envelope guarantees that whenever for the number density λ of events, $\lambda \leq \lambda^*$ is fulfilled, a specified minimum separation of length at least s will exist between the random events with a minimum probability p_0.

It is necessary to point out that a stochastic time separation can be achieved not only by reducing the number density of random demands and reducing the demand times but also by increasing the number of sources servicing the random demands.

The proposed idea of stochastic time and space separation forms the core of a new methodology for reliability analysis and setting reliability requirements based on minimum separation intervals between random events on a finite interval.

6.1.4 Stochastic Separation Based on the Probability of Overlapping of Random Events for More than a Single Source Servicing the Random Demands

Consider a finite time interval during which a fixed number of consumers place a demand for a particular service independently and randomly. The durations of the random demands come from a uniform distribution with mean $\mu = 140$ minutes, with a range $[\mu - 25, \mu + 25]$ minutes. Unsatisfied demand occurs if a random demand arrives while all sources are engaged in servicing other random demands. Suppose that the maximum tolerable probability of unsatisfied demand is $\alpha = 20\%$ and the random demands follow a homogeneous Poisson process on an operational time interval of 48 hours.

At a specified level of unsatisfied random demand ($\alpha = 20\%$), the maximum tolerable frequency of random demands that can be satisfied by a given number of sources can be determined, such that the probability of unsatisfied demand does not exceed the specified tolerable level.

This process will be referred to as *stochastic separation of random events based on a specified probability of overlapping*. In stochastic separation of random events based on the probability of overlapping, the absence of overlapping of the random events is guaranteed with a specified probability.

For random demands with number densities ranging from 0.01 hour^{-1} to 0.3 hour^{-1}, the simulated probability of unsatisfied demand has been plotted as shown in Figure 6.5. During the simulations, the actual number of random demands on the time interval is a random variable whose realisations were obtained by sampling the Poisson distribution.

The curves correspond to a single source, two sources, and three sources servicing the random demands. The intersections of the horizontal line corresponding to the maximum acceptable probability of unsatisfied demand with the curves representing the simulated probability of unsatisfied demand, determine the optimal number densities of random demands which are equal to 0.05 for a single source, 0.14 for two sources, and 0.265 for three sources. Number densities of demands larger than the optimal values lead to probability of unsatisfied demand larger than the maximum tolerable value of 20%. Number densities smaller than the optimal values, result in an inefficient use of the available sources and unnecessary restriction of the volume of random demands that can potentially be serviced. The obtained number densities provide not only a stochastic separation of

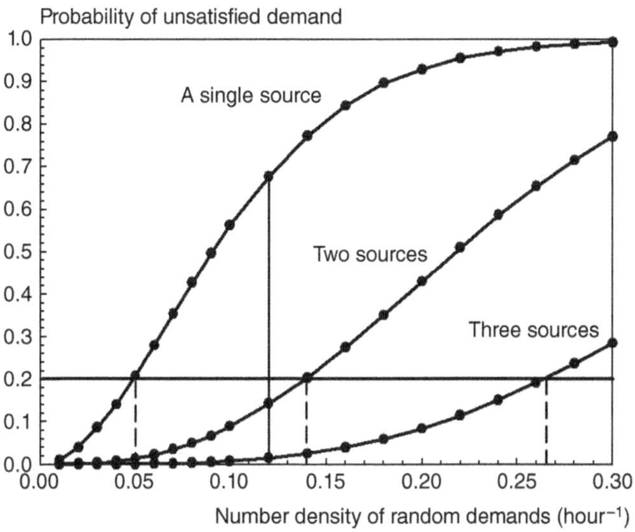

Figure 6.5 Probability of unsatisfied random demand as a function of the number density of random demands, for a different number of sources.

the random events at the maximum tolerable probability of overlapping of 20%; they also guarantee that the specified stochastic separation is characterised by a minimum cost. The optimal stochastic separation of random events has been obtained by using a Monte Carlo simulation algorithm and is characterised by an optimal balance between the probability of overlapping and the cost for reducing the probability of overlapping.

The results in Figure 6.5 also show that the inclusion of an extra source reduces drastically the probability of unsatisfied demand. Thus, for a density of random demands equal to 0.14, the probability of unsatisfied random demand for a single source is 0.77. The inclusion of a second source reduces the probability of unsatisfied demand to 0.2. Consequently, including a second source drastically improves the degree of stochastic separation of random events.

The results from Figure 6.5 can also be used to determine the number of sources servicing a specified volume of random demands which provide a degree of stochastic separation (probability of unsatisfied demand) of 20%.

Thus, for a density of random demands equal to 0.12 hour^{-1}, the optimum number of sources is two. This value is obtained from the closest intersection to the maximum tolerable probability level of 20% of the vertical line corresponding to a number density of 0.12 and the curves (Figure 6.5). A single source leads to an insufficient degree of stochastic separation because the probability of unsatisfied demand is equal to 0.677. Three sources yield too low probability of unsatisfied demand equal to 0.014, which is associated with unnecessary investment in extra sources. The optimal degree of stochastic separation corresponds to a probability of unsatisfied demand equal to 0.142 and is obtained from two sources (Figure 6.5).

The proper level of stochastic separation is the key to finding the optimal balance between the number of available sources and the risk of unsatisfied demand. More supplied sources than the optimal number is costly and undermines the profitability of the enterprise; fewer supplied sources than the optimal number increases the risk of unsatisfied demand.

6.1.5 Computer Simulation Algorithm Determining the Probability of Overlapping for More than a Single Source Servicing the Demands

The times of the start of the random demands are simulated by the product Lu_i, where u_i is a uniformly distributed random number in the interval (0,1) and L is the length of the operation interval $(0,L)$. The duration of the random demand is also a random variable, simulated by sampling the distribution of the random demands. Algorithms for sampling various standard distributions can, for example, be found in Todinov (2016a) and Ross (1997).

For each random demand, two events are generated: a 'start-of-demand' event; and an 'end-of-demand' event. The generated pair of events is 'stamped' with the time of their occurrence and inserted in a list, in ascending order of the time of their occurrence. The first event at the head of the list is always the event with the smallest time of occurrence. The variable 'remaining_sources' shows the number of remaining sources servicing the random demands. The list containing the 'start-of-demand' and 'end-of-demand' events is scanned, by starting with the earliest event.

The algorithm is based on the key observation that an overlap (simultaneous presence) of random demands of order $m+1$ can only appear if upon arrival of a 'start-of-demand' event, all available sources are engaged in servicing other random demands. Consequently, the check for an overlap of order $m+1$ or more (marking unsatisfied demand) is simply reduced to checking whether the number of available sources is equal to zero upon appearance of a 'start-of-demand' event.

In the case of a 'start-of-demand' event, if the current content of the variable 'remaining_sources' is zero, there is an overlap (simultaneous presence) of random demands of order $m+1$ which creates a constraint on the available sources servicing the random demands. In this case, the variable 'count' is incremented and the while-loop is exited with the statement 'break', after which a new simulation is initiated. The variable 'count' counts the number of simulations during which unsatisfied demand is present.

If the content of the variable 'remaining_sources' is greater than zero, the current random demand can be serviced. A resource is then allocated for servicing the random demand which is reflected by decreasing the content of the variable 'remaining_sources' by one.

In the case of an 'end-of-demand' event, a source is released, which is reflected by increasing the content of the variable 'remaining_sources' by one.

At the end of the simulation trials, the ratio of the variable 'count' and the number 'num_trials' of simulation trials gives the probability of unsatisfied demand.

Algorithm 6.1

```
count=0;
num_rand_demands=n;
for i=1 to num_trials do
{
   remaining_sources=m;
   For each random demand, generate a start-of-demand event and an
   end-of-demand event;
   Insert all generated events in a list, ordered according to the time of occurrence
    of the event;
  while (the list is not empty) do
  {
     Take the first event from the list;
     if (event.type='start of demand') then
          {
           if (remaining_sources=0) then
                {
                 count=count+1;
                 break;
                }
           else {
                remaining_sources=remaining_sources - 1;
                remove the current event from the list;
                }
          }
    else // a case of end-of-demand event
           {
           remaining_sources=remaining_sources+1;
           remove the current event from the list;
           }
  }
}
probability_of_unsatisfied_demand=count/num_trials;
```

6.2 Expected Time Fraction of Simultaneous Presence of Critical Events

The expected time fraction of simultaneous presence of critical events on a time interval can also be used as a powerful measure of the risk of simultaneous presence of random events. Consider the common case where the randomly appearing events are random demands for a particular resource/service. Suppose that m units of the resource are available, which can satisfy m simultaneous demands, but not $m+1$ or more simultaneous demands. The random demands have durations d_i and are represented by segments with lengths d_i. The operational time interval is represented by a segment with length L (Figure 6.6).

The risk of unsatisfied demand is estimated by the expected fraction of the time interval $(0,L)$ where more than m overlapping random demands are simultaneously present (Figure 6.6).

Before determining the expected time fraction of simultaneously present random demands, a key theorem will be proved. An overlapped region of order k is present if, for a particular point from that region, exactly k random demands are simultaneously present (an overlapping of order k is present). The duration fractions of the random demands are given by $\psi_i = d_i / L$, $i = 1, \ldots, n$.

Theorem 6.1 The expected overlapped fraction of order k ($k = 0, 1, \ldots, n$) from the time interval $(0,L)$, by n random events with duration fractions ψ_i, coming from a distribution with mean $\overline{\psi}$, is given by the k + 1 term of the binomial expansion $[(1-\overline{\psi})+\overline{\psi}]^n = \sum_{i=0}^{n} \binom{n}{i} \overline{\psi}^i (1-\overline{\psi})^{n-i} = 1$, where $1-\overline{\psi}$ is treated as a single variable $a = (1-\overline{\psi})$ and $\overline{\psi}$ is treated as another variable $b = \overline{\psi}$.

Proof: The expected fraction of the time interval $(0,L)$ covered by exactly m random events can be determined from the probability that a randomly selected point on the time interval $0,L$ will sample simultaneously exactly m overlapping random events. The probability that a randomly selected point on the time interval $(0,L)$ will sample simultaneously exactly m overlapping events is equal to the probability that a fixed point from the time interval $(0,L)$ will be covered exactly m times by randomly placed events in the time interval $(0,L)$. The probability $P(A_m)$ that a fixed point from the interval $(0,L)$ will be covered by exactly m random events out of n random events with duration time fractions $\psi_1, \psi_2, \ldots, \psi_n$ sampled from a distribution with mean $\overline{\psi}$ is given by

$$P(A_m) = \sum_{i1,\ldots,im} (\psi_{i1}\psi_{i2}\cdots\psi_{im}) \prod_{\substack{k=1 \\ k\neq i1;\ldots k\neq im}}^{n} (1-\psi_k) \tag{6.11}$$

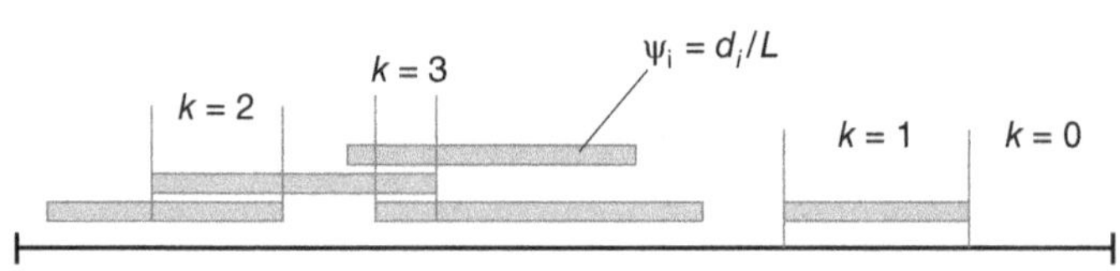

Figure 6.6 Overlapping of different order k for random demands on a time interval.

where $\sum_{i1,\ldots,im}$ is the sum over all distinct combinations of m indices $i1, i2, \ldots, im$ out of n indices $1, 2, \ldots, n$. The number of combinations of m out of n indices is $\binom{n}{m} = \frac{n!}{m!(n-m)!}$. Equation (6.11) gives the sum of the probabilities of $\frac{n!}{m!(n-m)!}$ mutually exclusive events, covering all the cases where exactly m random events cover the selected point and the rest of the random events $(n-m)$ do not.

Next, the expected values of the left- and right-hand sides of Eq. (6.11) are taken:

$$E[P(A_m)] = E\left[\sum_{i1,\ldots,im}\left(\psi_{i1}\psi_{i2}\cdots\psi_{im}\prod_{\substack{k=1\\k\neq i1;\ldots k\neq im}}^{n}(1-\psi_k)\right)\right] \tag{6.12}$$

Because the time fractions ψ_k of the overlapping random events are statistically independent random variables, according to a well-known result in statistics, the expectation of a product of statistically independent random variables is equal to the product of the expectations of the random variables. Consequently, Eq. (6.12) becomes:

$$\begin{aligned} E[P(A_m)] = \alpha &= \sum_{i1,i2,\ldots,im} E\left(\psi_{i1}\psi_{i2}\cdots\psi_{im}\times\prod_{\substack{k=1\\k\neq i1;\ldots;k\neq im}}^{n}(1-\psi_k)\right) \\ &= \sum_{i1,i2,\ldots,im} E[\psi_{i1}]E[\psi_{i2}]\ldots E[\psi_{im}]\times\prod_{\substack{k=1\\k\neq i1;\ldots;k\neq im}}^{n}(1-E[\psi_k]) \\ &= \binom{n}{m}\overline{\psi}^m(1-\overline{\psi})^{n-m} \end{aligned} \tag{6.13}$$

where α is the expected overlapped time fraction of order m and $\overline{\psi}$ is the mean of the distribution from which the events time fractions are sampled.

As a result, the expected overlapped time fractions of different order are given by the separate terms of the binomial expansion of

$$[(1-\overline{\psi})+\overline{\psi}]^n = 1 \tag{6.14}$$

where $(1-\overline{\psi})$ is treated as a single variable $a = (1-\overline{\psi})$ and $\overline{\psi}$ is treated as another variable $b = \overline{\psi}$.

$$[(1-\overline{\psi})+\overline{\psi}]^n = \sum_{i=0}^{n}\binom{n}{i}\overline{\psi}^i(1-\overline{\psi})^{n-i} = 1 \tag{6.15}$$

Suppose that the random events are random demands. Thus, the expected time fraction Δ from the interval $(0,L)$, covered by exactly m random demands, is given by

$$\Delta = \binom{n}{m}\overline{\psi}^m(1-\overline{\psi})^{n-m} \tag{6.16}$$

The expected covered fraction of order m, from the time interval (0,L), depends only on the mean of the distribution of the demand times and does not depend on the type of the distribution of the demand times or the variance of the demand times.

Now consider a case where n random demands come from a distribution with mean $\overline{d}$. The demand time fractions of the separate demands are $\psi_i = d_i/L, i = 1, \ldots, n$. The mean of the demand time fractions ψ_i is given by $\overline{\psi} = \overline{d}/L$. Suppose that a single source can service only a single random demand at a time. The next result then holds.

For n random demands and m sources servicing the demands, the expected time fraction during which $m+1$ *or more demands are simultaneously present on the time interval is given by the expression*

$$\Delta_{\geq m+1} = 1 - \sum_{i=0}^{m} \binom{n}{i} \overline{\psi}^i (1-\overline{\psi})^{n-i} \tag{6.17}$$

where $\overline{\psi}$ is the mean of the distribution of the duration fractions of the demand times. The expected time fraction of unsatisfied demand does not depend on the distribution of the demands or their variance.

This result follows directly from the previous theorem and its proof will be omitted.

If the time fraction of the durations of the random demands is constant: $\psi = d/L$, the expected (average) fraction of time during which exactly m random demands are simultaneously present ($m = 0, 1, \ldots, n$) is given by the terms $\frac{n(n-1)\ldots(n-m+1)}{1\times 2\times \ldots \times m}(1-\psi)^{n-m}\psi^m$, which come from the binomial expansion of the expression $[(1-\psi)+\psi]^n$.

Thus, $(1-\psi)^n$ is the expected fraction of time during which no random demand is present; $n(1-\psi)^{n-1}\psi^1$ is the expected fraction of time during which exactly one random demand is present; $\frac{n(n-1)}{1\times 2}(1-\psi)^{n-2}\psi^2$ is the expected fraction of time during which exactly two random demands are simultaneously present; …; ψ^n is the expected fraction of time during which exactly n random demands are simultaneously present. The sum of all expected fractions of time is equal to one ($[(1-\psi)+\psi]^n = 1$).

Because the sum of all expected fractions is equal to one, the total expected fraction of time during which there is unsatisfied demand ($m+1$ or more random demands are simultaneously present) is given by

$$\Delta_{\geq m+1} = 1 - \left((1-\psi)^n + n(1-\psi)^{n-1}\psi^1 + \frac{n(n-1)}{1\times 2}(1-\psi)^{n-2}\psi^2 + \ldots \right.$$
$$\left. + \frac{n(n-1)\ldots(n-m+1)}{1\times 2\times \ldots \times m}(1-\psi)^{n-m}\psi^m \right) \tag{6.18}$$

Another implication of the result given by Eq. (6.14) is that the fraction of unsatisfied demand is the same as long as the sum of the mean time fractions characterising the separate user demands is the same. To demonstrate this, consider a case study involving different number of users and demand time fractions characterising the separate users such that the sum always remains the same.

6.2.1 Case Study: Expected Fraction of Unsatisfied Demand at a Constant Sum of the Time Fractions of User Demands

Given there are $n = 30$ users, each of which places a random demand over a time interval with length 400 hours. The means of the demands characterising the users are 640 minutes. A single source is servicing the user demands. The sum of the means of all user demands is $S = 30 \times 640 = 19\,200$ minutes.

The time fraction $\overline{\psi}$ characterising each user is $\overline{\psi} = 640/(400 \times 60) = 0.0267$. Unsatisfied demand occurs only if two or more simultaneous random demands are present at a particular time. For a single source ($m = 1$) Eq. (6.18) yields:

$$\Delta_{\geq 2} = 1 - (1 - \overline{\psi})^n - n\overline{\psi} \times (1 - \overline{\psi})^{n-1} = 1 - (1 - 0.0267)^{30}$$
$$- 30 \times 0.0267 \times (1 - 0.0267)^{30-1} \approx 0.19$$

This value was confirmed by a Monte Carlo simulation.

Next, the number of users is doubled to $n = 60$ and the mean duration of the demands characterising the users is halved to 320 minutes so that the sum of the means of all user demands remains the same: $S = 60 \times 320 = 19\,200$ minutes.

The time fraction $\overline{\psi}$ characterising the users is now $\overline{\psi} = 320/(400 \times 60) = 0.0133$. The expected fraction of unsatisfied demand is:

$$\Delta_{\geq 2} = 1 - (1 - 0.0133)^{60} - 60 \times 0.0133 \times (1 - 0.0133)^{60-1} \approx 0.19$$

which is confirmed by a Monte Carlo simulation.

Next, the number of users is doubled to $n = 120$ and the mean duration of the demands characterising the users is halved to 160 minutes so that the sum of the means of all user demands remains the same again: $S = 120 \times 160 = 19\,200$ minutes.

The time fraction $\overline{\psi}$ characterising the users is now $\overline{\psi} = 160/(400 \times 60) = 0.0067$. The expected fraction of unsatisfied demand is:

$$\Delta_{\geq 2} = 1 - (1 - 0.0067)^{120} - 120 \times 0.0067 \times (1 - 0.0067)^{120-1} \approx 0.19$$

As can be verified, the expected fraction of unsatisfied demand remains the same as long as the sum of the mean durations of the users' demands remains the same.

Now consider a case where n consumers demand a particular resource, during an operating period with length L. Suppose that the demand time fractions from each consumer come from a separate distribution with mean $\overline{\psi}_i$ $(i = 1, \ldots, n)$. Denote $\alpha = \frac{1}{n}\sum_{i=1}^{n}\overline{\psi}_i$.

Theorem 6.2 The expected overlapped fractions of order 0, 1, 2,…, n, for n overlapping events on a time interval are given by the terms of the expansion $[(1 - \alpha) + \alpha]^n$ where $(1 - \alpha)$ is treated as a single variable and α is treated as another variable.

Proof: This theorem can be proved by observing that the demand time fractions effectively come from a distribution mixture. Each demand comes with the same probability $1/n$ from separate distributions with means $\overline{\psi}_i$, characterising the durations of the consumers' demands. The mean of the distribution mixture is $\alpha = \frac{1}{n}\sum_{i=1}^{n}\overline{\psi}_i$. Because the overlapped fraction of random demands does not depend on the distribution of random demands but on the mean of the random demands only, n distinct demand fractions with means $\overline{\psi}_i$ $(i = 1, \ldots, n)$, from n consumers, are equivalent to n demands characterised by a time fraction $\alpha = \frac{1}{n}\sum_{i=1}^{n}\overline{\psi}_i$ equal to the mean of the demand fractions characterising the n consumers. As a result, *the expected overlapped fractions of order* 0, 1, 2,…, n, are given by the terms of the binomial expansion of $[(1 - \alpha) + \alpha]^n$.

The theorems discussed earlier and the conclusion that the expected fraction of unsatisfied demand does not depend on the distribution of the demand times have been verified by Monte Carlo simulations involving direct measurement and accumulation of the overlapping regions.

6.2.2 Case Study: Servicing Random Demands from Ten Different Users, Each Characterised by a Distinct Demand Time Fraction

Ten different types of users issue random demands with certainty. The means of the time durations (in minutes) characterising each user are 530, 204, 350, 150, 190, 610, 375, 400, 220, 115, correspondingly. The length of the operational time interval is 300 hours.

If a single server is servicing the random demands, the question of interest is the expected fraction of unsatisfied demand.

According to Theorem 6.2, the expected overlapped fractions of order $0, 1, 2, \ldots, n$ is given by the separate terms of the binomial expansion of $[(1-\alpha)+\alpha]^n = 1$, where

$$\alpha = \frac{1}{n}\sum_{i=1}^{n}\overline{\psi}_i = \frac{1}{10}\sum_{i=1}^{10}(530+204+350+150+190+610+375+400+220+115)/(300\times 60) = 0.01747$$

The expected fraction of unsatisfied demand is:

$$\Delta_{\geq 2} = 1-(1-\alpha)^n - n\alpha\times(1-\alpha)^{n-1} = 1-(1-0.01747)^{10} - 10\times 0.01747 \times(1-0.01747)^{10-1} = 0.0125$$

This result was also confirmed by a Monte Carlo simulation.

6.3 Analytical Method for Determining the Expected Fraction of Unsatisfied Demand for Repair

Theorem 6.1 provides the powerful possibility to determine the degree of unsatisfied demand for repair of any complex system with statistically independent component failures, characterised by constant hazard rates. A constant hazard (failure) rate means that the probability of failure within a short time interval practically does not depend on the age of the component. Suppose that the system includes n components. For components characterised by constant hazard (failure) rates, the probability p_i that a particular component i will be found in a failed state at any given time is

$$p_i = \frac{MTTR_i}{MTTF_i + MTTR_i} \tag{6.19}$$

where $MTTR_i$ is the mean time to repair of the ith component and $MTTF_i$ is the mean time to failure (MTTF) of the ith component (Trivedi 2002). The variable p_i in Eq. (6.19) gives the unavailability of the ith component and is effectively the expected time fraction for repair associated with the ith component during the lifetime of the system. It is assumed that each repair unit is capable of handling only a single failed component at a time. Because the component failures are independent from one another, unsatisfied demand for repair is present when an overlapping of demands for repair is present of order greater than the number of available repair units.

Now consider a case of a system including k types of components. The number of components in each type are $n_1, n_2, \ldots, n_k$, correspondingly $(n_1 + n_2 + \ldots + n_k = n)$. A component from type i is characterised by unavailability p_i, $i = 1, 2, \ldots, k$. Denote $\overline{p} = \sum_{i=1}^{k} \frac{n_i}{n} p_i$.

Theorem 6.3 The expected overlapped fractions of order 0, 1, 2,..., n, for n overlapping repairs on a time interval, are given by the terms of the expansion $[(1 - \overline{p}) + \overline{p}]^n$ where $(1 - \overline{p})$ is treated as a single variable and $\overline{p}$ is treated as another variable.

Proof: This theorem can also be proved by observing that the repair time fractions characterising the components effectively form a distribution mixture. Each demand for repair comes with a probability n_i / n from the type i component, characterised by unavailability p_i, $i = 1, 2, \ldots, k$. The mean of the distribution mixture is $\overline{p} = \sum_{i=1}^{k} \frac{n_i}{n} p_i$. The overlapped fraction of repair times depends only on the mean of the repair times. Consequently, the overlapped time fraction from distinct repair time fractions with means p_i $(i = 1,\ldots,k)$, from n components, is equivalent to the overlapped time fraction from n demands for repair with time fractions equal to $\overline{p} = \frac{1}{n} \sum_{i=1}^{n} p_i$. As a result, *the expected overlapped fractions of order* 0, 1, 2, ..., n, are given by the binomial expansion of $[(1 - \overline{p}) + \overline{p}]^n$.

Given that m repair units are available, the time fraction of unsatisfied demand is given by

$$\Delta_{\geq m+1} = 1 - \left((1 - \overline{p})^n + n(1 - \overline{p})^{n-1} \overline{p}^1 + \frac{n(n-1)}{1 \times 2}(1 - \overline{p})^{n-2} \overline{p}^2 + \ldots \right.$$
$$\left. + \frac{n(n-1)\ldots(n-m+1)}{1 \times 2 \times \ldots \times m}(1 - \overline{p})^{n-m} \overline{p}^m \right) \quad (6.20)$$

6.3.1 Case Study: Servicing Random Repairs from a System Including Components of Three Different Types, Each Characterised by a Distinct Repair Time

A system with three different types of components is present: type 1, including $n_1 = 10$ components with constant hazard rate $\lambda_1 = 0.01\ \text{hour}^{-1}$ and 45 minutes average repair time; type 2, including $n_2 = 14$ components characterised by a constant hazard rate $\lambda_2 = 0.02\ \text{hour}^{-1}$ and 110 minutes average repair time and type 3, including $n_3 = 18$ components characterised by a constant hazard rate $\lambda_3 = 0.035\ \text{hour}^{-1}$ and 14 minutes average repair time. The length of the operational time interval is 450 hours. If a single source is servicing the repairs of failed components, the question of interest is the expected fraction of unsatisfied demand for repair.

The first step is to calculate the unavailabilities characterising the components from each type.

For a constant hazard rate, the link between the MTTF and the hazard rate λ is: $MTTF = 1/\lambda$. Accordingly, $MTTF_1 = 1/\lambda_1 = 1/0.01 = 100$ hours, $MTTF_2 = 1/\lambda_2 = 1/0.02 = 50$ hours, and $MTTF_3 = 1/\lambda_3 = 1/0.035 = 28.57$ hours.

The unavailabilities are determined next:

$$p_1 = \frac{MTTR_1}{MTTR_1 + MTTF_1} = \frac{45/60}{45/60 + 100} = 0.00744;$$

$$p_2 = \frac{MTTR_2}{MTTR_2 + MTTF_2} = \frac{110/60}{110/60 + 50} = 0.0354;$$

$$p_3 = \frac{MTTR_3}{MTTR_3 + MTTF_3} = \frac{14/60}{14/60 + 28.57} = 0.0081$$

According to Theorem 6.3, the expected overlapped fractions of order 0, 1, 2, …, n are given by the separate terms of the binomial expansion of $[(1-\overline{p}) + \overline{p}]^n$ where

$$\overline{p} = \sum_{i=1}^{3} \frac{n_i}{n} p_i = \frac{10}{42} \times 0.00744 + \frac{14}{42} \times 0.0354 + \frac{18}{42} \times 0.0081 = 0.017.$$

The expected fraction of unsatisfied demand is:

$$\begin{aligned}\Delta_{\geq 2} &= 1 - (1-\overline{p})^n - n\overline{p} \times (1-\overline{p})^{n-1} \\ &= 1 - (1-0.017)^{42} - 42 \times 0.017 \times (1-0.017)^{42-1} = 0.16\end{aligned}$$

This result was confirmed by a Monte Carlo simulation.

Theorem 6.3 creates the unique opportunity *to assess the expected time fraction of unsatisfied demand for repair, for any system with components failing independently from one another, without the need for computer simulation*. The expected time fraction of unsatisfied demand for repair is obtained analytically from a simple binomial expansion.

6.4 Expected Time Fraction of Simultaneous Presence of Critical Events that have been Initiated with Specified Probabilities

Now consider the important case of n consumers, each placing a random demand with probability γ, during the time interval $(0,L)$. The time fractions of the random demands come from a distribution with mean $\overline{\psi}$. The next theorem then holds.

Theorem 6.4 *The expected overlapped time fractions of different order, for n consumers each placing a random demand with probability γ, are given by the terms of the binomial expansion of $[(1-\gamma\overline{\psi}) + \gamma\overline{\psi}]^n$, where $1-\gamma\overline{\psi}$ is treated as a single variable $a = (1-\gamma\overline{\psi})$ and $\gamma\overline{\psi}$ is treated as another variable $b = \gamma\overline{\psi}$.*

Proof: Placing a random demand with probability $0 \leq \gamma \leq 1$ is equivalent to sampling the demand time fraction from a distribution mixture defined as follows: The time fractions of the random demands are sampled with probability γ from a distribution with mean $\overline{\psi}$

and with probability $1-\gamma$ from the constant zero. Indeed, sampling zero demand time with probability $1-\gamma$ means a random demand with zero time fraction ($\psi = 0$). A zero demand time fraction cannot possibly contribute to the expected overlapped time fraction of any order and is equivalent to not having a random demand with probability $1-\gamma$.

The mean of the as-defined distribution mixture is $\mu = \gamma \times \overline{\psi} + (1-\gamma) \times 0 = \gamma \times \overline{\psi}$. According to Theorem 6.1, the expected overlapped time fraction of a particular order is the same, irrespective of whether random demands with constant time fraction equal to $\gamma \times \overline{\psi}$ are present or the demand time fractions come from the as-defined distribution mixture with mean $\mu = \gamma \times \overline{\psi}$. Because for random demands with constant time fractions $\mu = \gamma\overline{\psi}$, the expected overlapped time fraction of any particular order is given by the terms of the binomial expansion $[(1-\gamma\overline{\psi}) + \gamma\overline{\psi}]^n$, the theorem has been proven.

The total expected fraction of time during which there is unsatisfied demand ($m+1$ or more random demands are simultaneously present) is then given by

$$\alpha_{\geq m+1} = 1 - \sum_{i=0}^{m} \binom{n}{i} (\gamma\overline{\psi})^i (1-\gamma\overline{\psi})^{n-i} \tag{6.21}$$

The theorem has been verified by Monte Carlo simulations.

From Eq. (6.21), for a fixed number of consumers initiating demands with certain probability γ, reducing the mean $\overline{\psi}$ of the demand times by a certain factor $k > 1$ and reducing the probability of initiating the random demands γ by the same factor $k > 1$ has the same effect on the degree of stochastic separation between the random demands. Alternatively, the degree of stochastic separation does not change if an increase of the mean $\overline{\psi}$ of the demand times by a factor $k > 1$ has been compensated by a decrease by the same factor of the probability γ of initiating a random demand: $(k\overline{\psi}) \times (\gamma/k) = \overline{\psi}\gamma$.

6.4.1 Case Study: Servicing Random Demands from Patients in a Hospital

For an operational time interval of $L = 58$ hours, a single piece of medical equipment is available for precise diagnosis of a particular medical condition developed by some of the $n = 150$ patients located in a surgical hospital unit. Only a single patient can be assessed at a time and the duration of the assessment can vary uniformly within the range (10:40 minutes) The probability with which any of the patients will develop the particular medical state and will need the equipment is $\gamma = 0.6$. The question of interest is the expected percentage of time where two or more patients will need the equipment at the same time.

The mean of the demands from the patients is $(10+40)/2 = 25$ minutes and the mean demand time fraction is $\overline{\psi} = 25/(58 \times 60) = 0.00718$). For $m = 1$ units of medical equipment servicing the random demands from the patients, Eq. (6.21) results in $\alpha_{\geq 2} = 1 - (1-\gamma\overline{\psi})^n - n(\gamma\overline{\psi}) \times (1-\gamma\overline{\psi})^{n-1}$ which, after the substitution of the numerical values, gives $\alpha_{\geq 2} = 0.137$ for the expected fraction of unsatisfied demand. This is the expected percentage of time during which two or more patients will need simultaneously the equipment and this value has been confirmed by a Monte Carlo simulation: $\alpha_{\geq 2} = 0.137$.

Next, the demand durations are assumed to be distributed uniformly in the following intervals: 25 ± 0, 25 ± 5, 25 ± 10, 25 ± 15, 25 ± 20 and 25 ± 25 minutes, all of which have the same mean of 25 minutes. The simulations with each of the intervals resulted in the same expected fraction of unsatisfied demand equal to 0.137. These results confirm the earlier conclusion that the expected fraction of unsatisfied demand is independent of the variance of the demand times.

Consider finally the case where there are m types of consumers $n_1, n_2, \ldots, n_m$ $(n_1, n_2, \ldots, n_m; \sum_{i=1}^{m} n_i = n)$. Each of the consumers from type i, initiates a demand with probability γ_i and the demand time fractions characterising the consumer from type i come from a distribution with mean $\overline{\psi}_i$ $(i = 1,\ldots,m)$. Denote $\beta = \sum_{i=1}^{m} \frac{n_i}{n}\gamma_i\overline{\psi}_i$. Theorem 6.2 can then be generalised to the next theorem.

Theorem 6.5 The expected overlapped fractions of order 0, 1, 2, …, n, for n overlapping events on a time interval are given by the separate terms of the expansion $[(1-\beta)+\beta]^n$ where $(1-\beta)$ is treated as a single variable and β is treated as another variable.

Proof: Each user from type i is characterised by a distribution with mean $\overline{\psi}_i$. Placing a random demand with probability $0 \leq \gamma_i \leq 1$ is equivalent to sampling the demand time fraction from a distribution mixture defined as follows: For type i, the time fractions of the random demands are sampled with probability γ_i from the distribution with mean $\overline{\psi}_i$ and with probability $(1-\gamma_i)$ from the constant zero. Indeed, sampling zero demand time with probability $1-\gamma_i$ means a random demand with zero time fraction $(\psi = 0)$. A zero demand time fraction cannot possibly contribute to the expected overlapped fraction of any order and is equivalent to not having a random demand, with probability $1-\gamma$. The mean of the time fraction from sampling type i user is $\mu_i = \gamma_i \times \overline{\psi}_i + (1-\gamma_i) \times 0 = \gamma_i \times \overline{\psi}_i$. Because a user from type i is selected with probability n_i/n, the mean of the time fraction from sampling all m types of users is $\beta = \sum_{i=1}^{m} \frac{n_i}{n}\gamma_i\overline{\psi}_i$.

Because the overlapped fraction of random demands does not depend on the distribution of the random demands but on the mean of the random demands only, n demand time fractions from all n users, are equivalent to n demand time fractions with means $\beta = \sum_{i=1}^{m} \frac{n_i}{n}\gamma_i\overline{\psi}_i$. As a result, *the expected overlapped fractions of order* 0, 1, 2, …, n, are given by the binomial expansion of $[(1-\beta)+\beta]^n$.

6.4.2 Case Study: Servicing Random Demands from Four Different Types of Users, Each Issuing a Demand with Certain Probability

Four different types of users issue random demands with probabilities $\gamma_1 = 0.3$, $\gamma_2 = 0.5$, $\gamma_3 = 0.7$, and $\gamma_4 = 0.9$. The number of users from each type is $n_1 = 15$, $n_2 = 40$, $n_3 = 20$, $n_4 = 25$, correspondingly. The means of the time durations characterising each type of user are 55, 89, 33, and 44 minutes, correspondingly. The length of the operational time interval is 68 hours; the expected time fractions characterising the separate types of users are $\overline{\psi}_1 = 55/(68 \times 60) = 0.0135$, $\overline{\psi}_2 = 89/(68 \times 60) = 0.0218$, $\overline{\psi}_3 = 33/(68 \times 60) = 0.0081$, and $\overline{\psi}_3 = 44/(68 \times 60) = 0.011$, correspondingly. The total number of users is $n = n_1 + n_2 + n_3 + n_4 = 100$.

If a single source is servicing the random demands, the question of interest is the expected fraction of unsatisfied demand.

According to Theorem 6.4, the expected overlapped fractions of order 0, 1, 2,..., n is given by the separate terms of the binomial expansion of $[(1-\beta)+\beta]^n$, where

$$\beta = \sum_{i=1}^{4} \frac{n_i}{n}\gamma_i \overline{\psi}_i = 0.15 \times 0.3 \times 0.0135 + 0.4 \times 0.5 \times 0.0218 + 0.2 \times 0.7 \times 0.0081 + 0.25 \times 0.9 \times 0.011 = 0.00857$$

The expected fraction of unsatisfied demand is:

$$\Delta_{\geq 2} = 1-(1-\beta)^n - n\beta \times (1-\beta)^{n-1} = 1-(1-0.00857)^{100} - 100 \times 0.00857 \times (1-0.00857)^{100-1} = 0.21$$

This result was confirmed by a Monte Carlo simulation.

6.5 Stochastic Separation Based on the Expected Fraction of Unsatisfied Demand

6.5.1 Fixed Number of Random Demands on a Time Interval

Consider a finite time interval during which a number of consumers place a single demand for a particular service independently and randomly, during a time interval 0,L. The duration times come from a distribution with mean d. The mean demand time fraction is therefore $\overline{\psi} = d/L$. Unsatisfied demand occurs if a random demand arrives while all m available sources are engaged in servicing random demands. Suppose that the maximum tolerable expected fraction of unsatisfied demand is α.

Solving Eq. (6.22) with respect to n then yields the maximum number of random demands that can be serviced at the expected fraction of unsatisfied demand α:

$$\alpha = 1 - \sum_{i=0}^{m} \binom{n}{i} \overline{\psi}^i (1-\overline{\psi})^{n-i} \tag{6.22}$$

This equation can be solved by a repeated bisection.

Figure 6.7 gives the expected fraction of unsatisfied demand as a function of the number n of random demands. The figure corresponds to a demand time fraction $\overline{\psi} = d/L = 0.15$ of an individual demand. Consider a maximum acceptable expected time fraction of unsatisfied demand $\alpha = 10\%$ and a number of available sources $m = 3$.

By using repeated bisection, it was determined that at most 12 users can be serviced by the sources without exceeding the specified level $\alpha = 10\%$ of expected time fraction of unsatisfied demand.

At another specified level of the expected time fraction of unsatisfied demand (for example, $\alpha = 20\%$), the maximum number n of consumers that can be serviced can be determined directly from the corresponding curve (Figure 6.7). As can be verified from the plots in Figure 6.7, increasing the number of sources m increases significantly the number of consumers that can be serviced without exceeding the tolerable risk of unsatisfied demand.

Suppose that the number of consumers n (each placing a single random demand on the time interval) and the maximum tolerable expected fraction of unsatisfied demand α have

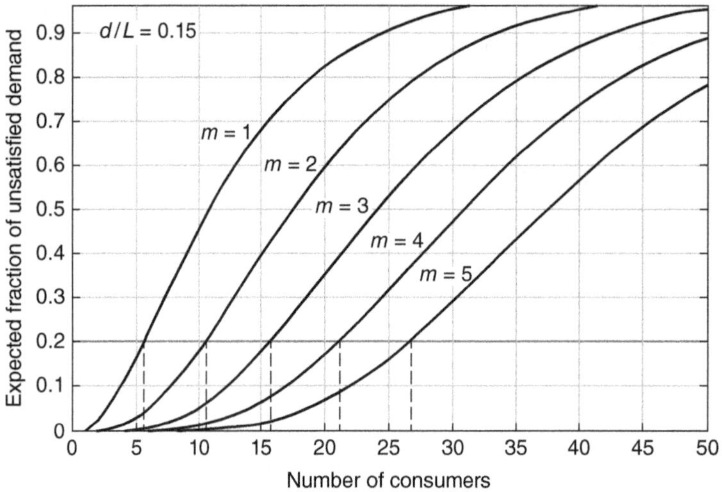

Figure 6.7 Expected fraction of unsatisfied random demand as a function of the number of consumers n, for different number of sources m servicing the demands.

been specified. Solving Eq. (6.22) with respect to m now yields the number of sources required to service n consumers such that the expected fraction of unsatisfied demand does not exceed α. The solutions complying with these requirements provide a *stochastic separation of random events based on the expected overlapped time fraction*. The stochastic separation, based on the expected overlapped time fraction, guarantees that the expected degree of overlapping will be equal to or below the specified tolerable level.

Finding the minimum number of sources which guarantees a degree of stochastic separation based on the expected overlapped time fraction is critical in striking the right balance between the risk of unsatisfied demand and costs. If the servicing sources are medical personnel, repairmen, or extra equipment, increasing the number of sources increases the degree of stochastic separation but also increases the salary costs and other investment, which undermines profit. Too few sources means increased risk of unsatisfied demand, risk of fatalities, damage to health, dissatisfied customers, etc. Again, Eq. (6.22) can be solved with respect to m, by keeping the number of consumers n and the demand time fraction $\overline{\psi} = d/L$ constant.

Consider a maximum acceptable level of the expected time fraction of unsatisfied demand $\alpha = 10\%$ and a number of consumers $n = 20$, each characterised by a demand time fraction $\overline{\psi} = d/L = 0.2$. For these input data, the repeated bisection algorithm determined that the minimum number of sources must be six, in order to achieve the 10% expected fraction of unsatisfied demand which is the specified degree of stochastic separation between the random events.

6.5.2 Random Demands Following a Poisson Process on a Time Interval

Suppose now that the random demands follow a homogeneous Poisson process with density λ, on a time interval with length L. The number of available sources is m. The durations of the random demands follow a particular distribution with mean $\overline{\psi}$. Again, unsatisfied demand is present only if the number of random demands is greater than the number of available servicing sources m. The expected fraction of unsatisfied demand α on the time interval (0,L) is equal to the probability that a randomly selected point on the time interval 0,L will sample simultaneously exactly $m+1$ or more overlapping random demands. The probability that a randomly selected point on the time interval (0,L) will sample $m+1$ or more overlapping demands is equal to the probability that a fixed point from the time interval (0,L) will be covered $m+1$ or more times by randomly placed demands in the time interval (0,L).

A selected point in the time interval (0,L) can be covered $m+1$ or more times in the following mutually exclusive ways: there are exactly $m+1$ random demands in the interval (0,L) and the selected point is covered by all of them; there are exactly $m+2$ random demands in the time interval (0,L) and the selected point is covered $m+1$ or more times; there are exactly $m+3$ random demands in the time interval (0,L) and the selected point is covered $m+1$ or more times; and so on. The probability α that the selected point will be covered $m+1$ or more times can then be determined as a sum of the probabilities of the following mutually exclusive events: (i) exactly $m+1$ demands are present in the time interval (event A_{m+1} characterised by a probability $P(A_{m+1})$) and the selected point is covered by $m+1$ or more random demands (event B, characterised by a conditional probability $P(B| m+1)$); (ii) exactly $m+2$ demands are present in the time interval (event A_{m+2}, characterised by a probability $P(A_{m+2})$) and the selected point is covered by $m+1$ or more random events (event B, characterised by a conditional probability $P(B| m+2)$ and so on). For the probability α of covering the selected point, the total probability theorem yields the expression:

$$\alpha = P(A_{m+1}) \times P(B|m+1) + P(A_{m+2}) \times P(B|m+2) + \dots \tag{6.23}$$

For random demands following a homogeneous Poisson process with density λ, on a time interval with length L, the probability that there will be exactly $m+1$ random demands on the time interval 0,L is given by $P(A_{m+1}) = \frac{e^{-\lambda L}}{(m+1)!}(\lambda L)^{m+1}$; the probability that there will be exactly $m+2$ random demands on the time interval 0,L is given by $P(A_{m+2}) = \frac{e^{-\lambda L}}{(m+2)!}(\lambda L)^{m+2}$ and so on. According to Section 6.2, the probability that the coverage of the selected point will be of order $m+1$ or higher, given that there are $n \geq m+1$ random demands on the time interval, is given by $P(B|n) = 1 - \sum_{i=0}^{m} \binom{n}{i} \overline{\psi}^i (1-\overline{\psi})^{n-i}$. For $n \leq m$, the probability that the coverage of the selected point will be of order $m+1$ or higher, is zero.

The substitution in Eq. (6.23) yields

$$\alpha = \sum_{n=m+1}^{\infty} \frac{e^{-\lambda L}}{n!}(\lambda L)^n \left[1 - \sum_{i=0}^{m} \binom{n}{i} \overline{\psi}^i (1-\overline{\psi})^{n-i}\right] \tag{6.24}$$

for the probability that the selected point will be covered: this is also the expected fraction α of unsatisfied demand.

For practical purposes, the computation of the sum (6.24) can be truncated with negligible error at $n = N$, which is, for example, 10 times larger than the expected number λL of demands in the time interval 0,L ($N = 10 \times (\lambda L)$). Equation (6.24) then becomes

$$\alpha = \sum_{n=m+1}^{N} \frac{e^{-\lambda L}}{n!}(\lambda L)^n \left[1 - \sum_{i=0}^{m} \binom{n}{i} \overline{\psi}^i (1-\overline{\psi})^{n-i}\right] \tag{6.25}$$

6.5.2.1 Case Study: Servicing Random Failures from Circular Knitting Machines by an Optimal Number of Repairmen

Consider demands for maintenance/servicing from circular knitting machines in a plant. The frequency of demands for maintenance from the machines is $\lambda = 7$ hour^{-1}, and the average duration of a single maintenance is $d = 15$ minutes. During maintenance, the operation of the knitting machine is stopped. The demands for maintenance follow a homogeneous Poisson process during the operational period with length $L = 68$ hours. The maximum tolerable fraction of unsatisfied demand for maintenance is 30% (the fraction of time during which one or more knitting machines have to wait to be fixed). The question of interest is the minimum number of repairmen necessary to guarantee the maximum tolerable fraction of unsatisfied demand for maintenance of 30%.

In the case of a single repairman conducting all maintenance, $m = 1$, Eq. (6.25) becomes

$$\alpha = \sum_{n=2}^{N} \frac{e^{-\lambda L}}{n!}(\lambda L)^n[1 - (1 - \overline{\psi})^n - n(1 - \overline{\psi})^{n-1}\overline{\psi}] \tag{6.26}$$

This equation yields $\alpha = 0.52$ as expected fraction of unsatisfied demand, which is unacceptable. (This value has been confirmed by a Monte Carlo simulation.)

For three repairmen servicing the random demands for maintenance ($m = 3$), Eq. (6.25) becomes

$$\alpha = \sum_{n=4}^{N} \frac{e^{-\lambda L}}{n!}(\lambda L)^n \times \left[1 - (1 - \overline{\psi})^n - n(1 - \overline{\psi})^{n-1}\overline{\psi} - \frac{n(n-1)}{2!}(1 - \overline{\psi})^{n-2}\overline{\psi}^2 - \frac{n(n-1)(n-2)}{3!}(1 - \overline{\psi})^{n-3}\overline{\psi}^3\right] \tag{6.27}$$

For the same input data, Eq. (6.27) yields $\alpha = 0.1$ expected fraction of unsatisfied demand for maintenance, which has also been confirmed by a Monte Carlo simulation. This level of unsatisfied demand is now acceptable but the number of repairmen (and the cost of salaries) is three times higher.

For two repairmen servicing the random demands for maintenance ($m = 2$), Eq. (6.25) becomes

$$\alpha = \sum_{n=3}^{N} \frac{e^{-\lambda L}}{n!}(\lambda L)^n \left[1 - (1 - \overline{\psi})^n - n(1 - \overline{\psi})^{n-1}\overline{\psi} - \frac{n(n-1)}{2}(1 - \overline{\psi})^{n-2}\overline{\psi}^2\right] \tag{6.28}$$

For the same input data, Eq. (6.28) yields $\alpha = 0.256$, which is below the maximum tolerable limit. (This value has also been confirmed by a Monte Carlo simulation.)

Two repairmen is the optimal choice. Increasing the number of repairmen to three increases the degree of stochastic separation but also significantly increases salary costs which undermines profit. Reducing the number of repairmen to one gives an unacceptably high fraction of unsatisfied demand for maintenance which slows down production and also undermines profit.

In the case of random failures following a homogeneous Poisson process, including additional repair personnel significantly reduces the expected fraction of unsatisfied demand. Finding the minimum number of repairmen which guarantees a degree of stochastic separation based on the maximum tolerable expected overlapped fraction is critical in striking the right balance between the risk of unsatisfied demand for maintenance and cost of maintenance.

Using the closed-form Expression (6.25) is significantly more efficient compared with estimating the expected fraction of unsatisfied demand by a direct Monte Carlo simulation. The closed-form expressions permit their embedding in fast optimisation loops thereby reducing dramatically the computation time. Furthermore, the closed-form expressions also permit important inferences to be made about the properties of a system consisting of consumers placing random demands on a time interval.

7

Improving Reliability and Reducing Risk by Segmentation

7.1 Segmentation as a Problem-Solving Strategy

The present chapter introduces an important domain-independent risk reduction method referred to as *the method of segmentation*. Segmentation is the act of dividing an entity (assembly, system, process, task, time, etc.) into a number of distinct parts.

Reducing the variation of returns by segmenting and diversifying an investment portfolio into many non-correlated stocks is a well-documented technique for reducing financial risk by segmentation. With increasing the number of non-correlated stocks, the variance (volatility) of the portfolio, which is a measure of the risk associated with the portfolio returns, is reduced significantly (Teall and Hasan 2002).

Micro-segmentation, aimed at improving the cyber security by isolating different applications and parts of computer networks, has been discussed in Mämmelä et al. (2016). The method of segmentation has been used as one of the principles for resolving technical contradictions in the development of TRIZ methodology for inventive problem solving (Altshuller 1984, 1996, 1999). However, the formulated principle of segmentation was primarily formulated as a tool for generating inventive solutions by resolving technical or physical contradictions and not as a tool for reliability improvement and risk reduction. Examples featuring segmentation as a tool for resolving technical contradictions have been discussed in Altshuller (1984, 1996, 1999), but no specific discussion has been provided related to the mechanisms through which segmentation reduces risk. No discussion regarding the mechanisms through which segmentation reduces risk exist in more recent literature related to TRIZ (Terninko et al. 1998; Savransky 2000; Orloff 2006, 2012; Rantanen and Domb 2008; Gadd 2011).

In addition, segmentation as a problem-solving tool in TRIZ has been introduced in a rather narrow context: primarily as size segmentation or time segmentation. However, a physical division of the size is not the only instance when segmentation is present. Segmentation is also present when no physical division is done but additional boundaries with different properties are introduced in the homogeneous component. Such is the case of welding stiffening rings around an underwater pipeline, at regular intervals. The purpose of these rings is to contain the eventual collapse of the pipeline between two welded rings, thereby minimising the extent of damage. Segmentation is present without a physical division of the whole object. In addition, TRIZ methodology does not consider *logical segmentation* where no physical division exists yet the system is effectively segmented. In a logical segmentation of a computer network into several distinct parts, for example,

Methods for Reliability Improvement and Risk Reduction, First Edition. Michael Todinov.

no reduction of size or complexity is present. In effect, during logical segmentation, the barriers set between the different parts of the network increase complexity.

Essentially, segmentation has been used as a very efficient problem-solving tool, long before the emergence of TRIZ methodology and any other methodology for creative problem solving. Thus, a powerful problem-solving strategy based on segmentation has been known for a long time – the divide-and-conquer strategy. The divide-and-conquer approach breaks a problem into simpler sub-problems for which solutions can be obtained with the available means. Next, the obtained solutions are merged until the solution of the initial problem is obtained. The divide-and-conquer approach is at the heart of the *mergesort algorithm* for sorting arrays of large size n, whose worst-case running time is $O(n\ln_2 n)$ (Sedgewick 1992).

Segmentation is at the heart of the decomposition method for system reliability analysis (Todinov 2016a) where the initial complex system is decomposed into four simpler systems depending on the four distinct states of two key components. The system reliability of the initial system is obtained by aggregating (combining) the system reliabilities of the resultant four simpler systems. Finally, the segmentation of a complex task into multiple simpler manageable sub-tasks has always been the cornerstone in managing the execution of projects. Evaluating a complex mathematical expression, by segmenting it into simpler terms which are evaluated separately, reduces the likelihood of an error during the evaluation.

Segmentation is also at the heart of one of the biggest inventions in mathematics – differential and integral calculus. The chain rule for differentiating a function of a function is effectively an application of the method of segmentation. Segmenting a complex domain into multiple simpler sub-domains permits determining its area/volume. Segmenting a large time interval into multiple very small time intervals permits presenting a complex motion by a series of simple motions and determining the travelled distance. In all these cases, segmentation works because it replaces a complex problem with a series of sub-problems with trivial solutions.

Indeed, if the continuous function $f(x)$ has been defined on a domain D with boundaries the vertical lines $x=a, x=b$ (Figure 7.1), the area S beneath the plot of the function $f(x)$ can be obtained as a sum of the areas $dS=f(x)dx$ of infinitesimally small space strips (segments) with lengths $f(x)$ and thickness dx.

The area S beneath the plot of the function $y=f(x)$ is determined by integration:

$$S = \int_a^b f(x)\,dx$$

which is effectively a summation of the areas $dS=f(x)dx$ of all space segments within the time interval $[a, b]$. The method of space segmentation works because the area of each elementary space segment has been approximated by a rectangle with area $dS=f(x)dx$. The complex shape of the initial domain has been approximated by a series of simple rectangles.

The double integration is also effectively done by segmentation. In determining the volume of an object with complex shape by double integration, the volume is essentially converted into segments with the same infinitesimal thickness and with cross-sectional area dependent on the position along one of the coordinate axes. In turn, the cross-sectional area of each slice is essentially converted into segments (strips) with heights dependent on the position of the strip along another coordinate axis. As a result, the evaluation of the complex volume is reduced to two sequential summations (integrations) involving segments.

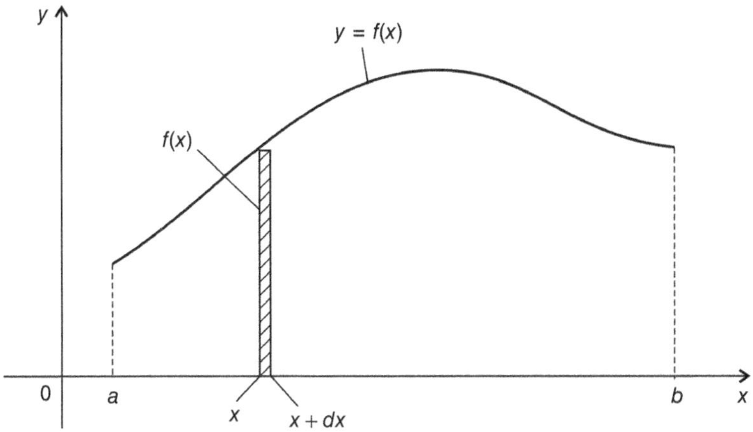

Figure 7.1 Segmentation during integration.

Techniques, which involve acts of segmentation, have already been used for reliability improvement and risk reduction. Despite the existence of a number of isolated applications, which involve segmentation, they have not been linked with the general method of segmentation and no systematic analysis of the various mechanisms through which segmentation improves reliability and reduces risk currently exists. In Todinov (2015), it has been suggested that reliability improvement and risk reduction can be achieved by the method of segmentation and this idea was subsequently further developed (Todinov 2017c).

7.2 Creating a Modular System by Segmentation

An important mechanism through which segmentation improves reliability and reduces risk is by creating self-contained modules, with clearly specified input and output, linked with other modules. The advantages of the modular system is that a fault or failure is confined in one of the modules which makes the identification of the error and its repair much easier compared with a structure that has not been segmented. The faulty module can be replaced/bypassed without the need of replacing the entire system.

Creating a modular system by segmentation is an efficient way of managing complexity in contemporary software systems without sacrificing transparency and reliability. Segmentation is the basis of nestedness. A module is composed of modules, which in turn are composed of modules, and so on. Nestedness permits defining and managing very complex structures including a very large number of elements. Thus, a network could be defined, consisting of nodes and connections among them. Each of the nodes can be a network itself consisting of nodes and links among them, and so on.

Nestedness also provides economy of thinking which permits a simple description of very complex algorithms thereby reducing the likelihood of errors. This is evident in recursive algorithms which make active use of nestedness to decompose a problem of size *n* into two sub-problems of size *n*/2, for example, through recursive calls. In turn, the recursive calls decompose each of the two sub-problems of size *n*/2 into two sub-problems of size *n*/4, and so on until trivial sub-problems are reached.

The modular structure makes it possible to successfully coordinate the work of many developers at different geographic locations and assemble a very complex software system with minimum amount of errors, in a very short time. In developing software with modular structure, many developers can work in parallel which permits a separation: using the best developer skills for each particular module, e.g. the best developers for the user interface, for the numerical routines, and for the database programming.

In software development, modularisation permits focussing on the interaction between modules rather than the content of each module which can be thought of as a 'black box'. This approach makes it easy to develop the algorithm correctly and choose the correct methods.

Furthermore, modular structures permit reusing standard components whose high reliability is guaranteed, instead of developing/manufacturing them. This is particularly relevant to designing reliable software. A number of libraries with high-quality reliable routines are already available (for example, for standard user interfaces, for solving algebraic and differential equations, for determining extrema of functions, etc.). They require only a specified format of the input. These have been developed by expert teams, extensively tested, and their high reliability is guaranteed. Using standard components instead of developing them from scratch drastically reduces the development time and the presence of bugs in the developed software.

A monolithic piece of code is highly interconnected, unreadable, and error-prone. The high degree of inter-connectedness creates the so called 'spaghetti type code', where a small modification of the code in one place often creates dangerous and difficult to isolate bugs.

Modules in software programs typically consist of a small number of statements. This makes their testing easy and drastically reduces the number of programming faults (bugs). Within a piece of software, a module can be tested without fully developing it, by making it return a constant value, for example. Later, the module can be fully developed by using a specific algorithm for evaluating the precise value. The errors can be easily traced to a particular module which can be debugged step-by-step, instead of debugging the code of a large interconnected program which may include thousands of statements.

Finally, a modular structure of a piece of software permits easy updating of the different functions with better/faster functions without disrupting the architecture of the system and the established connections between the separate functions.

The life of a mechanical component can also be extended by segmenting it into replaceable sections. This creates the possibility to replace a failed section without having to replace the entire component.

In mechanical systems, for example, this technique produces maximum benefit if zones subjected to intensive wear are identified and subsequently segmented into replaceable sections. Changing a worn out replaceable section avoids replacing the entire component.

The life of a conveyor belt carrying tiles for polishing, for example, can be increased significantly by applying this method. After some period of continuous operation of the production line, the surface of the conveyor belt will be eroded by the abrasive action of the tiles. To avoid the replacement of the entire conveyor belt, the surface zones of the conveyor belt, which are in direct contact with the polished tiles and undergo the most intensive wear, are designed as replaceable segments. Instead of replacing the expensive conveyor belt, only the surface segments are replaced or repaired (for example, by restoring the worn rubber coating on the steel plates). This technique has been widely used in replacing inserts for journal bearings and pads for friction brakes.

Segmenting a complex device into separate blocks greatly reduces the downtime for repair, because such a device is easy to disassemble and assemble. Identifying the block where failure originated is easy and the failed block can be replaced quickly, without having to dismantle the entire device. The reduced downtime for repair increases the availability and efficiency of the production process.

The logical segmentation of critical information into categories and classes, decreases significantly the time for retrieval of a key piece of information. This is of particular importance during troubleshooting, when a correct decision must be made very quickly in order to avert an accident or minimise the consequences from an accident.

Segmentation can also be used to resolve a major common technical contradiction – reducing the weight of designs. This can be done by splitting components into light segments connected with light joints. Segmenting a heavy beam into a light truce formed by light straight members connected through joints is a demonstration of this technique.

This technique can also be applied by segmenting components into working zones, linked with light bridges, which take directly the contact stresses and wear. This method is used in packaging of goods where, instead of heavy packaging boxes, only packaging segments along the edges of the packaged items are used, connected with light links. Segmentation is also used whenever a heavy monolithic foundation is replaced by multiple lighter foundation segments providing the required support.

7.3 Preventing Damage Accumulation and Limiting Damage Propagation by Segmentation

7.3.1 Creating Barriers Containing Damage

These are probably one of the most important mechanisms by which segmentation improves reliability and reduces risk: *(i) by creating barriers which prevent damage accumulation; (ii) by creating barriers which prevent the spread of damage; and (iii) by creating deliberate weak boundaries dissipating or deflecting damage.*

Since many failures are initiated when damage accumulates beyond a critical level (e.g. fatigue failure), segmentation can be used to delay the failure occurrence. Delaying damage accumulation essentially reduces the likelihood of reaching a threshold level of damage after which the damage starts propagating. The strong physical boundaries created by the segmentation also prevent the propagation of damage, which is another distinct mechanism through which segmentation reduces the risk of failure.

The strong boundaries created by segmentation can be used to block the pathways through which the damage accumulates or to delay the spread of damage. Segmentation can physically isolate and contain the consequences in case of failure. Space segmentation by introducing blast walls reduces damage from blast waves and is an efficient measure against domino-type failures of fuel tanks and pressure vessels built in close proximity. The boundaries created by segmentation provide passive protection against the spread of fire, radiation, toxic substances, or dangerous operating conditions.

The boundaries resulting from segmentation prevent damage from building to a critical level which precipitates failure. Thus, preventing the formation of large build-ups of snow, water, overheated water vapour, etc., by a physical segmentation, reduces the likelihood of accidental release of a large amount of potential energy.

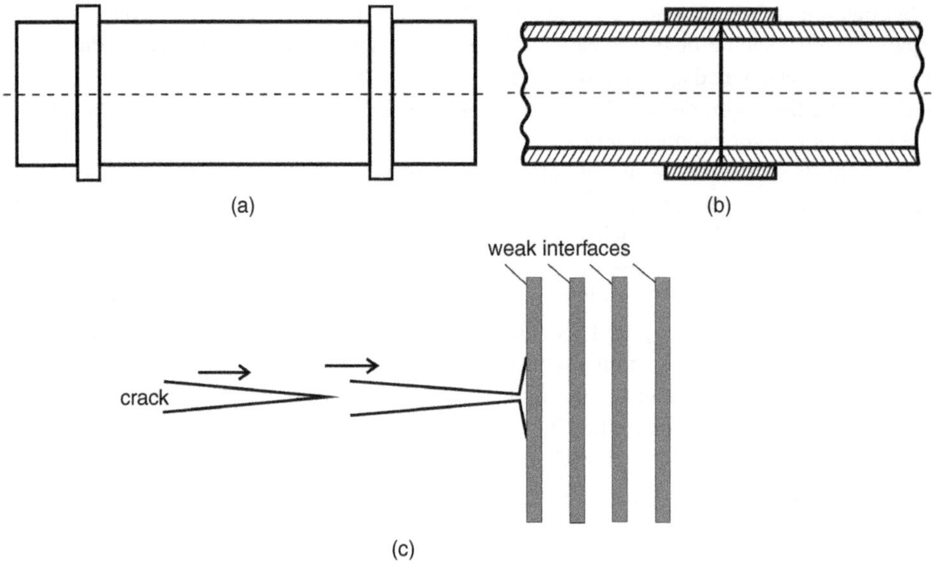

Figure 7.2 (a) A segmented pipeline with strong boundaries (stiffened welded rings); (b) a segmented pipe with weak boundaries (with only a contact between the two surfaces); and (c) a layered structure with weak interfaces deflecting damage.

Crack arrestors as strips or rings made of tougher material (Figure 7.2a) is an example of segmentation limiting the spread of damage by creating strong boundaries. The mechanism of crack arresting consists of reducing the strain energy flow to the crack tip upon encountering a tougher material strip.

Another example of segmentation by creating strong boundaries can be given with buckling of a pipeline subjected to a high external hydrostatic pressure. Buckling could be eliminated by increasing the thickness of the pipeline but this option is associated with significant costs. Control of buckling propagation achieved by using buckle arrestors is a cheaper and more preferable option. Buckle arrestors are thick steel rings welded or attached at regular intervals to the pipeline in order to halt the propagating buckle and confine damage to a relatively small section (Figure 7.2a). In this way, the losses from buckling are limited to the length of the section between two buckle arrestors. In case of failure, only the buckled section will be cut and replaced. The spacing between buckle arrestors can be optimised on the basis of a cost–benefit balance between the cost of installation of the arrestors and the expected cost of intervention and repair. In this case, risk is reduced by reducing the consequences of failure.

Segmentation limiting the spread of damage by creating boundaries can be used for reducing technical risk in a wide range of applications. It has been used effectively to increase the resistance of ships to flooding. The volume of the hull is divided into watertight compartments. If flooding is localised, only a single or a few compartments are affected, which allows the ship to retain buoyancy. For a long time, space segmentation with fireproof doors of the corridors in a building, has been used to delay the spread of fire. In these cases, risk is reduced by reducing the consequences of failure.

Segmenting large formations of people into smaller groups has been used to prevent the spread of infectious diseases. Valuable assets can be protected by reducing the likelihood of accessing them or by reducing the amount of accessible assets through segmentation.

Segmentation of computer networks limits the spread of damage and has clear security benefits. For a segmented computer network, accessing a computer in one segment does not automatically give the attacker easy access to other segments. A segmented network design can significantly slow down the speed with which an attacker moves towards the valuable service and provides more opportunities for a successful detection. In addition, securing each segment by firewalls makes accessing the valuable service much more difficult because numerous security walls must be breached before access could be gained. The result is a significantly reduced likelihood of unauthorised access.

7.3.2 Creating Weak Interfaces Dissipating or Deflecting Damage

Segmentation also works through the mechanism of weak interfaces. The weak boundaries resulting from segmentation often help dissipate the damage escalation given that failure has occurred. In Figure 7.2b, the crack is arrested at the edge of the pipeline section because, at the edge, the two segments are only placed in contact. This is effectively a weak interface which dissipates the energy of the crack and prevents the crack from spreading into the neighbouring segment. As a result, segmenting a pipe into many parts with weak interfaces (Figure 7.2b) confines the damage within a single segment only, which reduces significantly the consequences from failure. The weak boundaries provided by the segmentation of composite material with multiple layers deflect propagating cracks (Figure 7.2c). As a result, the propagation of the crack into the next layer is delayed and the fatigue life is increased.

7.3.3 Reducing Deformations and Stresses by Segmentation

Segmentation improves the flexibility of assemblies which improves their capability to align and accommodate strains without excessive elastic and plastic deformations and stresses.

Thermal deformations and thermal stresses, for example, can be reduced by a geometrical segmentation which consists of segments separated by expansion gaps. Thus, the thermal expansion gaps left between rail segments and between building panels helps to accommodate thermal expansion strains and reduce the thermal stresses whose magnitudes are otherwise capable of destroying the structures.

Damage appearing as excessive deformation can be reduced by segmenting a process associated with a large amount of energy into intermediate processes, each associated with a smaller amount of energy. The result is greater control, minimal intermediate deformations, minimal total deformation, and minimal risk of failure.

For example, instead of applying the full thermal load in a single weld bead, the weld bead can be segmented into several smaller multi-pass welds. Each of the smaller welds introduces a smaller amount of thermal energy and the result is a significantly reduced geometric distortion of the welded part compared with welding with a single weld bead, associated with a large amount of input energy.

7.3.4 Reducing Hazard Potential by Segmentation

Segmentation of hazardous substances can be applied with success to limit the amount of energy locked in the substance and its potential to cause harm. Processing very small (segmented) volumes of toxic substances at a time, for example, significantly reduces the hazard potential of the handled substance and eliminates the risk of poisoning in the case of accidental spillage. Segmentation of hazardous substances avoids steps that could potentially cause harm and is fully in line with the 'precautionary principle' in risk management (Sunstein 2002). What makes the segmentation method particularly appealing is that it does not require external resources to achieve risk reduction. Risk reduction is achieved within the current system, without the involvement of auxiliary resources. Normally, the segmentation method does not involve big investment to reduce risks and leads to economical risk reduction solutions. Handling only a limited quantity of harmful substance at a time does not involve substantial investment but it does guarantee that, in the case of accident, the potential harm will be small. Unlike some other risk reduction measures, segmentation is a low-cost risk reduction method and does not lead to indefensibly big expenses exhausting safety budgets.

7.3.5 Reducing the Likelihood of Errors by Segmenting Operations

Segmentation of an operation or design into smaller incremental parts can be used with success for reducing the possibility of errors. A design with clearly defined small steps, each of which is assessed and tested thoroughly before moving to the next stage, permits control over each step and easy identification of the source of errors. Upon identification of an error, the operation or design can be easily rolled back to the previous stage where the error was missing. Adding finer incremental parts one-by-one permits the exact location of the step at which the error was introduced. This is the underlying idea behind incremental design. It is particularly efficient in the design of complex software algorithms.

The principle of segmentation is also efficient in reducing the likelihood of errors in evaluating complex expressions. The complex expression is segmented by using temporary variables. For example, to reduce the likelihood of errors in evaluating the complex expression

$$y = \ln \sqrt{t + \ln \sqrt{3 - (\sin^3 x + 2\cos^3 x)^2}} \tag{7.1}$$

the expression is segmented into simpler expressions by using the temporary variables a, b, and c: $a = \sin^3 x + 2\cos^3 x$; $b = 3 - a^2$; and $c = t + \ln \sqrt{b}$.

The evaluation of expression (7.1) reduces to the evaluation of $y = \ln \sqrt{c}$.

7.3.6 Limiting the Presence of Flaws by Segmentation

Segmentation to a very small size (e.g. spinning very thin glass fibres) does not allow an imperfection to be present and this is another mechanism through which segmentation prevents accumulation of damage and decreases the likelihood of failure. This is one of the reasons for the exceptional strength of glass fibres and carbon fibres, for example.

Glass fibres are spun from molten glass and are with diameters between 10 μm and 100 μm. The absence of imperfections, because of the very small cross sections, contributes to their exceptional tensile strength. Because of this, they are used for reinforcement in glass fibre polymers.

Carbon fibres are very thin (<10 μm in diameter) and also possess exceptional tensile strength due to very small sections and the absence of imperfections. They are commonly woven into textiles and used as reinforcement in polymer, metal or carbon matrices (Ashby and Jones 2002).

7.4 Improving Fault Tolerance and Reducing Vulnerability to a Single Failure by Segmentation

A common failure mechanism is locally initiated damage which spreads to cause a total collapse of a component or structure. Segmentation can be used to improve the resistance to local damage which subsequently spreads and causes total collapse.

A monolithic glass panel will shatter totally if hit by an object because the initial crack from the projectile spreads through the entire panel. Segmenting the glass panel into small glass segments makes the panel resistant to local damage. A glass panel made of small glass segments will suffer only local damage but will not shatter totally. The segmentation prevents a crack appearing in one segment from penetrating another segment and extending. As result, the local failure of a glass segment does not transcend into a failure of the panel. This solution can be improved if objects made of brittle material are more likely to be hit in a particular area only. Only the regions that are likely to be hit and damaged can then be segmented, without segmenting the entire object. This case can be illustrated by a brittle prismatic object falling to the ground. It can be shown that the likelihood that a prismatic object falling to the ground will contact the ground with a corner is significant. As a result, corners are easily damaged and a crack appearing at a corner quickly propagates into the rest of the object. If the corners of the prismatic object are segmented, a crack appearing at a corner will not be capable of spreading through the rest of the object and the damage will be limited.

A similar effect is present for walls built with stones and bricks.

A theoretical justification that segmentation indeed improves fault tolerance has been given in Todinov (2017c). The conclusion was that in a segmented structure, the maximum tolerable number density of the flaws in the material can be significantly higher than the maximum tolerable number density of the flaws in the monolithic structure. Segmentation increases the fault tolerance significantly.

7.4.1 Case Study: Improving Fault Tolerance of a Column Loaded in Compression by Segmentation

To demonstrate theoretically that segmentation indeed increases fault tolerance, suppose that $\lambda_1(\sigma)$ is the number density of flaws in the material of a monolithic cylindrical column with volume V, subjected to a compression (Figure 7.3a). A *critical flaw* is a flaw which, if present in the column, will initiate an unstable crack which propagates through the material of the column. In the case of a monolithic column, a critical flaw, if present, will cause failure of the column.

Suppose that the flaws follow a homogeneous Poisson process in the material of the column and the probability that a flaw will be critical is $F_c(\sigma)$. For example, the probability $F_c(\sigma)$ that a flaw will be critical can be thought of as the probability $F_c(\sigma) = P(D \geq d_c(\sigma))$ that the diameter D of the flaw will be greater than a particular critical value $d_c(\sigma)$, dependent on the loading stress σ in the material, caused by the distributed force F (Figure 7.3a).

The probability of failure p_f of the column, at a loading stress σ, is now equal to the probability that at least a single critical flaw will be present in the stressed volume of the loaded column. According to an equation derived in Todinov (2005), this probability is given by

$$p_f = 1 - \exp[-\lambda_1(\sigma)\, F_c(\sigma)\, V] \tag{7.2}$$

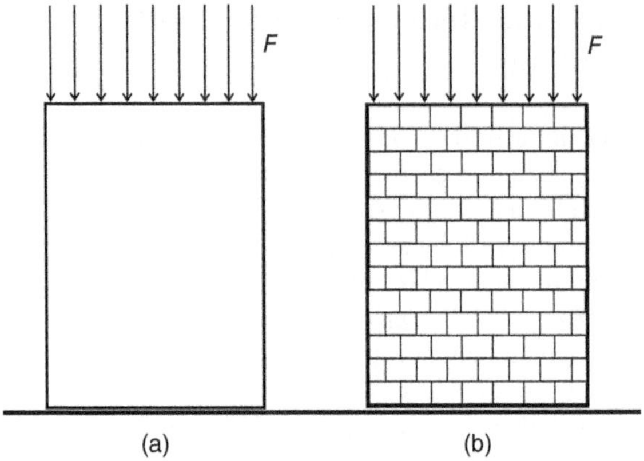

Figure 7.3 Segmentation improves the fault tolerance of a column loaded in compression.

Note that the product $\lambda_{cr,1}(\sigma) = \lambda_1(\sigma) F_c(\sigma)$ in Eq. (7.2) denotes the number density of the *critical flaws*. These are flaws which, if present in the material of the column or a brick, cause failure with certainty at the loading stress σ.

Eq. (7.2) can then be rewritten as

$$p_f = 1 - \exp[-\lambda_{cr1}(\sigma)\, V] \tag{7.3}$$

If the maximum acceptable probability of failure of the column has been specified to be p_f, the maximum acceptable number density of the critical flaws in the material of the monolithic column can be determined:

$$\lambda_{cr,1}(\sigma) = -\frac{1}{V}\ln(1 - p_f) \tag{7.4}$$

Now suppose that the column has been segmented into n bricks with the same volume V_b and made of the same material (Figure 7.3b). Suppose that the brick failures are not linked. This means that because of the brick boundaries, a crack starting from a critical flaw in any of the bricks cannot spread through the neighbouring bricks and cause the collapse of the column. A crack appearing in a particular brick causes only failure of the brick where the initiating flaw resided. After the collapse of a brick, there will be a redistribution of the stress and some local stress concentration effect. Assuming a very large number of bricks n in the column, these effects, although important, can be ignored in the first approximation and the loading stress σ before and after the collapse of a brick will be considered to be constant and uniform throughout the column.

Because of the segmentation of the column, and the large number of bricks, suppose that a collapse of the column occurs only if a certain fraction p_b of failed (cracked) bricks are present in the column. Denote the absolute number of failed bricks by n_b. Because of the large number n of bricks in the column, the maximum tolerable probability of failure of a brick in the column is approximately equal to $p_b \approx n_b/n$.

Assume that for each brick, the probability that a flaw in the brick will be critical (will cause failure of the brick) is equal to $F_c(\sigma)$. Again, the probability that a flaw in the brick will be critical can be thought of as the probability $F_c(\sigma) = P(D \geq d_c(\sigma))$ that the diameter D of the flaw will be greater than a particular critical value $d_c(\sigma)$, dependent on the loading stress σ. The maximum acceptable probability of failure of a brick can now be related to the maximum acceptable number density of flaws $\lambda_{cr,2}(\sigma)$ in the bricks:

$$p_b = 1 - \exp[-\lambda_{cr,2}(\sigma)\, V_b] \tag{7.5}$$

Solving this equation with respect to $\lambda_2(\sigma)$ gives:

$$\lambda_{cr,2}(\sigma) = -\frac{1}{V_b}\ln(1 - p_b) \tag{7.6}$$

for the maximum acceptable number density of the flaws in the material.

Taking the ratio of Eqs. (7.6) and (7.4) gives:

$$\frac{\lambda_{cr,2}(\sigma)}{\lambda_{cr,1}(\sigma)} = \frac{V}{V_b} \times \frac{\ln(1 - p_b)}{\ln(1 - p_f)} = n\frac{\ln(1 - p_b)}{\ln(1 - p_f)} \tag{7.7}$$

where $n = V/V_b$ is the number of the bricks in the column. For the specific values $p_f = 0.001$, $p_b = 0.03$ and $n = 300$ bricks,

$$\frac{\lambda_{cr,2}(\sigma)}{\lambda_{cr,1}(\sigma)} = 300 \times \frac{\ln(1 - 0.03)}{\ln(1 - 0.001)} \approx 9 \times 10^3$$

As a result, in the case of a segmented column, the maximum tolerable number density of critical flaws in the material of the bricks can be orders of magnitude higher than the maximum tolerable number density of the critical flaws in the material of the monolithic column. Segmentation increased fault tolerance significantly. This result is rather counter-intuitive to many experts reducing risk in specific domains which emphasises the importance of domain-independent methods of risk reduction transcending the boundaries of the narrow specific domains.

Now, suppose that the monolithic column has been loaded to a stress σ_1 while the brick column has been loaded to stress σ_2. With increasing the stress from σ_1 to σ_2, the number density of the critical flaws increases monotonically. This has been illustrated in Figure 7.4.

From Eq. (7.7), it follows that at stress σ_1, the segmented column can tolerate significantly more critical flaws than the monolithic column: $\frac{\lambda_{cr,2}(\sigma_1)}{\lambda_{cr,1}(\sigma_1)} = n\frac{\ln(1-p_b)}{\ln(1-p_f)} > 1$. The stress in the segmented column can then be increased to σ_2 which corresponds to a concentration of the critical flaws $\lambda_{cr}(\sigma_2) = \lambda_{cr,2}(\sigma_1)$ equal to the number density $\lambda_{cr,2}(\sigma_1)$ of critical faults the segmented column can tolerate at stress σ_1. As a result, segmentation increases the capability of carrying out larger loading stress without increasing the specified probability of failure of the column.

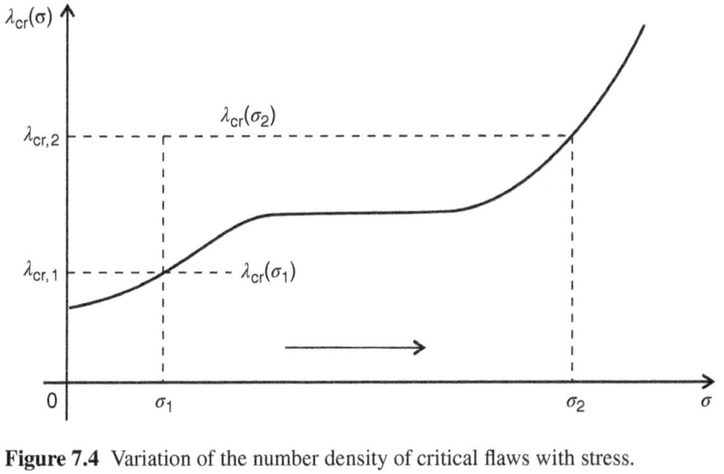

Figure 7.4 Variation of the number density of critical flaws with stress.

Note that it cannot be assumed that the number density of the critical flaws is linked through a power law $\lambda_{cr}(\sigma)=k\sigma^m$ with the loading stress σ, where k and m are constants. This assumption is incorrect. The power law states that the number density of critical flaws is a strictly increasing function of the stress which is not confirmed by the experiments and simple logic. If no new critical flaws are created within a particular stress range (σ_a, σ_b), there is no reason for the number density of the critical flaws to increase in this stress range. In this case, the dependence $\lambda_{cr}(\sigma)$ will remain flat in the region σ_a, σ_b. This is the fundamental reason why the Weibull model is an incorrect model for the probability of failure of brittle components, triggered by flaws (Todinov 2009b, 2010).

7.4.2 Reducing the Vulnerability to a Single Failure by Segmentation

A similar argument related to the fault tolerance of segmented entities can be present for components loaded in tension (Figure 7.5a,b). Ropes and cables are typical segmented structures built of multiple twisted strands or wires. The solid rod with cross section S in Figure 7.5a has been segmented into n parallel wires with cross sections s (Figure 7.5b). The sum of the cross sections $S=n\times s$ of the wires is equal to the cross section S of the monolithic rod.

While a crack initiated from a flaw in the monolithic rod propagates through the rod and causes failure, the segmented design in Figure 7.5b does not allow a local fracture of a single wire to penetrate into the neighbouring wires. Even though the material and the cross-sectional areas are the same, the fault tolerance of the segmented rod is higher. The argument is similar to the one presented for a column loaded in compression.

Segmentation replaces a single critical failure occurring at a macro level with non-critical failures occurring at a micro level. Suppose that a single wire is used for resisting a force F (Figure 7.5c). Failure of the wire causes delays and lost production. Suppose that the single wire has been segmented into four thinner wires resisting the force F, and the force F is still resisted even if a single wire or two wires fail. Segmentation reduces risk by replacing a single critical failure at a macro-level with non-critical failures at a micro-level. This essentially replaces a sudden failure at a macro-level with gradual deterioration of the system at a micro-level through many non-critical failures.

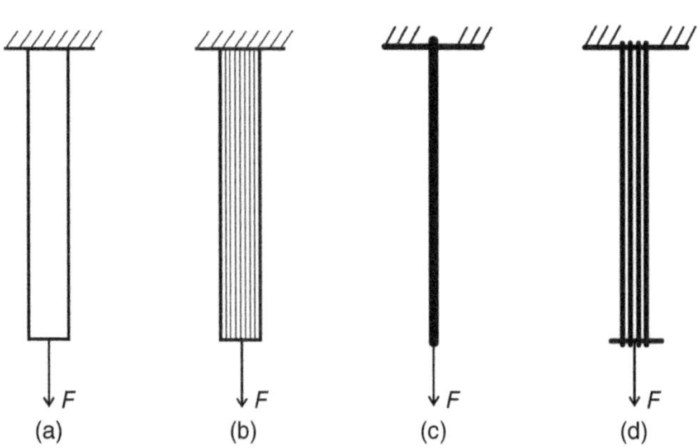

Figure 7.5 (a,b) Segmentation reduces the vulnerability to a single failure; (c,d) segmentation improves the fault tolerance of a rod loaded in tension.

Analytical justification of this technique will be given with a single wire used for resisting the force F in Figure 7.5c. Because the single wire could contain a material flaw which causes accelerated corrosion and failure of the faulty wire, suppose that the reliability of the wire associated with one year of continuous operation is $r_0 = 0.75$.

Suppose also that a segmentation of the wire has been made by replacing it with four thinner parallel wires (Figure 7.5d). Each of the thinner wires also could contain a material flaw which causes accelerated corrosion and fast premature failure of the faulty wire. As a result, each of the thinner wires is characterised by a lower reliability $r_1 = 0.70$, associated with the operational time interval of one year. Any two of the thinner wires are sufficient to resist the force F.

The probability that the segmented wire will still be resisting the force after one year of continuous operation, is now equal to the probability that at least two (two, three, or all four) wires survive the operational interval of one year. If the thinner wires fail due to a presence of a material flaw, independently from one another, this probability is given by

$$R = r_1^4 + \frac{4!}{3! \times (4-3)!} r_1^3 (1-r_1)^1 + \frac{4!}{2!(4-2)!} r_1^2 (1-r_1)^2 = 0.916$$

As a result of the segmentation, the reliability of the segmented wire increased significantly compared with the single wire, even though the reliability of the thinner wire segments was inferior to the reliability of the thicker wire.

Similarly, for a flange with very few fasteners, failure of any single fastener will cause a loss of containment. In other words, every single failure of a fastener is critical. Consider now a flange with many fasteners, each of which has a lower reliability and lower load-carrying capacity compared with the original fasteners. This assembly will not be vulnerable to a single failure of a fastener or even to failure of several fasteners. The single critical failures in the initial assembly have been replaced by non-critical failures in the segmented assembly. The sudden failure has been replaced by gradual deterioration.

The critical failure of the solid rod in Figure 7.5a is replaced by non-critical failures of the wires, on a micro-level. Instead of a sudden failure, the segmented rod will experience gradual deterioration through the non-critical failures of the separate wires. When inspection of the segmented rod discovers that the number of failed single wires at the surface becomes greater than a specified maximum tolerable quantity, the rod will be replaced and the catastrophic consequences from sudden failure will be avoided.

The segmentation method also greatly enhances the reliability of devices obtaining a signal from a sensor. The sensor triggers a particular action/alarm if the signal indicates a dangerous concentration of a particular chemical, dangerous magnitude of a force, torque, pressure, temperature, humidity, etc. Segmentation of a single sensor into multiple sensors, even with inferior reliability, makes the device less vulnerable to malfunction of a sensor or even to simultaneous failures of several sensors.

Segmenting an assembly into a number of identical sections makes the assembly easily reconfigurable if a section is damaged. For example, a long chute on a building site, made of separate small segments which fit into each other, can be easily damaged by the transported building debris. The segmented chute however, can be reconfigured easily into a fully functioning chute by simply discarding the damaged segment. There exist also segmented robotic systems that reconfigure themselves automatically upon failure of any of their building segments. This is done by bypassing the failed segment, with insignificant loss of functionality (Paley 2010).

Decreasing vulnerability and increasing the capability to reconfigure by segmentation increase the resilience of the system.

Segmentation reducing the risk of unauthorised access to a valuable asset is practiced whenever the valuable assets are divided (segmented) into smaller parts, each of which is subsequently stored in a different place. Accidental access to any part of the asset will not result in an automatic loss of the entire asset. Segmenting a single command centre into a mesh-type network of interconnected parts significantly reduces the vulnerability to a single strike.

7.5 Reducing Loading Stresses by Segmentation

7.5.1 Improving Load Distribution by Segmentation

Segmenting a component into several smaller components increases the contact area and often brings a significant reduction of the loading stresses. This segmentation mechanism increases reliability by distributing load over many load-carrying units.

In cases where torque is transmitted from a rotating shaft to a gear, sprocket or pulley or vice versa, retention devices such as keys (Figure 7.6a) or splines (Figure 7.6b) are used to prevent the relative rotation of the gear, sprocket, or pulley. For a specified torque, the stresses acting on a single key are high because the load is concentrated over a relatively small area. Splines can effectively be regarded as segmented keys, uniformly spaced around the shaft. The segmentation in this case, distributes the load over a larger area, decreases the stresses and ensures higher resistance of the connection to failure in the case of overload.

In another example, consider a flange with very few fasteners. A flange connection with a very small number of fasteners leads to excessive stresses in some of the fasteners. Increasing the number of fasteners, whose combined tensile strength is equivalent to the tensile

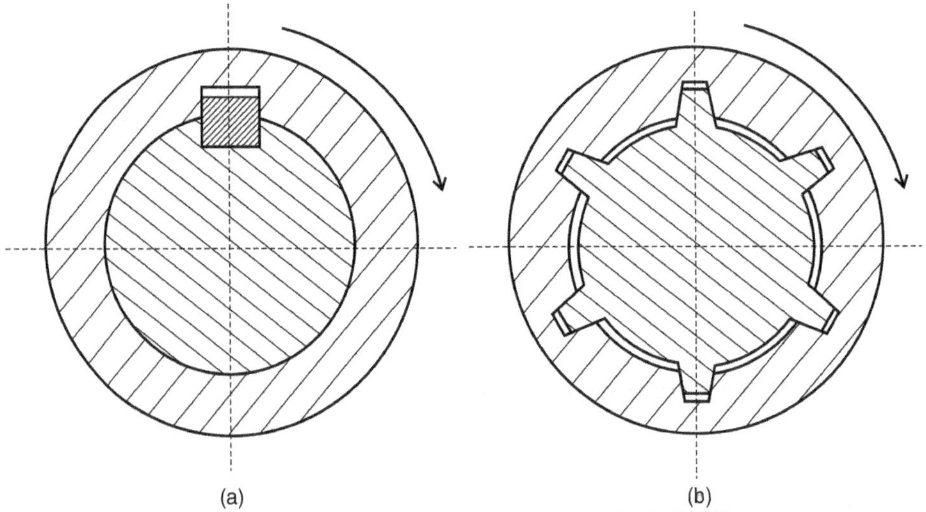

Figure 7.6 Splines can effectively be regarded as segmented keys uniformly spaced around the shaft.

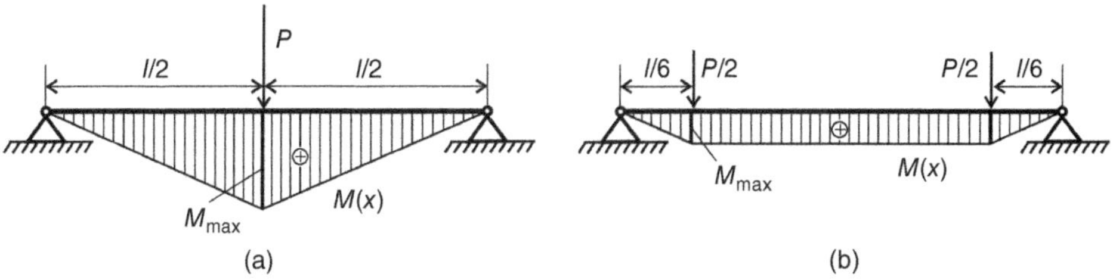

Figure 7.7 Segmenting the concentrated load *P* significantly reduced the bending moment and the stresses from bending.

strength of the very few fasteners, improves the load distribution and improves the reliability of the assembly.

Segmentation of a component often results in increased contact area and better conforming contact with other components which reduces contact stresses significantly. Reduced contact stresses result in reduced wear and increased resistance to fracture in the case of overload. For example, segmentation consisting of several rows of ball bearings, instead of a single row, significantly reduces the contact stresses and wear of the ball bearings.

Instead of applying the load into a single contact point, segmenting the load often decreases the bending moments and the stresses from bending. This is illustrated by the example in Figure 7.7. The maximum value of the bending moment in the loading from Figure 7.7a is $M_{max} = Pl/4$. After segmenting the load *P* into two loads with magnitudes $P/2$ applied at a distance $l/6$ from the supports (Figure 7.7b), the maximum bending moment becomes $M_{max} = Pl/12$. The maximum bending moment and the bending stresses resulting from it have been decreased three times by segmenting the concentrated load.

Often, the improved load distribution achieved by segmentation is used to achieve a greater level of balancing. The increased level of balancing results in a greater stability, smaller vibration amplitudes, and inertia forces which enhances reliability. Such is the case of introducing multiple cylinders in an internal combustion engine, multiple blades in a turbine, etc.

7.5.2 Improving Heat Dissipation by Segmentation

The increase of the surface to volume ratio by this mechanism increases the total surface area of the segmented parts, which helps heat dissipation. Heat dissipation is an important aspect of the reliability of many devices, and electronic devices in particular. Fast heat dissipation and equilibrium temperature within the acceptable limits is a necessary condition for fault-free operation of many electronic components. For a given volume of the component, segmentation increases the surface area through which heat is lost and the component is cooled. An increased heat transfer means a low equilibrium temperature and a more reliable operation. If segmentation of the heated components is not practicable, highly segmented radiators attached to the components are used for fast heat dissipation.

Heat dissipation is also an important aspect of the reliability of mechanical devices. In mechanical systems, an example can be given with a single *V*-belt, segmented into multiple parallel *V*-belts. Because of the segmentation, the heat released due to hysteresis losses is better dissipated by multiple parallel *V*-belts. Better heat dissipation leads to a lower equilibrium temperature of the belt and enhanced reliability.

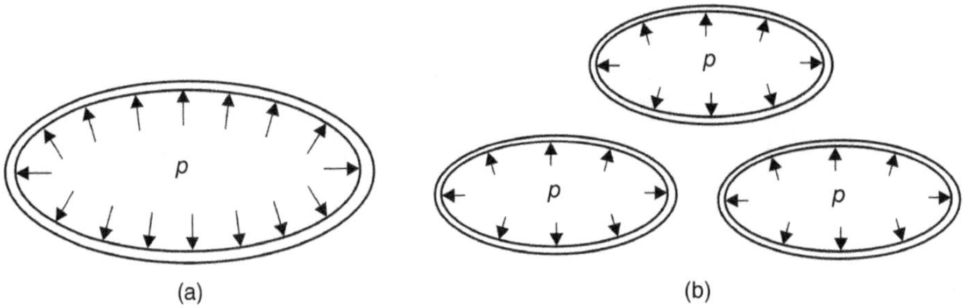

Figure 7.8 Segmenting the pressure vessel into several smaller pressure vessels increases the perimeter to cross-sectional area ratio and significantly reduces the loading stress acting in the shell.

7.5.3 Case Study: Reducing Stress by Increasing the Perimeter to Cross-Sectional Area Ratio Through Segmentation

Segmentation increases the perimeter to cross-sectional area ratio. This segmentation mechanism can be used to reduce the stresses acting in the shell of pressure vessels. This rather subtle yet very powerful reliability improvement mechanism has been depicted in Figure 7.8a featuring a convex pressure vessel with volume V, and thickness t of the shell. A convex pressure vessel is defined as follows: for any two points from the interior of the pressure vessel, each point on the segment connecting the two points also belongs to the interior of the pressure vessel.

The pressure vessel contains fluid exerting pressure p on the inside of the shell (Figure 7.8a). Segmenting the vessel into m smaller and similar in shape (convex) pressure vessels with thickness of the shell t and volumes V_1, $V_2,\ldots,V_m$ (Figure 7.8b) ($V = V_1 + V_2 + \ldots + V_m$), reduces significantly the maximum stress acting in the shell of each of the smaller vessels.

Indeed, an expression for the stress in the shell of the pressure vessel can be derived from the equilibrium of elementary forces. A slice has been taken from the pressure vessel (Figure 7.9). The z-axis is perpendicular to the x and y axes.

The pressure p is always perpendicular to the wall of the pressure vessel (Figure 7.9). The elementary force created by the pressure p, on an elementary surface area ds on the

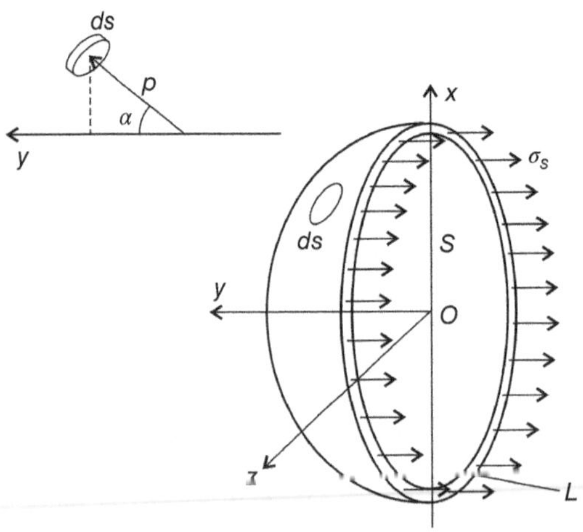

Figure 7.9 Derivation of the magnitude of the stress acting in the shell of a vessel under pressure with magnitude p.

inside wall of the vessel, is equal to $p\,ds$. The component of this elementary force along the y-axis is $(pds) \times \cos\alpha$ where α is the angle which the normal to the elementary surface element subtends with the y-axis (Figure 7.9). Note that the projection of the elementary force can also be written as a product of the pressure p and the projection of the elementary area $ds \cos\alpha$ on the (x,z)-plane. The resultant force, which is a sum of the components of all elementary forces along the y-axis, due to the internal pressure p, is therefore given by the product pS of the pressure p and the total projected area S of the inner surface of the pressure vessel on the (x,z)-plane. This resultant force must be counterbalanced by the force $tL\sigma_S$ created by the stress σ_S acting in the shell of the pressure vessel, where t is the thickness of the shell and L is the perimeter of the sliced part of the pressure vessel along which the counteracting stresses σ_S act (Figure 7.9). From the equilibrium equation $tL\sigma_{S0} = pS$, the expression

$$\sigma_{S0} = \frac{pS}{tL} \tag{7.8}$$

is obtained immediately for the stress acting in the shell of the initial pressure vessel.

Next, denote by s and l the projected area and the perimeter of the similar cross section from the smaller pressure vessel.

Similarly, for the stress σ_{S1} acting in the shell of the smaller pressure vessel

$$\sigma_{S1} = \frac{ps}{tl} \tag{7.9}$$

is obtained. With reducing the size of the pressure vessel by segmentation, the ratio of the cross-sectional area and the perimeter of the cross section decreases:

$$\frac{s}{l} < \frac{S}{L} \tag{7.10}$$

Since the pressure p and the thickness t of the wall have not been altered by the segmentation, from Eqs. (7.8)–(7.10), it follows that $\sigma_{S1} < \sigma_{S0}$. Consequently, segmenting a pressure vessel with arbitrary shape into smaller similar pressure vessels reduces the loading stress in the shell.

The degree of reduction of the loading stress will be illustrated by a simple special case of a spherical pressure vessel with inner diameter D and thickness of the shell t, subjected to a pressure with magnitude p (Figure 7.10).

The cross-sectional area of a section across the centre of the spherical vessel is $S = \pi D^2/4$ and the perimeter of the cross section is $L = \pi D$. If the initial pressure vessel is segmented into a number of pressure vessels with inner diameter d, for the ratio of the loading stresses in the wall, the expression

$$\sigma_{S1} = \sigma_{S0}\frac{d}{D} \tag{7.11}$$

is obtained from Eqs. (7.8) and (7.9). If the initial pressure vessel with diameter D is segmented into pressure vessels with twice as small diameter d, the loading stress σ_{S1} in the shell, due to the pressure p, is halved.

This example demonstrates that by increasing the perimeter to cross-sectional area ratio, segmentation reduces the magnitude of the stresses in pressure vessels significantly.

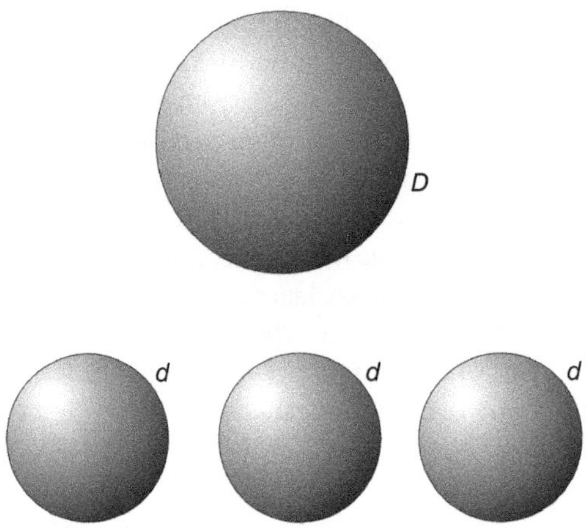

Figure 7.10 Segmentation of a spherical pressure vessel.

7.6 Reducing the Probability of a Loss/Error by Segmentation

7.6.1 Reducing the Likelihood of a Loss by Segmenting Opportunity Bets

Segmentation can also be used with success to produce a low-cost reduction of the risk of a loss from a risk-reward bet. Risk-reward events/bets can materialise as benefit or loss. An investment in a particular enterprise is a typical example of a risk-reward bet. A successful investment is associated with returns (benefits) while an unsuccessful investment is associated with losses.

Suppose that $0 \le p_s \le 1$ is the probability that the risk-reward bet will be a 'success' and $p_f = 1 - p_s$ is the probability that the risk-reward bet will be a 'loss'.

The expected values of the benefit and the loss given that the risk-reward event has materialised are denoted by $\overline{B}_s$ and $\overline{C}_f$, respectively.

The expected profit $\overline{G}$ from a risk-reward bet is then given by:

$$\overline{G} = p_s \times \overline{B}_s + p_f \times \overline{C}_f, \tag{7.12}$$

where p_s is the probability of a beneficial outcome with expected magnitude $\overline{B}_s$ and $p_f = 1 - p_s$ is the probability of a loss with expected magnitude $\overline{C}_f$ (the loss $\overline{C}_f$ has been taken with a negative sign). If $\overline{G} > 0$, the risk-reward bet will be referred to as an *opportunity bet.*

Traditionally, the *maximum expected profit criterion* is used for making an optimal choice among risky prospects containing risk-reward bets (Moore 1983; Denardo 2002). According to this criterion, a rational decision maker compares the expected profits from a number of risky prospects and selects the prospect with the largest expected profit.

The maximum expected profit criterion however, does not account for the significant impact of the actual number of risk-reward events/bets in a risky prospect. The critical dependence of the choice of a risky prospect on the number of risk-reward bets in it has not been discussed in studies related to ranking risky alternatives (Starmer 2000; Nielsen and Jaffray 2006; Richardson and Outlaw 2008). Even in a recent publication, probably the most comprehensive treatise on the theory of betting (Epstein 2009), no discussion has been provided on the impact of the limited number of risk-reward bets on the choice of a risky prospect. However, the number of risk-reward bets in a risky prospect has a crucial impact on the choice of a risky prospect and cannot be ignored.

The potential profit G from a risk-reward bet is a random variable following a Bernoulli distribution with parameter p_s. For constant values of the benefit given success $\overline{B}_s$ and the loss given failure $\overline{C}_f$, the probability distribution of the potential profit G is given by $P(G = \overline{B}_s) = p_s$ and $P(G = \overline{C}_f) = p_f$. The probability distribution of the potential profit G can be considered to be a distribution mixture including two distributions with means $\overline{B}_s$ and $\overline{C}_f$ and variances $V_1 = 0$ and $V_2 = 0$, sampled with probabilities $p = p_s$ and $1 - p = p_f$. Consequently, from the theory of the distribution mixtures (Todinov 2002a), for the variance of the potential profit we have

$$Var(G) = p_s p_f (\overline{B}_s - \overline{C}_f)^2 \tag{7.13}$$

7.6.1.1 Case Study: Reducing the Risk of a Loss from a Risky Prospect Involving a Single Opportunity Bet

This case study involves a risky prospect containing a single opportunity bet with parameters: $p_s = 0.3$, $\overline{B}_s = 300$, $p_f = 0.7$, $\overline{C}_f = -90$ and expected profit $E(G) = 0.3 \times 300 - 0.7 \times 90 = 27$. The probability of a net loss from this risky prospect is 70%.

The risk of a net loss from this risky prospect can be reduced by segmenting the opportunity bet. This creates a second risky prospect which now contains three opportunity bets with the same probability of success and failure but with three times smaller magnitudes for the benefit given success and the loss given failure: $p_s = 0.3, \overline{B}_s = 300/3 = 100, p_f = 0.7, \overline{C}_f = -90/3 = -30$. The expected profit *from the risky prospect containing the three segmented opportunity bets* is $E(G_{123}) = 3 \times (0.3 \times 100 - 0.7 \times 30) = 27$. Because a net loss from the second risky prospect can be generated only if a loss is generated from every single segmented bet, the probability of a net loss from the segmented bets is $p_{f,123} = 0.7^3 \approx 0.34$.

Clearly, the second risky prospect involving segmented opportunity bets is to be preferred to the first risky prospect. Segmenting the opportunity bet in the initial risky prospect significantly reduced the risk.

This example shows that the risk associated with an opportunity bet can be reduced significantly if the opportunity bet is segmented (split) into several opportunity bets characterised by the same probability of success and failure as the original bet but with proportionally smaller benefit and loss. Indeed, consider a risky prospect containing a single opportunity bet, characterised by a probability of success p_s, benefit given success $\overline{B}_s$, probability of failure p_f, and loss given failure $\overline{C}_f$. This opportunity bet can be segmented into m opportunity sub-bets, each characterised with the same probability of success and failure p_s and p_f but with m times smaller expected benefit and loss $\overline{B}_s/m$ and $\overline{C}_f/m$. The expected profit from the segmented bets is:

$$E(G_{1,\ldots,m}) = m \times (p_s\overline{B}_s/m + p_f\overline{C}_f/m) = E(G) \tag{7.14}$$

is equal to the expected profit from the original bet. Considering Eq. (7.13), the variance

$$V(G_{1,\ldots,m}) = \sum_{i=1}^{m} V_i = \sum_{i=1}^{m} p_s p_f (\overline{B}_s/m - \overline{C}_f/m)^2 = \frac{1}{m} p_s p_f (\overline{B}_s - \overline{C}_f)^2 \tag{7.15}$$

of the profit from the risky prospect with m opportunity bets is m times smaller than the variance $p_s p_f(\overline{B}_s - \overline{C}_f)^2$ of the profit characterising the initial opportunity bet.

Segmenting the initial opportunity bet significantly reduced the risk of a loss. This example also shows that the number of the risk-reward bets should be a key consideration in selecting a risky prospect.

7.6.2 Reducing the Likelihood of a Loss by Segmenting an Investment Portfolio

Risk of failure can be reduced by segmenting the investment into smaller investments in many unrelated sectors whose returns are non-correlated. While the expected returns are not affected by the segmentation, the variance (volatility) of the return from the whole portfolio of non-correlated stocks is reduced significantly. Investment funds reduce risk by buying the shares of many companies. Even small investors can obtain a segmented and well diversified portfolio with low transaction costs by investing, for example, in a unit trust.

The principle of diversification can be illustrated by an example involving a portfolio with n securities. For such a portfolio, the variance σ_p^2 of the returns is given by (Teall and Hasan 2002):

$$\sigma_p^2 = \sum_{i=1}^{n} w_i^2 \sigma_i^2 + 2 \sum_{i<j} w_i w_j \sigma_i \sigma_j \rho_{ij} \tag{7.16}$$

where σ_i is the standard deviation of the returns from security i, and ρ_{ij} is the linear correlation coefficient between the ith and jth security. The weight w_i shows how much money is invested in security i relative to the total amount invested in the entire portfolio $\left(\sum_{i=1}^{n} w_i = 1\right)$. If the portfolio is based on three securities only, Eq. (7.16) becomes

$$\sigma_p^2 = w_1^2\sigma_1^2 + w_2^2\sigma_2^2 + w_3^2\sigma_3^2 + 2w_1w_2\sigma_1\sigma_2\rho_{12} + 2w_2w_3\sigma_2\sigma_3\rho_{23} + 2w_1w_3\sigma_1\sigma_3\rho_{13} \tag{7.17}$$

For the purposes of the illustration, let us assume equal weights $w_1 = w_2 = \ldots = w_n = 1/n$, variances $\sigma_1 = \sigma_2 = \ldots = \sigma_n = \sigma$, and correlation coefficients $\rho_{ij} = \rho$, $i = 1, n; j = 1, n; i \neq j$. Eq. (7.16) then becomes

$$\sigma_p^2 = \sigma^2/n + \rho\sigma^2(1 - 1/n) \tag{7.18}$$

In the case of a portfolio based on two securities only whose returns are perfectly negatively correlated ($\rho = -1$), substituting $n = 2$ in Eq. (7.18) gives $\sigma_p^2 = 0$. In other words, such a portfolio is associated with no risk.

In the case of a portfolio based on a large number n of securities, from Eq. (7.18), for the portfolio variance,

$$\sigma_p^2 \approx \rho\sigma^2 \tag{7.19}$$

is obtained. In other words, with increasing segmentation (the number of the securities), the volatility of the portfolio, which is a measure of the risk associated with the portfolio returns, has been reduced to $\rho\sigma^2$. If the returns from the different securities are non-correlated ($\rho_{ij} = \rho = 0$), the portfolio volatility (the risk) becomes zero ($\sigma_p^2 = 0$).

By using optimisation techniques, the weights w_i which yield the minimum volatility (risk) of the portfolio specified by Eq. (7.16) can be determined. A number of additional constraints must also be satisfied. Such is the constraint

$$\overline{R}_{\min} = \overline{R}_1 w_1 + \overline{R}_2 w_2 + \ldots + \overline{R}_n w_n \tag{7.20}$$

where $\overline{R}_1, \ldots, \overline{R}_n$ are the expected returns from the individual securities and $\overline{R}_{\min}$ is the minimum expected return from the portfolio required by the investor. Additional constraints are represented by Eqs. (7.21) and (7.22)

$$w_1 + w_2 + \ldots + w_n = 1 \tag{7.21}$$

$$0 \le w_i \le 1, \; i = 1, n \tag{7.22}$$

7.6.3 Reducing the Likelihood of Erroneous Conclusion from Imperfect Tests by Segmentation

Another important application of the segmentation method can be found in a low-cost reduction of the likelihood of erroneous conclusion from an imperfect test.

Suppose that a test results either in a positive or negative outcome about the presence of a particular attribute in the test substance. The test is not perfect and produces an error with probability p. However, the probability of erroneous conclusion can be reduced by segmenting the available test substance and conducting separate tests with the segmented parts, instead of performing a single test with the entire available substance (Figure 7.11).

Suppose that each test with the segmented test substance results in an error with probability p. After all outputs from the separate tests become available, the presence of the attribute is decided by the prevalent outcome (positive or negative) from the individual tests with the segmented substance. In order for the segmented test to produce an erroneous conclusion, more than half of the tests with segmented substance must result in an error. For the special case of $n = 2k + 1$ identical tests with the segmented substance, at least $k + 1$ test results must be erroneous. Because the tests are statistically independent and the probability of an error in each test is constant, the conditions for a *binomial experiment* are fulfilled and the number of erroneous outputs X follows the *binomial distribution*. The probability that the number or erroneous outputs will be greater than or equal to $k + 1$ is given by:

$$P(X \ge k + 1) = \sum_{x=k+1}^{n} \frac{n!}{x!(n-x)!} p^x (1-p)^{n-x} \tag{7.23}$$

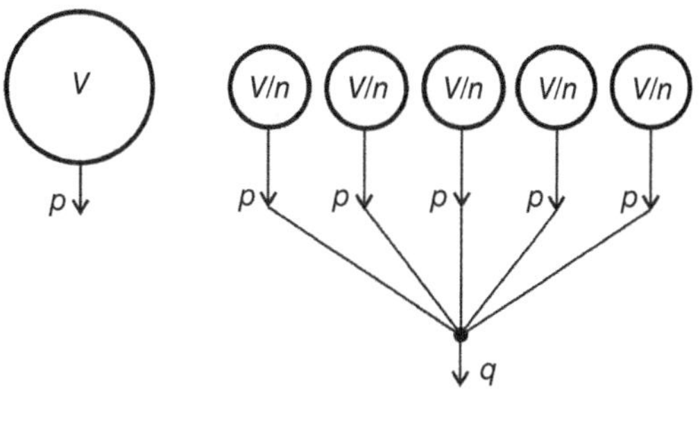

Figure 7.11 Segmenting the available substance with volume V into smaller volumes V/n and conducting multiple parallel tests with the segmented parts significantly reduces the probability of erroneous conclusion from imperfect tests.

For $n = 11$ and $p = 0.1$, for example, the probability of erroneous output from the segmented tests is

$$P(X \geq 6) = \sum_{x=6}^{11} \frac{11!}{x!(11-x)!} 0.1^x (1-0.1)^{11-x} = 0.0003 \tag{7.24}$$

As a result, the relatively high probability of erroneous output of $p = 0.1$, characterising the initial single test, has been decreased 333 times by segmenting the test.

The same segmentation technique can be applied in taking the correct decision about approving a particular decision, transaction, etc. If the process of decision making is split between a number of independent decision makers, the likelihood of making a wrong decision can be reduced significantly.

For an expert in a specific domain (for example, a food or toxicology expert) it is not at all obvious that the risk of erroneous conclusion could be decreased to such an extent simply by segmenting the tested substance. This is also a low-cost risk reduction because no investment is made into more precise test equipment associated with a smaller error from a single test.

7.7 Decreasing the Variation of Properties by Segmentation

It is a well-known fact from statistics that replicating a single measurement characterised by a variance σ^2 with the average from n measurements, each of which is characterised by a variance σ^2, decreases the variance of the averaged result to σ^2/n.

A coarse microstructure composed of two microstructural constituents, A and B (Figure 7.12a) is associated with big variation of the properties of the sample (transect) T.

If the segmentation of the microstructure from Figure 7.12a is increased (Figure 7.12b) by fragmenting the microstructural constituent B, the variation of properties of the sample (transect) T is decreased.

Since the variation of properties is largely determined by the variation of the intercepts from the microstructural constituents, the empirical distribution of the intercepts and the intercept variance are essential factors in determining the distribution of properties. They

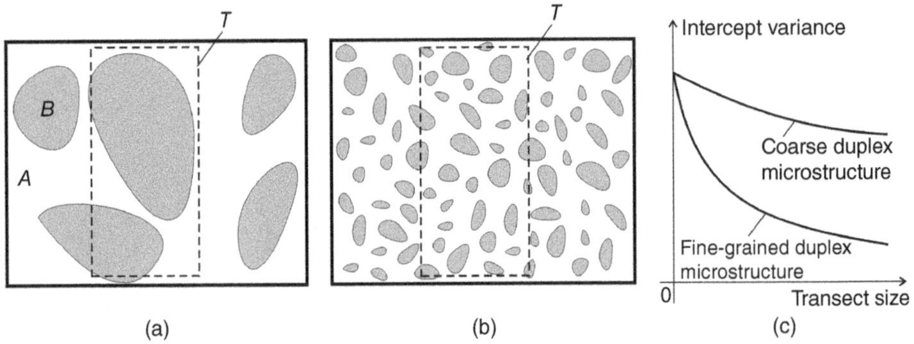

Figure 7.12 Decrease of the intercept variance and the variation of the sample properties with increasing the microstructural segmentation.

are important fingerprints of inhomogeneous materials and contain valuable information regarding the risk of poor properties (Todinov 2002b).

One of the reasons for the decrease in the variation of the properties of the sample with increasing the microstructural segmentation is the decreased variation of the intercepted volume fraction from microstructural constituent *B* with increasing the segmentation of the microstructural constituent *B*. Figure 7.12c depicts the decrease of the variance of the intercepted volume fraction from microstructural constituent *B*, with increasing the size of the transect (sample).

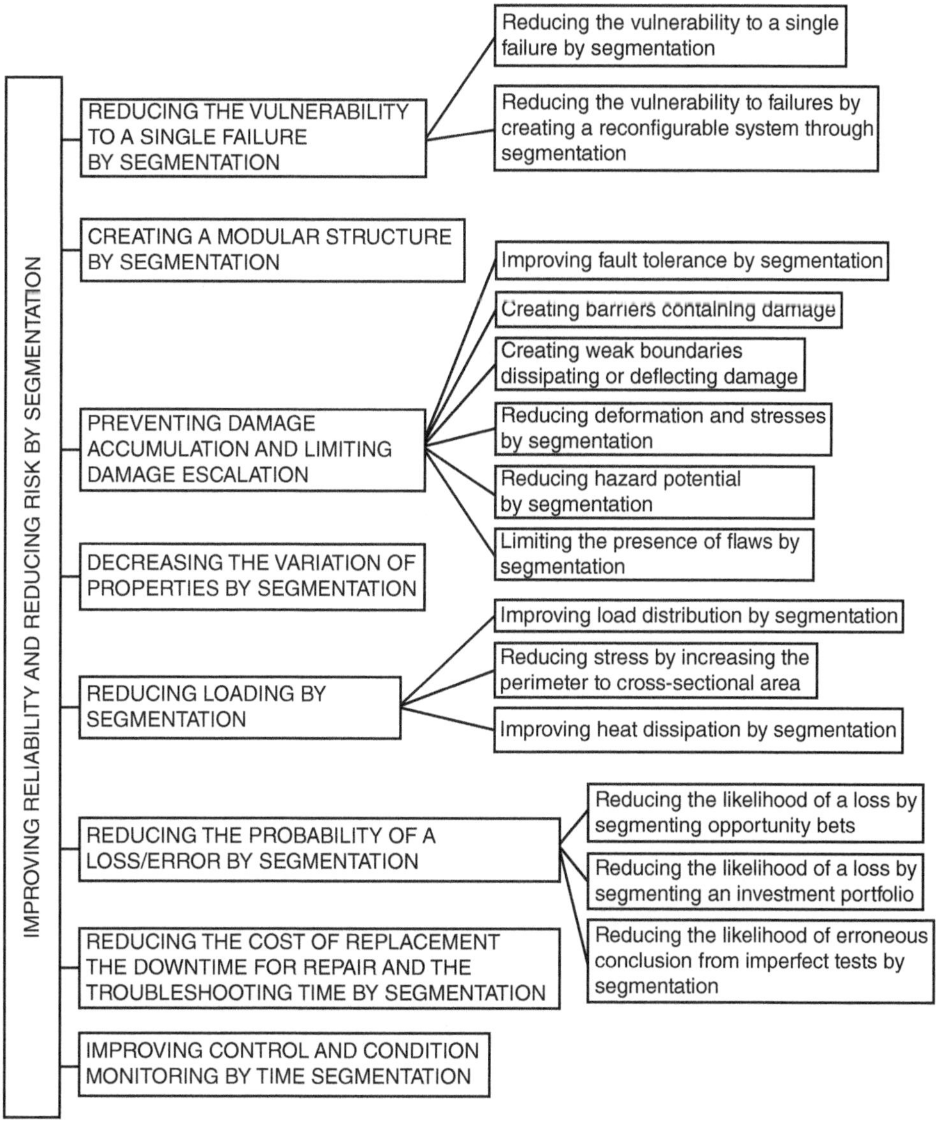

Figure 7.13 Various mechanisms for improving reliability and reducing risk by segmentation.

7.8 Improved Control and Condition Monitoring by Time Segmentation

Time discretisation (segmentation) has been used with success for real-time control of various risks, for example the risk of flooding. At regular time slices (segments), a data acquisition system based on sensors, serial interface and computer, monitors key risk parameters like on-site water level, flow, rainfall, etc., and after post-processing, estimates the risk of flooding (Zhang 2012).

Time segmentation can also be used for analysing complex motion even if the laws of motion are not known in their integral form. Such is the case of tracing the trajectory of flying objects in the presence of air resistance. Time segmentation with a very small time step is introduced, followed by updating the direction and magnitude of the air-resistance force and the gravity force. Next, an updated value for the current acceleration and velocity of the object are calculated after which the values of the new coordinates of the flying object are obtained. Continuing this process yields a very precise approximation of the actual trajectory of the object. Decreasing the size of the time segment (time slice) increases the precision of the approximation.

Time segmentation combining sampling of key parameters and corrective feedback is at the heart of the discrete-time controllers of missiles. Time segmentation is also central to the feedback control of mechatronic systems such as industrial manipulators, robots, etc. (Awrejcewicz et al. 2017).

A classification of the discussed mechanisms of improving reliability and reducing risk by segmentation is presented in Figure 7.13.

8

Improving Reliability and Reducing Risk by Inversion

8.1 The Method of Inversion

The reverse strategy for reliability improvement and risk reduction, which is the focus of this chapter, has always been an important technique in problem solving. It was already known to Pappus of Alexandria in about 320 CE, who described it as 'beginning with the desired outcome and working backwards until something known is reached'. The famous algebraist Carl Gustav Jacobi, whose favourite maxim was 'invert, always invert', used the reverse strategy with great success to solve many hard mathematical problems. An important example of the reverse strategy in problem solving is recursion, widely used for solving difficult combinatorial problems (Wirth 1976). A typical approach in recursion is reducing the initial problem of size n to two or more problems of size $n-1$, which in turn are reduced to problems of size $n-2$, and so on, until problems with trivial solutions are obtained. Starting from the trivial solutions and moving backwards helps to assemble the solution.

It is then no surprise that reverse strategy features in texts related to creativity and inventive problem solving. The reverse strategy is at the heart of many inventions and Altshuller's TRIZ framework (Altshuller 1984, 1996, 1999) identified a number of useful applications of this technique.

As a problem-solving tool, recently, the method of inversion has been applied for maximising the throughput flow in networks in (Todinov 2013a,b). The classical *saturation algorithms* for maximising the throughput flow start from an empty network and continue by gradually saturating the edges with flow. The method of inversion applied to maximising the throughput flow starts from fully saturated with flow edges and works backwards, by redistributing and draining flow from the network, until a maximum throughput flow is reached (Todinov 2013b). For networks with a few imbalanced nodes, in the case where only the maximum throughput flow is of interest, the inverse method, starting from fully saturated flow network and working backwards by draining the excess flow will outperform many classical saturation algorithms for determining the maximum throughput flow.

Reverse strategy is often used in planning projects, operations, or processes where success is critically dependent on whether people, equipment, and resources are available when and where needed. This is particularly relevant to cases where critical deadlines need to be met. In order to reduce the risk of failure, the planning of the project/process is often started from the desired outcome and progressed backwards. Starting from the desired end result and planning backwards helps to determine what resources are necessary to achieve the goal and to ensure how to make the necessary resources available at the right place and at the right time.

Methods for Reliability Improvement and Risk Reduction, First Edition. Michael Todinov.

Despite the progress made in using the reverse strategy for problem solving and planning, no treatment currently exists on the application of the reverse strategy for reliability improvement and risk reduction. Isolated applications of reliability improvement and risk reduction based on inversion do exist but these have not been linked with the generic method of inversion.

The reverse strategy is a powerful method for improving reliability and reducing risk and will be referred to as *the method of inversion*. The systematic application of the method of inversion for improving reliability and reducing risk has been presented in (Todinov 2015, 2017b).

Inverse states in the form of compressive residual stresses introduced by shot peening and their role in improving the reliability of automotive suspension springs have been discussed in Todinov (1999, 2000). Recently, Fu et al. (2015) presented a comprehensive review of methods for introducing compressive residual stresses by cold expansion, to improve the fatigue life of connection holes in aircraft structures.

The method of inversion achieves risk reduction mainly by eliminating dangerous failure modes or by providing a different frame of thinking which helps to see easily the risk-reducing solution. In what follows, the different mechanisms through which the method of inversion reduces risk are discussed in detail.

There are several major lines through which the method of inversion achieves reliability improvement and risk reduction: (i) by inverting functions, relative position, and motion; (ii) by inverting properties and geometry; (iii) by introducing inverse states; and (iv) by inverse thinking.

8.2 Improving Reliability by Inverting Functions, Relative Position, and Motion

Most of the processes in engineering are reversible. The basic function does not normally change if the relative position, motion or features of interacting objects are reversed. *By reversing functions, relative position, motion or features, the basic required functions are preserved whilst a number of critical failure modes disappear or are mitigated.* This is an important idea behind the method of inversion. Inverting motion, relative position of objects, and the location of features often eliminates failure modes, simplifies the interaction, and results in enhanced performance. An important feature of the method of inversion is that by applying this method, often, reliability of a product is improved at no extra cost.

The method of inversion works because between two states A and its inverse $\overline{A}$, with equivalent functions, one of them will probably be with greater resistance against a particular dangerous failure mode and superior overall reliability.

An example of eliminating failure modes by inverting a function can be given with the fail-safe air breaks of trucks. Instead of air pressure energising the breaks when these are needed, the function of air pressure is inverted. The air pressure keeps the brakes released which permits the truck to move. In the case of low pressure in the air line due to a puncture or failure of the compressor, the brakes are applied securing the truck. In this way, a dangerous failure mode is avoided: loss of air pressure and inability to apply breaks when needed.

A simple illustration of the method of inversion can also be given by making a bolt connection with threaded blind holes. Inverting upside down the orientation of the threaded blind holes prevents the collection of dirt, abrasive particles, and moisture in the threaded holes, because gravity helps to clean the holes. The result is reduced corrosion and wear and extended life of the bolt connection. The method of inversion works because the inverse orientation of the blind hole is more resistant to the collection of dirt, abrasive particles, and moisture. A very similar situation is present in drilling vertical blind holes in components, by a robot, on a manufacturing line. This operation is associated with the failure mode: 'metal chips stuck at the bottom of the blind hole because of gravity'. However, if the hole is drilled on a component positioned upside down, the failure mode is eliminated because gravity now helps to clean the hole.

Often, making a moving object stationary while moving the stationary object in the opposite direction (inverting the direction of motion) eliminates major failure modes and results in a significantly improved performance. This idea underlies the Cosworth® sand casting process (Campbel 2015) where the molten metal is never poured down into the sand mould as is the case in the classical sand casting process. The molten metal flows in the opposite direction (uphill) into the mould. This type of metal flow eliminates turbulence and mitigates a major failure mode: 'trapping oxides into the metal' which reduces significantly the fatigue strength of the cast component.

Moving parts exhibit more failures compared with stationary parts. This is usually due to increased kinetic energy, wear, fatigue, vibration, heat generation, and erosion associated with moving parts.

Reducing the number of moving parts is an efficient way of improving the reliability of a system and inverting the motion can be used directly to reduce the number of moving parts.

Suppose that a required function is guaranteed by the relative motion of three components with respect to each other. If an extra uniform motion with opposite direction to the motion of the first component is imparted to all three components, the first component will become stationary and the motion of the rest of the components will become a composition of their motion before the inversion and the motion of the first component taken with a negative sign. The relative motion of the components will remain the same but the reliability of the system will be improved because the number of moving parts has been decreased by one.

There are numerous cases where inverting the relative location of features eliminates failure modes, enhances reliability and reduces risk.

A simple example is the flanged wheel of a railroad train. Flanged wheels are needed to prevent the train from sliding off the track. Initially, the railroad tracks were flanged instead of the wheels and thousands of miles of unsafe railroad tracks were manufactured, with thousands of miles of unnecessary flanges. This situation changed dramatically when inversion was applied to the location of the flange. The location of the flange was shifted from the railway track to the wheel. In this way, a number of failure modes were eliminated while preserving the basic function of keeping the train from sliding off the track. As a result, railway safety improved significantly.

This inversion method will be illustrated by a case study involving an alarm circuit.

8.2.1 Case Study: Eliminating Failure Modes of an Alarm Circuit by Inversion of Functions

Consider the alarm circuit in Figure 8.1, activated by a normally open switch *S*. Pressing the switch *S* energises the latching relay *R* which closes its normally open contact *K* and the latching contact *L*. As a result, the alarm remains activated even after releasing the switch *S* (Figure 8.1b).

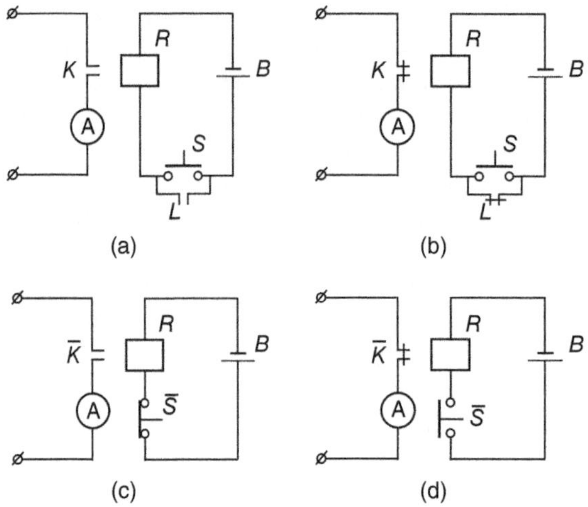

Figure 8.1 Improving the reliability on demand of an alarm system by inversion of functions.

There is a major drawback associated with the solution in Figure 8.1a. The alarm system in Figure 8.1a is vulnerable to an open circuit failure mode and failure of the power source *B*. If the control circuit fails open or the wires in the circuit of the normally open switch *S* have been cut or the power source *B* is not available (e.g. if the battery *B* is flat), pressing the switch *S* will fail to activate the alarm when it is needed (Figure 8.1b). This failure mode can be eliminated by inverting the normally open switch *S* and the normally open reed relay *K* to a normally closed switch $\overline{S}$ and normally closed reed relay $\overline{K}$ (Figure 8.1c). The relay in the circuit is a high-coil resistance reed relay.

When the normally closed switch $\overline{S}$ is not open, the coil *R* of the reed relay is energised, the normally closed contact $\overline{K}$ of the relay stays open and the alarm is not activated. If the normally closed switch $\overline{S}$ is opened it stays open, the reed contact $\overline{K}$ closes and the alarm is activated (Figure 8.1d). There will be an activated alarm in each of the following cases: (i) open circuit of the switch $\overline{S}$ due to a fault; (ii) open circuit of the switch $\overline{S}$ due to malicious cutting of any of the wires; and (iii) a flat battery *B*. A false alarm is a more benign outcome compared with not having an alarm when it is needed. The false alarm is actually beneficial because it is an indication that the alarm circuit is faulty and needs repair. The high resistance of the reed relay coil *R* (in the order of several kiloohms) makes the power consumption negligible for keeping the reed relay contact open.

In this example, by applying the method of inversion, security has been improved at no extra cost.

8.2.2 Improving Reliability by Inverting the Relative Position of Objects

The next major mechanism through which inversion improves reliability is that *inversion of relative position often increases the response to loading stresses.*

Inversion can be used to achieve a self-reinforcing response. Self-reinforcement is present *when external/internal forces or states are channelled towards inducing counterforces strengthening proportionally the system's response.* An important feature of self-reinforcement is that increasing the external/internal forces intensifies the system's response against these forces. Such a self-reinforcement application is given in Figure 8.2, in the design of a cover for containers under pressure.

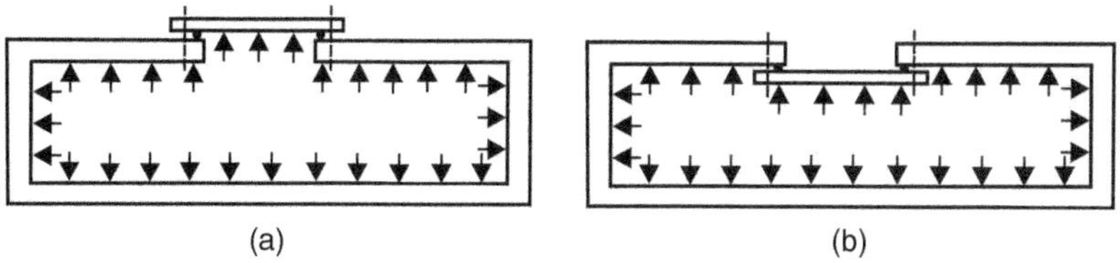

Figure 8.2 Improving reliability by inversion of the relative position of a cover on a container under pressure.

For the design in Figure 8.2a, the loading stresses in the screws fixing the cover can be reduced and the reliability of the seal can be increased significantly by *inverting the position of the cover with respect to the pressure vessel.* The design in Figure 8.2b channels the force induced by the internal pressure into a self-reinforcing force which helps to form a reliable seal and prevent leakage. Increasing pressure increases the magnitude of the self-reinforcing response, strengthens the seal and increases the resistance to increased pressure. In this example, by applying the method of inversion reliability has also been improved at no extra cost.

A similar effect is present for injection pistons with self-reinforcing seals. The elastic material of the seal is pressed against the walls of the cylinder. The higher the pressure, the tighter the seal formed between the piston and the cylinder.

8.2.2.1 Case Study: Inverting the Position of an Object with Respect to its Support to Improve Reliability

A reinforcing effect, achieved by inverting the position of an object with respect to its support, is given in Figure 8.3a. The brackets are supporting a part subjected to a varying force F (Figure 8.3a). The resultant effective stress range $\Delta\sigma_1$ present in the supporting bracket is shown in Figure 8.3a. The stress range $\Delta\sigma_1$ can be reduced to the stress range $\Delta\sigma_2$ in Figure 8.3b, if the relative location of the supporting brackets and the component is inverted as it is shown in Figure 8.3b. In the inverted location, the weight of the component diminishes the stress range and increases fatigue life (Todinov 2017b).

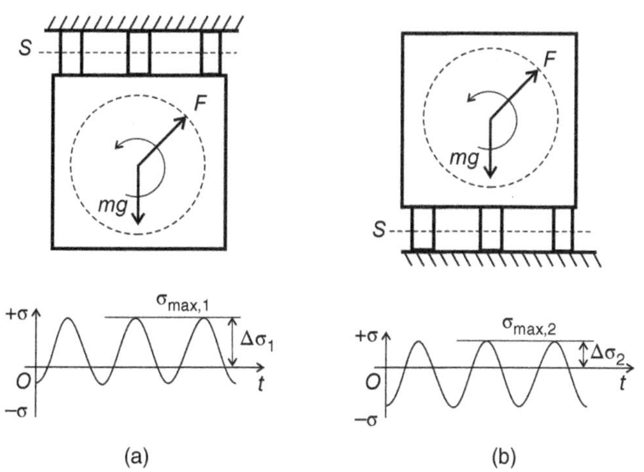

Figure 8.3 Improving reliability by inverting the position of an object with respect to its support.

Indeed, the fatigue life of the bracket is strongly dependent on the rate of fatigue crack growth. The fatigue crack growth rate da/dN is commonly estimated from the Paris power law (Paris et al. 1961; Paris and Erdogan 1963):

$$da/dN = C\,\Delta K^m \tag{8.1}$$

where a is the crack size, N is the number of load cycles, C and m are constants depending on the material, environment, test temperature, and load ratio $R=\sigma_{min}/\sigma_{max}$; $\Delta K = K_{max} - K_{min}$ is the range of the stress intensity factor. $K_{max} = \sigma_{max} Y\sqrt{\pi a}$ and $K_{min} = \sigma_{min} Y\sqrt{\pi a}$ are the stress intensity factors which correspond to the maximum (σ_{max}) and minimum (σ_{min}) value of the uniform tensile stress perpendicular to the crack plane.

Consequently, $\Delta K = \Delta\sigma\, Y\sqrt{\pi a}$ where $\Delta\sigma = \sigma_{max} - \sigma_{min}$ is the stress range. Because fatigue cracks do not propagate if the minimum loading stress is compressive, in calculating the stress range, the minimum loading stress should be set to zero ($\sigma_{min} = 0$) and the stress range becomes $\Delta\sigma = \sigma_{max,2}$. Y is a dimensionless factor that depends on the geometry of the loaded crack.

For relatively short cracks, it can be assumed that the geometry factor Y is independent of the crack length a and Eq. (8.1) becomes

$$da/dN = C \times (Y\,\Delta\sigma\,\sqrt{\pi a})^m \tag{8.2}$$

where $\Delta\sigma = \sigma_{max} - \sigma_{min}$ is the range of the uniform tensile stress perpendicular to the crack plane. Equation (8.2) can be integrated by separating the variables. In the usual case where $m > 2$, integrating the equation results in N_f load cycles to failure, given by:

$$N_f = \frac{2}{(m-2)C(Y\Delta\sigma\sqrt{\pi})^m}\left(\frac{1}{a_0^{(m-2)/2}} - \frac{1}{a_f^{(m-2)/2}}\right) \tag{8.3}$$

where a_0 is the initial crack size and a_f is the final crack size at which fracture is initiated. Assuming a fast fracture failure mode, the final crack size at which fracture will be initiated can be determined from

$$Y\sigma_{max}\sqrt{\pi a_f} = K_{Ic} \tag{8.4}$$

which governs the onset of fast fracture. In Eq. (8.4), K_{Ic} is the fracture toughness of the material and Y is the geometry factor. Equation (8.4) essentially states that fast fracture occurs when the stress intensity factor $Y\sigma_{max}\sqrt{\pi a_f}$ becomes equal to the fracture toughness K_{Ic} of the material. From this equation, for the final crack size a_f, we have:

$$a_f = \frac{1}{\pi}\left(\frac{K_{Ic}}{Y\sigma_{max}}\right)^2 \tag{8.5}$$

In other words, the larger the maximum stress σ_{max} from the loading cycle, the smaller the final crack size at which fast fracture occurs. In other words, for the final crack sizes characterising the two designs 'a' (1) and 'b' (2) (Figure 8.3), we have $a_{f1} < a_{f2}$.

For the two competing designs, the effective stress ranges from the fatigue loading are different $\Delta\sigma_1 = \sigma_{max,1} > \Delta\sigma_2 = \sigma_{max,2}$ (Figure 8.3).

The different stress ranges characterising the two designs result in different fatigue lives. According to Eq. (8.3), the ratio of the fatigue lives characterising the two designs is

$$\frac{N_{f1}}{N_{f2}} = \left(\frac{\Delta\sigma_2}{\Delta\sigma_1}\right)^m \left(\frac{A_1}{A_2}\right) \tag{8.6}$$

where $A_1 = \left(\frac{1}{a_0^{(m-2)/2}} - \frac{1}{a_{f1}^{(m-2)/2}} \right)$, $A_2 = \left(\frac{1}{a_0^{(m-2)/2}} - \frac{1}{a_{f2}^{(m-2)/2}} \right)$. Because $a_{f1} < a_{f2}$ it follows that $A_1 < A_2$. From this result and also from $\Delta\sigma_2 < \Delta\sigma_1$, it follows that $\left(\frac{\Delta\sigma_2}{\Delta\sigma_1}\right)^m \left(\frac{A_1}{A_2}\right) < 1$. As a result, $\frac{N_{f1}}{N_{f2}} < 1$, which means that the second design is characterised by a larger fatigue life.

With increasing the weight of the component, the stress range decreases and the fatigue life increases. As a result, inverting the location of the component and the supporting brackets, results in a reinforced design characterised by an increased fatigue life and improved reliability. The larger the weight, the larger is the reduction of the stress range, and the larger is the increase of the fatigue life. The reliability has been improved at no extra investment.

8.3 Improving Reliability by Inverting Properties and Geometry

Inversion of mechanical properties often eliminates failure of valuable components. Suppose that two components *A* and *B* are working in contact and component *A* is significantly more expensive than component *B*. If component *A* is the softer component, the damage will accumulate in the more expensive component. Inverting mechanical properties (hardness) by making component *B* softer than component *A* may not alter the basic required function but will replace a highly undesirable accumulation of wear damage in component *A* with accumulation of wear damage in the cheaper component *B*. Component *B* can be replaced at a much lower cost than component *A*.

An invariant is a quantity that does not change despite the changes in some parameters of the system. Well-known examples are: (i) the total momentum of an isolated system; (ii) the total energy of an isolated system; and (iii) the relative distances between several objects after an uniform motion is imparted to each of the objects.

This technique is introduced through the following basic applications: (i) improving the reliability of screw joints; and (ii) reducing the risk of catastrophic failure of pressure vessels.

8.3.1 Case Study: Improving Reliability by Inverting Mechanical Properties Whilst Maintaining an Invariant

For a bolt and flange joint where the load on the bolt fluctuates, the force–deformation diagram for a stiff bolt and elastic flange is shown in Figure 8.4a. The elastic constants of the bolt and the flange are $k_b = \tan(\alpha)$ and $k_f = \tan(\beta)$, respectively. The bolted joint is subjected to a required preload (clamping) force of magnitude *P*. This is the invariant in the assembly which will be preserved. Because of a variation of the load on the bolt, the bolted joint is subjected to a pulsating external force with magnitude *F*, which causes a pulsating force of magnitude F_b in the bolt (Figure 8.4a).

The range of the pulsating force in the bolt can be reduced by inverting the elastic constants of the bolt and the flange whilst keeping the required clamping force *P* constant. The large elastic constant of the bolt is reduced, and the small elastic constant of the flange is increased (Figure 8.4b) whilst preserving the magnitude of the clamping force *P* constant.

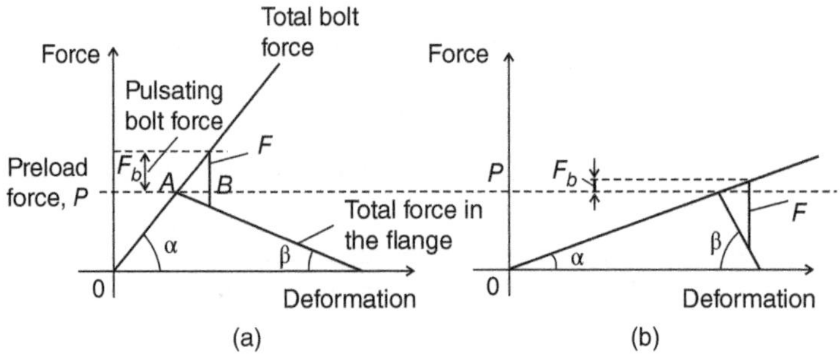

Figure 8.4 Range F_b of the pulsating force acting in the bolt of a preloaded bolted joint: (a) a stiff bolt and elastic flange; (b) elastic bolt and a stiff flange.

The net effect is a smaller range of the pulsating force F_b in the bolt (Figure 8.4b) which results in a larger fatigue life (reliability).

This effect can be quantified through the amplitude of the pulsating force F_b, by expressing the length L_{AB} of the segment *AB*, in two different ways (Figure 8.4a):

$$L_{AB} = \frac{F_b}{\tan(\alpha)} = \frac{F - F_b}{\tan(\beta)} \tag{8.7}$$

Considering that $k_b = \tan(\alpha)$ and $k_f = \tan(\beta)$, from (8.7) we have:

$$F_b = \frac{k_b}{k_b + k_f} F \tag{8.8}$$

Equation (8.8) shows that inverting the elastic properties of the bolt and the flange (reducing the elastic constant of the bolt and increasing the elastic constant of the flange), reduces the magnitude of the pulsating force F_b in the bolt and increases its fatigue life. The invariant property (specified preload force *P*) has been preserved during the inversion. The elastic constant of the bolt is given by $k_b = EA/L$, where *L* is the length of the bolt, *A* is the cross-sectional area and *E* is the Young's modulus of the material. Reducing the elastic constant of the bolt can be done by selecting a bolt with a larger length *L* or reducing the cross-sectional area *A* of the bolt by thinning down the shank diameter of the bolt.

8.3.2 Case Study: Improving Reliability by Inverting Geometry Whilst Maintaining an Invariant

Consider the cylindrical pressure vessel with semispherical ends in Figure 8.5a. The pressure vessel has a volume *V*, diameter *D*, length *L*, thickness *t* of the shell and contains fluid exerting pressure *p* on the inside of the shell (Figure 8.5a). The shape of the pressure vessel can be modified by inverting the dimensions *D* and *L* while maintaining the same volume and internal pressure *p*. The relatively large initial diameter *D* is reduced to a smaller diameter *d* (Figure 8.5b) while the length *L* is increased to a larger length *l* (Figure 8.5b). During this inversion, the volume is kept constant, equal to the required volume of the pressure vessel:

$$V = V_1 = \pi D^2(D/6 + L/4) = V_2 = \pi d^2(d/6 + l/4)$$

The largest tensile stress acting in the pressure vessel from Figure 8.5a is the hoop stress σ_{H1}, determined from $\sigma_{H1} = pD/(2t)$. The largest tensile stress acting in the wall of

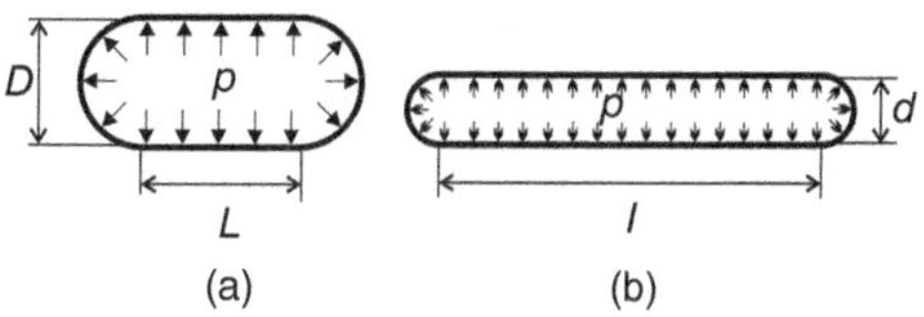

Figure 8.5 Inverting the dimensions by preserving an invariant volume significantly reduces the hoop stress.

the modified pressure vessel from Figure 8.5b is the hoop stress σ_{H2}, determined from $\sigma_{H2} = pd/(2t)$.

Even though the volume and the internal pressure remained unchanged, the inversion of the dimensions reduced significantly the hoop stress, which is the largest principal tensile stress acting in the shell of the pressure vessel. The axial principal tensile stress is also reduced.

The reduced hoop stress in the second design, makes it even possible to reduce the shell thickness t and obtain not only a safer design but also a design with a significantly reduced weight.

Indeed, suppose that an alloy with fracture toughness of $K_{Ic} = 112\,\text{MPa}\sqrt{m}$ and yield stress $\sigma_s = 1100\,\text{MPa}$ has been used for the design of a cylindrical pressure vessel with semi-spherical ends, containing gas with pressure 20 MPa. The dimensions of the pressure vessel (in mm) are according to Figure 8.6a. The radius of the spherical cap is 350 mm and the length L is 418 mm. The volume is $V = 0.34\,\text{m}^3$. The thickness of the shell for the initial design is $t_1 = 12\,\text{mm}$.

By applying inversion of the geometric parameters preserving the volume, the thickness of the shell is reduced to $t_2 = 5$ mm, the diameter of the pressure vessel is reduced to 400 mm and the length of the vessel is increased to 2439 mm. As a result, a modified design is obtained with the same volume $V = 0.34\,\text{m}^3$ (Figure 8.6b).

The hoop stress in the shell of the pressure vessel is given by $\sigma = pd/(2t)$, where p is the internal pressure and d is the internal diameter of the vessel, and t is the thickness of the

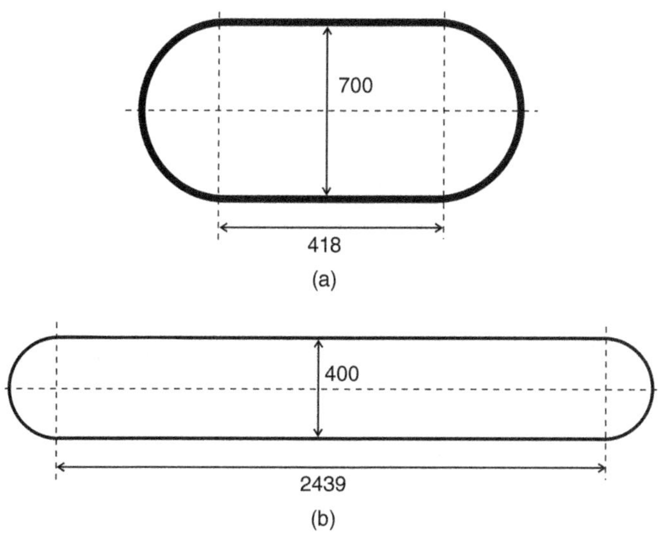

Figure 8.6 Modifying the shape of the pressure vessel by inverting the geometric parameters, while keeping constant volume improves reliability and reduces the weight of the design.

shell. For a through crack oriented in the worst possible direction (perpendicular to the hoop stress), the stress intensity factor is: $K_I = \sigma\sqrt{\pi a} = \frac{pd}{2t}\sqrt{\pi a}$. A *leak before break condition* must be present for the pressure vessel to be safe. This means that the shell of the presssure vessel must be able to support a through crack with length $2a = 2t$. As a result, the leak before leak condition becomes

$$\frac{pd}{2t}\sqrt{\pi t} < K_{Ic}$$

where K_{Ic} is the fracture toughness of the material of the shell.

The hoop stress characterising the design in Figure 8.6a is $\frac{pd_1}{2t_1} = \frac{20 \times 10^6 \times 0.7}{2 \times 0.012} = 583.3$ MPa.

The hoop stress characterising the design in Figure 8.6b is $\frac{pd_2}{2t_2} = \frac{20 \times 10^6 \times 0.4}{2 \times 0.005} = 800$ MPa and both hoop stresses are smaller than the yield stress of the material. However, the leak-before-break condition for the design in Figure 8.6a:

$$\frac{p_1 d_1}{2t_1}\sqrt{\pi t_1} = 583.3 \times 10^6 \sqrt{\pi \times 0.012} = 113.26\ \text{MPa}\sqrt{m} > K_{Ic} = 112\ \text{MPa}\sqrt{m}$$

is not fulfilled. The pressure vessel from Figure 8.6a is not safe.

The leak-before-break condition for the design in Figure 8.6b:

$$\begin{aligned}\frac{p_2 d_2}{2t_2}\sqrt{\pi t} &= 800 \times 10^6 \sqrt{\pi \times 0.005} \\ &= 100.27\ \text{MPa}\sqrt{m} < K_{Ic} = 112\ \text{MPa}\sqrt{m}\end{aligned}$$

is fulfilled and the pressure vessel is safe. The volume of material required by the design in Figure 8.6a is

$$\begin{aligned}V_{M1} &= 2\pi(d_1/2) \times L_1 \times t_1 + 4\pi(d_1/2)^2 t_1 \\ &= (2\pi \times 0.35 \times 0.418 + 4\pi \times 0.35^2) \times 0.012 \approx 0.0295\ \text{m}^3\end{aligned}$$

while the volume of material required by the design in Figure 8.6b is

$$\begin{aligned}V_{M2} &= 2\pi(d_2/2) \times L_2 \times t_2 + 4\pi(d_2/2)^2 t_2 \\ &= (2\pi \times 0.2 \times 2.439 + 4\pi \times 0.2^2) \times 0.005 \approx 0.0178\ \text{m}^3\end{aligned}$$

As a result of the inversion of the geometric parameters preserving an invariant volume, the safety of the design in Figure 8.6a has been significantly improved and, simultaneously, the weight of the design has been reduced by 40%.

8.4 Improving Reliability and Reducing Risk by Introducing Inverse States

8.4.1 Inverse States Cancelling Anticipated Undesirable Effects

A deliberate state, inverse to the anticipated state can be introduced to cancel the undesirable effect. The two states superpose and the result is the absence of an undesirable effect or a significantly attenuated undesirable effect. In acoustics, this principle works in noise-cancellation headphones. A sound wave is emitted with the same amplitude but with inverted phase to the noise. The result is a significant attenuation of the harmful noise and reduced risk of hearing damage.

This principle also underlies active methods of controlling vibration. The active vibration control involves suitable vibration sensors (e.g. accelerometers), controllers, and actuators for vibration control. The signal from the vibration sensors is fed to a controller and through an actuator, a spectrum of cancellation vibrations are generated in response. The advances in sensor, actuator, and computer technology made active methods of control cost-effective and affordable.

Deliberate inverse states cancelling undesirable anticipated effects are at the heart of many temperature-compensation circuits designed to mitigate the impact of generated heat on the parameters of the electronic devices. For example, creating inverse states through a Wheatstone bridge cancels the undesirable impact of temperature and is the foundation of various measurement techniques characterised by a very small error.

In mechanical engineering, a typical example is the compensation clock pendulum in which the temperature elongation of the pendulum rod is counteracted by an opposite expansion so that the period of oscillations remains the same. A typical application of an inverse state as a counterbalancing force are the counterweights in cranes which reduce the loading on the lifting motor and improve the balance and stability of the crane. Another example is the gate fail-safe valve which is maintained open by hydraulic pressure acting against a counterbalancing compression spring. Upon failure of the hydraulic system, the counterbalancing spring expands and returns the valve to the closed (safe) position.

8.4.2 Inverse States Buffering Anticipated Undesirable Effects

Inverse states are often created for buffering anticipated states with undesirable effect. This technique underlies reducing the risk of failure of zones generating heat. Components working in close contact (e.g. piston–cylinder) and moving relative to each other generate heat which, if not dissipated, causes intensive wear, reduced strength, and deformations. The risk of failure of such an assembly is reduced significantly if one of the parts (e.g. the cylinder) is cooled to dissipate the released heat. The introduced negative state (cooling) reduces friction and wear thereby improving reliability.

Cold expansion, used in aviation for creating compressive stresses at the surface of fastener holes (Figure 8.7), is an example of an inverse state buffering an anticipated undesirable effect (Fu et al. 2015).

Cold expansion is done by passing a tapered mandrel through the hole. The inverse state created in the vicinity of the hole (compressive residual stress field), counters the tensile loading stresses during operation and impedes the formation of fatigue cracks at the edge of the hole and their propagation which reduces the risk of fatigue failure.

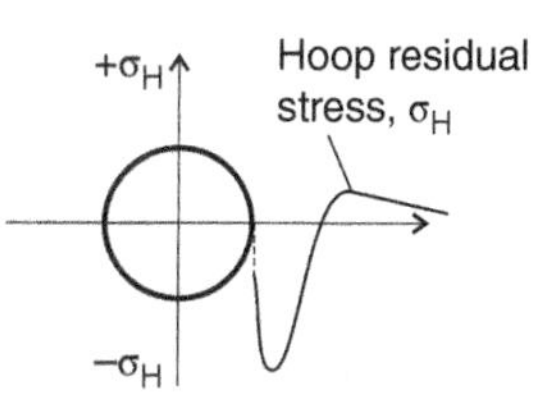

Figure 8.7 Countering the stress-concentration effect of a hole by creating compressive stresses through cold expansion.

In order to counter the tensile stresses from loading at the surface and improve fatigue resistance, *shot peening*, which introduces compressive stresses at the surface, has been used as an important element in manufacturing technology (Niku-Lari 1981; Bird and Saynor 1984). For compression springs, for example, shot peening induces compressive residual stress at the surface of the spring wire (Todinov 2000). During loading, the compressive residual stress at the spring surface is subtracted from the tensile stress from loading. The result is a smaller stress range and increased fatigue life of the shot-peened spring. As a result of this operation, the fatigue life of leaf springs, for example, can be increased up to 10 times.

In the construction industry, pre-tensioning of building components is a typical example of buffering a negative effect by introducing an inverse state. Thus, tensile stresses from bending of concrete beams can be reduced if pre-loaded in tension tendons (steel cables or rods) are inserted in the beam to provide a clamping load. After the concrete sets, the beam is pre-loaded in compression. The compressive stress from pre-loading is an inverse state which compensates tensile loading stresses. Since the tensile stresses from bending superpose with the compressive residual stresses, the net stress during service is compressive or a tensile stress with significantly reduced magnitude. Pre-stressed concrete is the main material for floors in high-rise buildings. In addition, pre-stressing makes it possible to construct large spans in bridges and buildings with large column-free spaces.

An inverse state of compressive residual stresses at the surface, acting as a buffer compensating the tensile service stresses from loading, can also be created by a special heat- and thermochemical treatment such as *case-hardening, gas-carburising*, and *gas-nitriding*.

The *corrosion, erosion*, and *wear allowances* added to the computed sections of components (e.g. pipes, support structures exposed to wind and rain, cutting tools, etc.) are other examples of inverse states anticipating the loss of wall thickness. They act as buffers compensating for the loss of wall thickness and decrease significantly the risk of failure.

The use of inverse states as buffers has a wide application in many other areas of human activity. In project management, providing time buffers for certain critical tasks reduces the risk of delay, should particular risks materialise. Similarly, in managing stock in the presence of random demands, increasing the reserve of a particular safety-critical stock (e.g. a particular life-saving medicine) reduces the risk of running out of stock in the case of clustering of random demands.

Increasing the financial reserves of a bank or a company makes it less vulnerable to depleting its reserves due to materialised credit and market risks.

Increasing the number of prepared specimens protects against the risk of producing an insufficient number of functional specimens. The increased number of prepared specimens makes allowance for faults such as lack of appropriate preparation, accidental damage of specimens during storage and transportation, accidental damage of specimens during tests, flawed material, etc.

8.4.3 Inverse States Reducing the Likelihood of an Erroneous Action

A typical inverse state introduced to reduce the likelihood of an erroneous action can be found in railway signalling. Instead of signalling in the case of danger, the signals are set by default into 'danger'. In order for the train to pass, it is necessary to clear the danger signal. If, for some reason, the person responsible for issuing the signal is incapacitated or the signalling system fails to be activated, a train will not be exposed to danger. In this way, by introducing an inverse state a dangerous failure mode is eliminated.

Another example is given with people compiling a set of m out of n options, where m is very close to the available number of options n. (e.g. $m = n - 1$, $m = n - 2$) To reduce the risk of compiling a wrong set of options, instead of building the total number of options by progressing from zero to m, the options are selected by offering initially a selection of all n available options and reducing their number to m by deleting unwanted options. The smaller number of steps involved in the inverse approach reduces the possibility of errors.

8.5 Improving Reliability and Reducing Risk by Inverse Thinking

The third major mechanism for improving reliability and reducing risk is by inverse thinking.

8.5.1 Inverting the Problem Related to Reliability Improvement and Risk Reduction

Instead of focusing on how to improve the reliability of a product; the focus is inverted on how to make the product fail. Instead of focusing on how to produce a reliable design the focus is on how to precipitate failure modes. The benefit from inverting the thinking about the problem is that the problem is effectively reframed; another perspective is set which breaks the usual pattern of thought and brings new avenues of exploration.

The original problem is often significantly simplified by restating the opposite problem for which it is easier to find a solution. *Restating the problem from 'how to improve reliability' to 'how to precipitate failure' often reveals unexpected resources for reliability improvement*. Improving reliability is often difficult to frame and quantify. Inverting the problem to 'how to induce failure', simplifies the problem and helps to find a solution.

Furthermore, in assessing a product, there is always an element of denial about the shortcomings of the product and an element of exaggeration of the product safety. This mindset blinds the analyst from seeing insidious failure modes which compromise the product safety. Focusing on what makes the component/system fail (instead of what makes it more reliable), breaks the standard perception about the safety of the company's own product and channels thinking proactively into how to make the product fail. Once an answer to the inverted question 'how to precipitate failure' is produced, the contributing factors, circumstances, and their interaction precipitating failure can be clearly seen. This strategy often makes it possible to identify and eliminate critical failure modes and yields important ideas for reliability enhancement and risk reduction. It brings the playing devil's advocate technique to the extreme, by asking not only questions about what could possibly go wrong but also how to make the system/component malfunction.

A simple example of risk reduction achieved by inverse thinking is improving the defence against unauthorised access to a valuable service through a computer program controlling the access (e.g. controlling the access to a bank account). An important approach to improving the defence against unauthorised access is to invert the problem to 'how to compromise the computer program controlling the access and make it fail'. Invariably, the inversion reveals a number of software vulnerabilities which could be exploited to gain unauthorised access. Common vulnerabilities discovered by using this technique are: 'intercepting the control over the program execution after induced division by zero', 'intercepting the program execution after induced buffer overflow', stealing passwords because of failure to clear the keyboard buffer, inadequate protection against coordinated simultaneous attacks exhausting possible password combinations, etc. Focusing on how to induce failure of the program controlling the access *is a very important route to eliminating software vulnerabilities and reducing risk.*

The *subversion analysis technique* and the *anticipatory failure determination approach* described in Kaplan et al. (1999) are largely an application of the method of inverse thinking. The focus of these approaches is on how to invent failures by using the available resources. Even though the anticipatory failure determination approach is a useful technique for identifying rare and unexpected failure modes, it restricts thinking to the available resources only. However, the method of inverse thinking does not pose such a restriction. It looks beyond the available resources and helps discover failure modes which also depend on particular stacking of external random events and circumstances. Thus, a fatal outcome from a critical condition of a patient can be precipitated not only by considering various problems caused by the faults in the available lifesaving resources (e.g. lifesaving equipment and medical personnel). The fatal outcome can also be caused by clustering of simultaneous demands for lifesaving equipment from a number of patients in a similar critical condition such that all life-saving resources are engaged when a new demand arrives. In the next case study, the quality of management can be improved significantly by inverse thinking.

8.5.1.1 Case Study: Reducing the Risk of High Employee Turnover

Reducing the risk to an organisation by improving the quality of management is a task which is difficult to frame and solve. Inverting the problem to 'how to be a bad manager' helps to identify negative actions and behaviour which have a markedly negative impact on employee satisfaction, increase employee turnover and the risk to the organisation.

Some of the identified negative actions and behaviour are as follows:

- Increase the workload to employees. Increase their work duties and responsibilities beyond what has been agreed and planned.
- Do not pay the actual number of their working hours or their full expenses.
- Distribute the workload non-uniformly. Increase the workload to some employees significantly more than the average workload.
- Do not praise employees. Do not acknowledge their contributions. Stay silent and do not celebrate their successes.
- Assign repetitive mechanical tasks which do not challenge employees intellectually and do not allow them to grow.
- Allocate work roles for which employees do not have the educational background, training or desire to do.
- Promote and hire the wrong people.
- Restrict communication. Do not seek advice from employees; do not consult them about key changes that affect them.
- Surround with a favourite small circle of employees and ignore the rest.
- Compare employees with their colleagues and discuss their inadequacy or lack of competence; criticise employees frequently and discuss their mistakes and shortcomings.

- Make employees fearful and stressed. Manipulate, bully and threaten them.
- Do not honour promises for promotion, staff development, salary rises, etc.
- Do not allow employees to pursue their passion and express themselves creatively in their work.
- Do not allow initiative.
- Do not trust employees, micromanage them.

These negative actions (the list is not exhaustive), are easy to identify, yet their elimination improves significantly the quality of management, reduces employee turnover and ultimately reduces the risk to the organisation.

8.5.2 Improving Reliability and Reducing Risk by Inverting the Focus

8.5.2.1 Shifting the Focus from the Components to the System

The inductive approach which is at the heart of the *failure modes and effects analysis* (FMEA; MIL-STD-1629A 1977), has been an important basic technique for identifying possible failure modes at a system level during the FMEA exercise. It consists of considering sequentially the failure modes of all parts and components building the system and tracking their effect on the system's performance.

The method for revealing failure modes provided by FMEA is very useful but, unfortunately, it cannot reveal all failure modes. For example, for a lubrication system, the failure modes 'too small size of the sump' and 'too large clearances between lubricated contact surfaces' cannot be discovered by the inductive approach.

Inverting the focus from the components to the system is a very useful approach in identifying new failure modes and is a valuable addition to FMEA. It consists of considering an already identified failure mode at a system level and investigating what kind of low-level failure could cause the same failure mode or contribute to it. For example, answering the question what could possibly contribute to the system failure mode 'too low oil pressure' identifies the important failure mode 'too large clearances between lubricated contact surfaces due to wear'. Inverting the focus from components to the system also helps to discover the low-level failure modes 'leaks from seals and gaskets' and 'inappropriate oil with high viscosity being used'.

Inverting the focus and asking the question what could possibly contribute to the system failure mode 'too high oil pressure', leads to the low-level failure mode 'incorrect design of the oil galleries'. Asking the question what could possibly contribute to a too high oil temperature, identifies the cause 'a small amount of circulating oil in the system' which helps to discover the low-level failure modes 'too low oil level' and 'too small size of the sump'. Undersized sumps lead to a high oil temperature which constitutes a failure mode at the system level.

8.5.2.2 Starting from the Desired Ideal End Result

Formulating the desired ideal end result is often an important step in finding a reliability enhancing solution.

A sprinkler system for putting out fires must be activated if the temperature reaches a critical threshold. Adding temperature sensors and a sprinkler activation system is an expensive option. In addition, this is a high-risk solution because the sprinkler activation system may fail and become non-operational when needed. An ideal sprinkler system operates

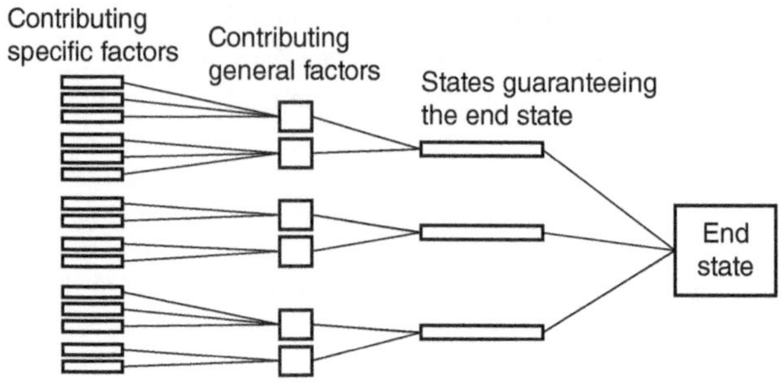

Figure 8.8 Identifying failure modes by starting from the end state and moving backwards to the contributing factors.

automatically if the temperature reaches a critical threshold and is not operational if the temperature is below the threshold. Starting from the desired ideal sprinkler system suggests using an alloy with melting temperature equal to the critical threshold temperature activating the sprinklers. This solution yields a sprinkler system that operates whenever it is needed, avoids the use of unreliable activation systems, and has very few components. The result is a significantly improved reliability and reduced risk of malfunctioning.

8.5.2.3 Focusing on Events that are Missing

Improving reliability and reducing risk often benefits significantly from changing the focus from appearance of reliability-critical events on the operational time interval to events that are missing during the time of operation. Thus, lack of scheduled maintenance events points to likely problems with components which require frequent alignment, cleaning, lubrication, control testing, etc., in order to guarantee reliable operation. Lack of events of replacement of components undergoing intensive damage accumulation, points at increased likelihood of failure. Lack of failures of deliberate weaknesses (e.g. pressure membranes, shear pins, etc.) upon overloading poses very serious questions about the appropriate design of protective measures and the safety of personnel and expensive equipment.

8.5.3 Improving Reliability and Reducing Risk by Moving Backwards to Contributing Factors

The principle of inversion has applications transcending mechanical engineering. In what follows, risk management applications of the principle of inversion are featured which have applications in project risk management.

One of the simplest yet very efficient risk management technique based on the method of inversion is 'moving backwards to the contributing factors' and 'changing the focus to the opposite'.

Starting from an end state and moving backwards towards the general and specific contributing factors is a powerful technique for reliability improvement and risk reduction (Figure 8.8). Moving backwards to the contributing factors creates conceptual anchors for the thinking and generates a full range of alternative reliability improvement measures. In this way, the likelihood of missing a relevant reliability improvement measure is reduced significantly. This process is illustrated in Figure 8.8.

In the common case of a loaded component, high reliability is brought by the condition 'low probability that the load will exceed strength'. The general factors contributing to this condition are: (i) 'decreased load'; and (ii) 'increased strength'.

Contributing specific reliability improvement measures to the general factor 'decreased load' are: decreasing the mean value of the load by derating; decreasing the variability of the load by smooth loading; and eliminating the high values of the load by stress limiters.

Contributing specific reliability improvement measures to the general factor 'increased strength' are: increasing the mean value of the strength by selecting appropriate materials; increasing the mean value of the strength by increasing the load carrying sections; decreasing the strength variability by better technological processes, manufacturing control and inspection control; eliminating the lower tail of the strength distribution by a burn-in operation and environmental testing.

In some cases, the failure modes at a system level are obtained by negating basic required system functions. Moving back to the general contributing factors and the specific contributing factors is a powerful technique for discovering and eliminating failure modes. The next case study uses the method of inversion to identify failure modes of a lubrication system.

8.5.3.1 Case Study: Identifying Failure Modes of a Lubrication System by Moving Backwards to Contributing Factors

The purpose of a lubrication system is: (i) reducing wear; (ii) removing heat from friction zones and cooling the contact surfaces; (iii) cleaning the contact surfaces from abrasion particles and dirt; and (iv) protecting the lubricated parts from corrosion.

Consequently, the required functions from a generic lubrication system is *to supply constantly clean oil at a specified pressure, temperature, flow rate, composition, and viscosity to contacting and moving parts*. Not fulfilling any of the required system functions constitutes a system failure. Negating 'clean oil' results in 'unclean oil'. Moving back to the contributing general factors to unclean oil reveals the general contributing factors 'oil degradation' and 'oil contamination'.

Specific factors leading to oil degradation, for example, are 'depletion of the oil additives' and 'oxidation of the oil', 'inappropriate oil selected', and 'inappropriate viscosity of the selected oil'. Specific factors leading to the general contributing factor 'oil contamination' are, for example, 'oil with suspended particles due to a damaged filter', and 'oil with moisture'.

Negating 'oil with specified pressure' results in 'too low pressure of the supplied oil' or 'too high pressure of the supplied oil'. Moving back from 'too low pressure of the supplied oil' reveals the contributing general factors 'due to the oil used', 'due to component failures', 'due to excessive wear', and 'due to faulty design'.

Specific factors leading to the general factor 'too low pressure due to the oil used' are 'oil with too large viscosity' and 'too small quantity of circulating oil in the lubricating system'. Specific factors leading to the general factor 'too low oil pressure due to component failures' are 'failure of the oil pump', 'blockage of the oil filter', 'blockage of the cooler', and 'failure of the pressure relief valve'.

A specific factor leading to the general contributing factor 'too low oil pressure due to excessive wear' is 'excessive clearances in the journal bearings due to wear' while a specific factor leading to the general factor 'too low oil pressure due to faulty design' is 'inappropriately designed oil galleries'.

Moving back from 'too high pressure of the supplied oil' reveals the contributing general factors 'pressure relief valve stuck in closed position' and 'oil with inappropriate viscosity'.

Negating 'oil with specified temperature' results in 'too high temperature' of the supplied oil. Moving backwards reveals the contributing general factors 'too high temperature due to inappropriate cooling' and 'too high temperature due to the oil used'.

Specific factors leading to the general factor 'too high oil temperature due to inappropriate cooling' are 'failure of the cooling circuit' and 'cooler clogged with debris or lined with plaques'. Specific factors leading to the general factor 'too high temperature due to the oil used' are 'insufficient quantity of the circulating oil' and 'oil degradation'.

If factors are contributing to a more general factor, it is necessary to move back to a lower level in order to reveal the more specific contributing factors. Thus, moving back from the factor 'insufficient quantity of the circulating oil' reveals the more specific contributing factors 'incorrectly sized sump' and 'too low oil level'.

8.5.4 Inverse Thinking in Mathematical Models Evaluating or Reducing Risk

An example of inverse thinking in mathematical models evaluating or reducing risk can be given by calculating the reliability of a system with components logically arranged in parallel, working independently from one another (Figure 8.9a). The system is in working state at the end of a specified time interval if at least a single component is in working state. Let the events $C_1, C_2, \ldots, C_n$ stand for the first component is in working state at the end of the operational time interval, the second component is in working state at the end of the operational time interval, and so on (Figure 8.9b). The system will be in working state if at least a single component is in working state at the end of the specified operational interval. Consequently, the reliability of the system R_{sys} is equal to the probability of the union $R_{sys} = P(C_1 \cup C_2 \cup \ldots \cup C_n)$ of n statistically independent events.

The probability of the union of n statistically independent events (the reliability of the system) is determined from:

$$R_{sys} = P\left(\bigcup_{i=1}^{n} C_i\right) = \sum_{i=1}^{n} P(C_i) - \sum_{i<j}\sum P(C_i \cap C_j) + \sum_{i<j<k}\sum\sum P(C_i \cap C_j \cap C_k) - \ldots + (-1)^{n+1} P(C_1 \cap C_2 \cap \ldots \cap C_n) \quad (8.9)$$

This expression is also known as *the inclusion–exclusion expansion*.

The probabilities that the separate components will be in working state at the end of the operational interval are $r_1 = P(C_1)$, $r_2 = P(C_2)$, …, $r_n = P(C_n)$, respectively. These are

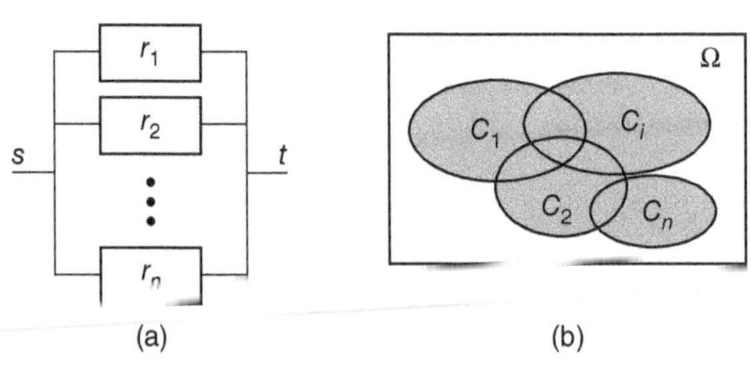

Figure 8.9 (a) A system with components logically arranged in parallel; (b) a union of n non-disjoint events.

effectively the reliabilities of the components with respect to the specified operational time interval. Equation (8.9) regarding the reliability of the system becomes

$$R_{sys} = \sum_{i=1}^{n} r_i - \sum_{i<j}\sum r_i r_j + \sum_{i<j<k}\sum\sum r_i r_j r_k - \ldots + (-1)^{n+1} r_1 r_2 \ldots r_n \tag{8.10}$$

According to the inclusion–exclusion expansion formula, the probabilities of all single events are first added. However, this means that the probability of the intersections of any pair of events has been added twice and should therefore be subtracted. Next, the probabilities of all twofold intersections are subtracted from the previous result. During these operations however, the contribution of the intersection of any three events has been added three times by adding the probabilities of the single events and subsequently subtracted three times by subtracting the probabilities of the twofold intersections. Consequently, the probabilities of all threefold intersections must be added. In short, terms with odd number of intersections are added while terms with even number of intersections are subtracted. For n statistically independent components, the probability of the union contains $2^n - 1$ terms which all must be evaluated in order to determine the probability of the union.

With increasing the number of components n, the number of terms in the inclusion-exclusion expansion increases exponentially, which makes it impossible to evaluate the system reliability even for a moderate number of components.

This difficulty can be easily avoided by applying the method of inversion. Instead of focusing on the reliability of the system, the problem is inverted, and the focus is placed on the probability of failure of the system. The system fails only if every single component is in a failed state at the end of the time interval. The probability that none of the components will be in working state is given by the simple expression:

$$P\left(\bigcap_{i=1}^{n} \overline{C}_i\right) = [1 - P(C_1)] \times [1 - P(C_2)] \times \ldots \times [1 - P(C_n)]$$

The reliability of the system (the probability of surviving the operational time interval) is then given by

$$\begin{aligned} R_{sys} &= P\left(\bigcup_{i=1}^{n} C_i\right) = 1 - [1 - P(C_1)] \times [1 - P(C_2)] \times \ldots \times [1 - P(C_n)] \\ &= 1 - (1 - r_1)(1 - r_2) \ldots . (1 - r_n) \end{aligned} \tag{8.11}$$

This expression, after expanding, leads to Eq. (8.10). The significant difference is that expression (8.11) can now be evaluated with n multiplications and $n + 1$ additions, only. Inverting the problem simplified the problem significantly and, as a result, the running time of the computation dropped from exponential $O(2^n)$ to linear $O(n)$.

8.5.4.1 Case Study: Using the Method of Inversion for Fast Evaluation of the Production Availability of a Complex System

The method of inversion can be used for simplifying mathematical models evaluating reliability/availability. This method is at the heart of the stochastic pruning method for fast evaluation of the production availability of complex repairable flow networks (Todinov 2016b).

Production availability is a key performance measure of repairable flow networks. It is defined by the ratio:

$$\psi = \overline{Q}_T / Q_{T0}$$

where $\overline{Q}_T$ is the total expected output of the network in the presence of component failures over a specified time interval and Q_{T0} is the total output that could be obtained in the absence of component failures over the specified time interval.

Traditionally, the production availability of repairable flow networks is revealed by using discrete-event simulators tracking the flow output, event-by-event. To reveal the variation of the total output, hundreds of thousands of failure-repair histories must be generated, each of which may contain thousands of failure-repair events ($s \approx 100000$, Figure 8.10a).

The drawback of the discrete-event simulators consists of the circumstance that the computational time increases enormously with increasing the size of the network, the length of the time interval and the failure frequencies of components.

The low efficiency of the discrete-event simulators for large systems is due to the event-by-event tracking of the output flow, for each failure-repair simulation history, $j = 1, 2, \ldots, s$ (Figure 8.10a), where s stands for the total number of failure-repair histories. The low efficiency of discrete-event simulators does not permit embedding them in a loop performing topology optimisation of the networks.

This drawback can be resolved if the operations in determining the expected output flow in the presence of failures are reversed. The discrete-event simulators track event-by-event the output flow in each simulation history followed by averaging the calculated output flows (Figure 8.10a).

Mathematically, the expected throughput flow in the presence of component failures, for s failure-repair histories (Figure 8.10a) is given by

$$\overline{Q}_T = \frac{1}{s} \sum_{j=1}^{s} \int_0^a Q_j^P(t)dt \tag{8.12}$$

where $Q_j^P(t)$ is the momentary throughput flow in the jth failure-repair history, at time t.

Inverting the problem suggests tracking first the average momentary output flow from all simulation histories at a given point in time, followed by integrating the averaged output

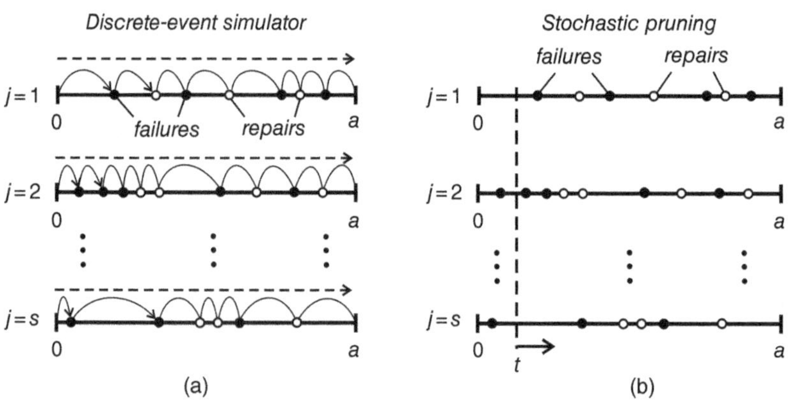

Figure 8.10 Applying the method of inversion simplifies greatly the task of determining the production availability of complex repairable flow networks.

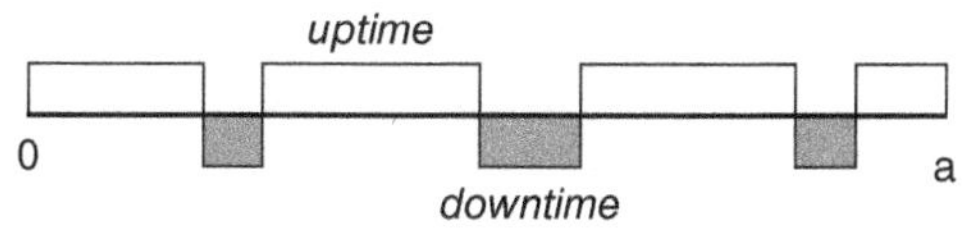

Figure 8.11 Each component in the network is either in working state (uptime) or in a state of repair (downtime).

flow over the time interval (Figure 8.10b). Tracking the average momentary output flow from all simulation histories at a given point in time is equivalent to tracking the momentary output flow from stochastically pruned networks: the networks from which edges are missing with probabilities equal to the unavailabilities of the components (edges). Unavailability is the probability that a particular component from the network will be in a failed state at a random point in time (Figure 8.11).

The method of inversion effectively reverses the integration and the summation in Eq. (8.12). It determines the expected momentary throughput flow for all stochastically pruned flow networks in all failure-repair histories at a time t and subsequently integrates this average throughput flow within limits (0,a):

$$\overline{Q'}_T = \int_0^a \left(\frac{1}{s} \sum_{j=1}^{s} Q_j^P(t) \right) dt \tag{8.13}$$

where $\overline{Q}_P = \frac{1}{s} \sum_{j=1}^{s} Q_j^P(t)$ is the expected momentary throughput flow from the stochastically pruned networks at time t. The momentary throughput flow $Q_j^P(t)$ is a non-negative function of the time t over the interval (0,a). According to the *Tonelli theorem* (Tonelli 1909), for non-negative functions, the integration and the summation can always be reversed without altering the result. Therefore, the expressions (8.12) and (8.13) are equal: $\overline{Q}_T = \overline{Q'}_T$. The throughput flow in the presence of failures becomes

$$\overline{Q}_T = \int_0^a \left(\frac{1}{s} \sum_{j=1}^{s} Q_j^P(t) \right) dt \tag{8.14}$$

This expression can be evaluated by a Monte Carlo integration.

$$\overline{Q}_T = \frac{1}{s} \int_0^a (Q_1^P(t) + Q_2^P(t) + \ldots + Q_s^P(t))dt = \frac{a}{s \times N} \left(\sum_{r=1}^{s\times N} Q_r^P \right) \tag{8.15}$$

where N is a sufficiently large number of samples over the time interval (0,a) and Q_r^P is the monetary throughput flow from the rth stochastically pruned network.

The production availability is determined from:

$$\psi = \overline{Q}_T / Q_{T0} = \frac{a}{s \times N} \left(\sum_{r=1}^{s\times N} Q_r^P \right) / (aQ_0) = \frac{1}{s \times N} \left(\sum_{r=1}^{s\times N} Q_r^P \right) / Q_0 \tag{8.16}$$

where Q_0 is the momentary output (throughput) flow from the original network, in the absence of any failures.

The stochastic pruning method works for both constant failure rates and time-dependent failure rates of the components.

As a result, the inversion made it possible to replace the direct integration with a Monte Carlo integration. The Monte Carlo integration samples the space stochastically pruned networks – time.

By inverting the problem, the task of determining the production availability was greatly simplified. There is no longer a need to generate and track, event-by-event, the failure repair histories. The result is a new method (the stochastic pruning method) for determining the production availability. Because no failure-repair histories are generated and tracked, the running time of the proposed algorithm is independent of the failure frequencies of the components. This reduces the computational time by many orders of magnitudes compared with the conventional discrete-event simulators (Todinov 2016b).

8.5.4.2 Case Study: Repeated Inversion for Evaluating the Risk of Collision of Ships

Combining inverse thinking with repeated inversion while maintaining an invariant property can be used with success for reducing the risk of collision of ships. The risk of collision of ships is an important problem already discussed widely in the relevant literature (Lebkowski et al. 2005; Oh et al. 2015). This problem is particularly acute for a large volume of sea traffic, high number of shipping lanes, increasing speeds and dimensions of newly built ships (Lebkowski et al. 2005). The ability to make fast decisions about the likelihood of collision of ships is a safety-critical activity.

The risk of a dangerous approach to a travelling ship can be assessed through the value r_{sd} where r_{sd} is the radius of the safety domain surrounding the ship. This radius makes a safety zone for the moving ship which cannot be entered by any other navigational object without a collision threat. A comprehensive discussion related to the safety domain and its definition has been provided in Lebkowski et al. (2005).

The method of inversion provides a very simple solution to this problem, yet it has never been applied to this problem before. This only confirms that without training in the domain-independent method of inversion, it is not easy to come to a solution based on the method of inversion.

The inversion method applied for assessing the risk of collision uses the fact that the relative distances between a number of navigating vessels (Figure 8.12a) does not change if the same uniform velocity is imparted to each vessel, in a particular common direction. In other words, the relative distances between the ships are invariant with respect to a common uniform velocity imparted to all ships.

If the common velocity is selected to be equal in magnitude and opposite to the velocity of one of the ships (e.g. the first ship), this ship will become stationary ($\mathbf{v}'_1 = \mathbf{v}_1 - \mathbf{v}_1 = \mathbf{0}$) and the rest of the ships will move with altered velocities equal to the vector sum of their old velocity and the inverted velocity of the first ship. In Figure 8.12b, the altered velocities of the ships become: $\mathbf{v}'_2 = \mathbf{v}_2 - \mathbf{v}_1$, $\mathbf{v}'_3 = \mathbf{v}_3 - \mathbf{v}_1$, and $\mathbf{v}'_4 = \mathbf{v}_4 - \mathbf{v}_1$. The relative distances between the ships however, will not change.

Essentially, the problem is more complex because the ships actually start and end their journeys at different times. This time dependence is presented in Figure 8.13 for two ships which start their travel at times ts_1 and ts_2, and end it at times te_1 and te_2, correspondingly. It is assumed that if a ship is located safely in port, no collision with other ships could possibly occur. Consequently, a collision between two navigating vessels can only occur during the time interval of parallel travel (in Figure 8.13, during the time interval ts_2, te_1). This observation significantly simplifies the problem, because the time intervals during which one of the ships is in port are excluded from consideration.

Another complication is that during their travel, the ships traverse not infinite straight lines but finite segments. Determining the smallest distances between the first ship S_1 and the second ship S_2 (Figure 8.14a) during their parallel time interval of navigation can be done by applying the method of inversion.

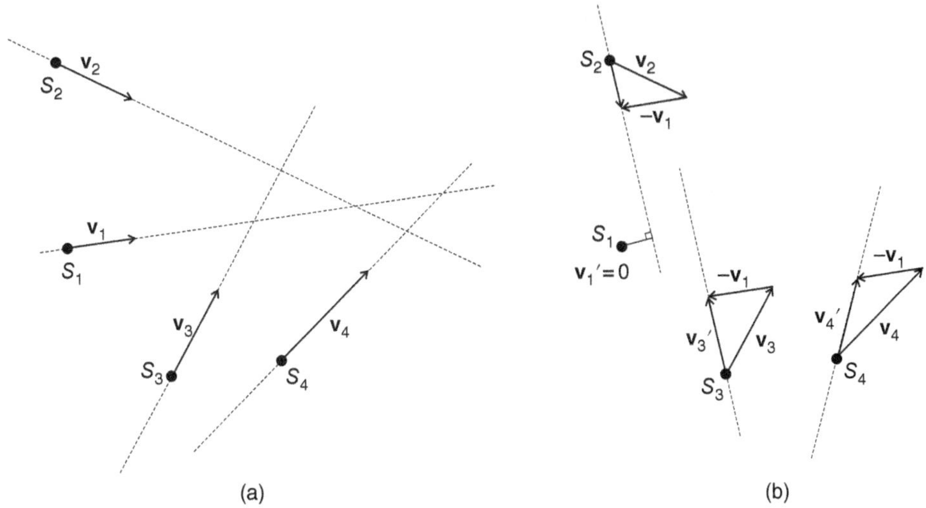

Figure 8.12 Determining the closest distance among ships travelling with constant speed, by using a repeated inversion of velocities.

Figure 8.13 Collision between two navigating vessels can only occur during the time interval of parallel travel.

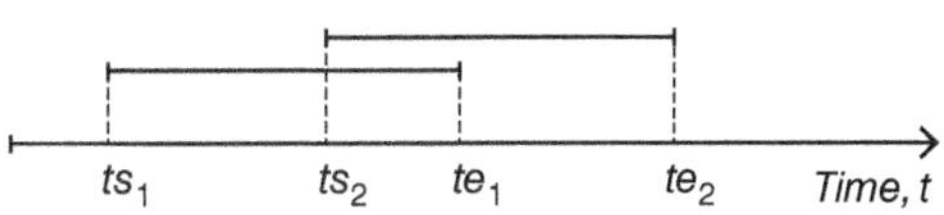

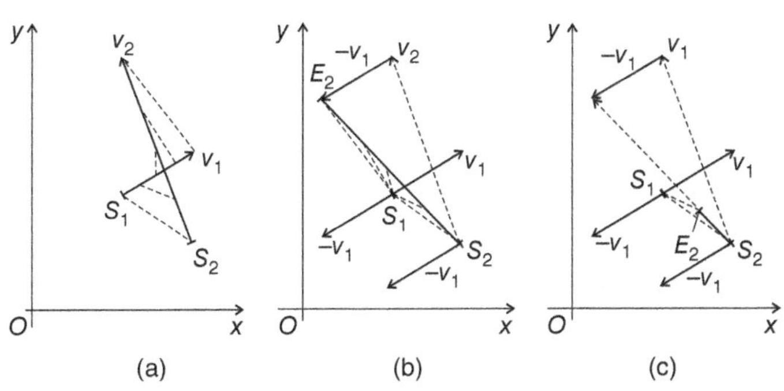

Figure 8.14 Determining the closest distance among ships travelling with constant speed, by using a repeated inversion of velocities.

A uniform velocity $-v_1$ is imparted to the first ship (Figure 8.14b). As a result, the first ship becomes stationary and the second ship S_2 starts moving with a new velocity $\mathbf{v}_2' = \mathbf{v}_2 - \mathbf{v}_1$ along a modified segment S_2E_2. The modified segment S_2E_2 travelled by the second ship, with the altered velocity, is oriented along the new trajectory defined by the vector $\mathbf{v}_2' = \mathbf{v}_2 - \mathbf{v}_1$ and the length of the travelled modified segment is equal to $|\mathbf{v}_2 - \mathbf{v}_1|\, t_{p12}$, where $|\mathbf{v}_2 - \mathbf{v}_1|$ is the magnitude of the modified velocity and t_{p12} is the time of parallel travel for ships S_1 and S_2 (Figure 8.14b). Next, the smallest distance between the first and second ship is determined. It is the shortest distance between the point S_1 and the modified segment S_2E_2 (Figure 8.14b).

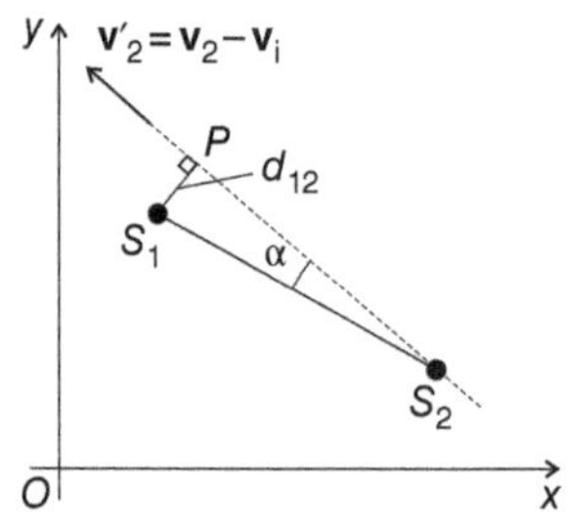

Figure 8.15 Determining the shortest distance to the modified travelled segment if both angles α and β are acute.

If both angles $\alpha \equiv S_1S_2E_2$ and $\beta \equiv S_1E_2S_2$ are acute (Figure 8.14b), the shortest distance is the length of the perpendicular to S_2E_2 (Figure 8.15).

Denote by (x_1, y_1) the coordinates of the ship S_1, by (x_2, y_2) the coordinates of the ship S_2, by (vx_1, vy_1) the components of the velocity $\mathbf{v_1}$ and by (vx_2, vy_2) the components of the velocity $\mathbf{v_2}$.

Because the components of the modified velocity vector $\mathbf{v}' = \mathbf{v_2} - \mathbf{v_1}$ of the second ship and the components of the vector $\mathbf{S_2S_1}$ are known, the angle α (Figure 8.15) is determined from

$$\cos\alpha = \frac{\mathbf{S_2S_1} \cdot \mathbf{v_2}'}{\|\mathbf{S_2S_1}\| \cdot \|\mathbf{v_2}'\|} \tag{8.17}$$

where $\mathbf{S_2S_1} \cdot \mathbf{v'}_2 = (x_1 - x_2)(vx_2 - vx_1) + (y_1 - y_2)(vy_2 - vy_1)$ is the scalar product of the vectors $\mathbf{S_2S_1}$ and $\mathbf{v'}_2$, $\|\mathbf{S_2S_1}\| = \sqrt{(x_1 - x_2)^2 + (y_1 - y_2)^2}$ is the magnitude of the vector $\mathbf{S_2S_1}$ and $\|\mathbf{v}_2'\| = \sqrt{(vx_2 - vx_1)^2 + (vy_2 - vy_1)^2}$ is the magnitude of the altered velocity $\mathbf{v'}_2$.

In a similar fashion, the cosine of angle $\beta \equiv S_1E_2S_2$ $(\cos\beta)$ is determined. If both $0 \leq \cos\alpha$ and $0 \leq \cos\beta$, the shortest distance is the perpendicular S_1P to the modified segment S_2E_2.

From expression (8.17), $\sin\alpha = \sqrt{1 - \cos^2\alpha}$ can be determined and the length of the perpendicular S_1P becomes:

$$S_1P \equiv d_{12} = \|\mathbf{S_2S_1}\| \sin\alpha = \frac{1}{\|\mathbf{v_2}'\|}\sqrt{\|\mathbf{S_2S_1}\|^2 \cdot \|\mathbf{v'}_2\|^2 - (\mathbf{S_2S_1}\cdot\mathbf{v'}_2)^2} \tag{8.18}$$

If $0 \leq \cos\alpha$ is not fulfilled, the angle α is obtuse and the shortest distance is the length S_2S_1; if $0 \leq \cos\beta$ is not fulfilled, the angle β is obtuse and the shortest distance is the length S_1E_2 (Figure 8.14c).

Once the smallest distances between the first ship and the remaining ships have been determined, the same procedure is applied to the second ship and the remaining ships. To each of the remaining ships (except the first ship whose distances to the rest of the ships have been determined), a uniform velocity of $-\mathbf{v_2}$ is imparted. As a result, the velocity of the second ship becomes zero ($\mathbf{v'}_2 = \mathbf{v_2} - \mathbf{v_2} = 0$) and the velocities of the rest of the ships become $\mathbf{v'}_3 = \mathbf{v_3} - \mathbf{v_2}$, $\mathbf{v'}_4 = \mathbf{v_4} - \mathbf{v_2}$, and so on. Now, the smallest distances between the second ship and the rest of the ships can be determined. This procedure is extended to the rest of the ships.

Finally, comparing the as-determined individual shortest distances d_{ij} $i<j$ between the ships reveals the absolute shortest distance between a pair of ships. The smallest absolute distance $d_{\min} = \min\{d_{ij}\}$ is determined and the possibility of collision is checked by comparing it with the radius r_{sd} of the safety zone. On the basis of the described method, an algorithm can be developed for determining the smallest distance among a specified number of ships moving with different speeds, in different directions.

Algorithm 8.1

```
//n - specified number of ships
dmin= Big_number; ind1=0; ind2=0;

for i=1 to n do
  for j=i+1 to n do
  {

   Determine the time interval of parallel travel
   between ships i and j;

  if (there is an interval of parallel travel)
    {Stop ship i by imparting a velocity with the same
     magnitude and opposite in direction to its current
     velocity.
       Determine the modified velocity of the jth ship by
       subtracting from its velocity vj, the velocity vi
       of the ith ship;
       Determine the length of the modified segment SjEj
       travelled by the jth ship;

     Evaluate the angles alfa and beta;

     if (alfa is an obtuse angle) d = SiSj;
     else if (beta is an obtuse angle) d=SiEi;
     else {
           Determine the length d=SiP of the perpendicular
           SiP;
          }

    if (d<dmin) then {dmin=d; ind1=i; ind2=j;}
    }
  }
 }
```

Initially, the smallest distance dmin between the travelling ships is initialised with a very big number. As each smallest distance *d* between a pair of ships is calculated sequentially, it is compared with the current minimum distance dmin. As a result, at the end of the calculations, the absolute minimum distance remains in the variable dmin and the indices of the corresponding ships remain in the variables 'ind1' and 'ind2'. The smallest distance *dmin* is determined after $n(n-1)/2$ comparisons. The method of inversion by maintaining an invariant yields a simple algorithm for assessing the risk of collision among ships.

The problem can be generalised in the case where the ships change their velocity magnitudes and directions with time $\mathbf{v_i} = \mathbf{v_i}(\mathbf{t})$ (Figure 8.16a). In this case, when considering

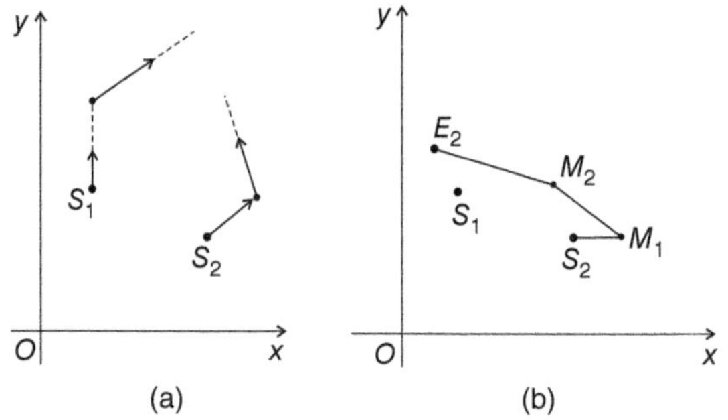

Figure 8.16 Determining the smallest distance in the case of ships travelling with variable velocities.

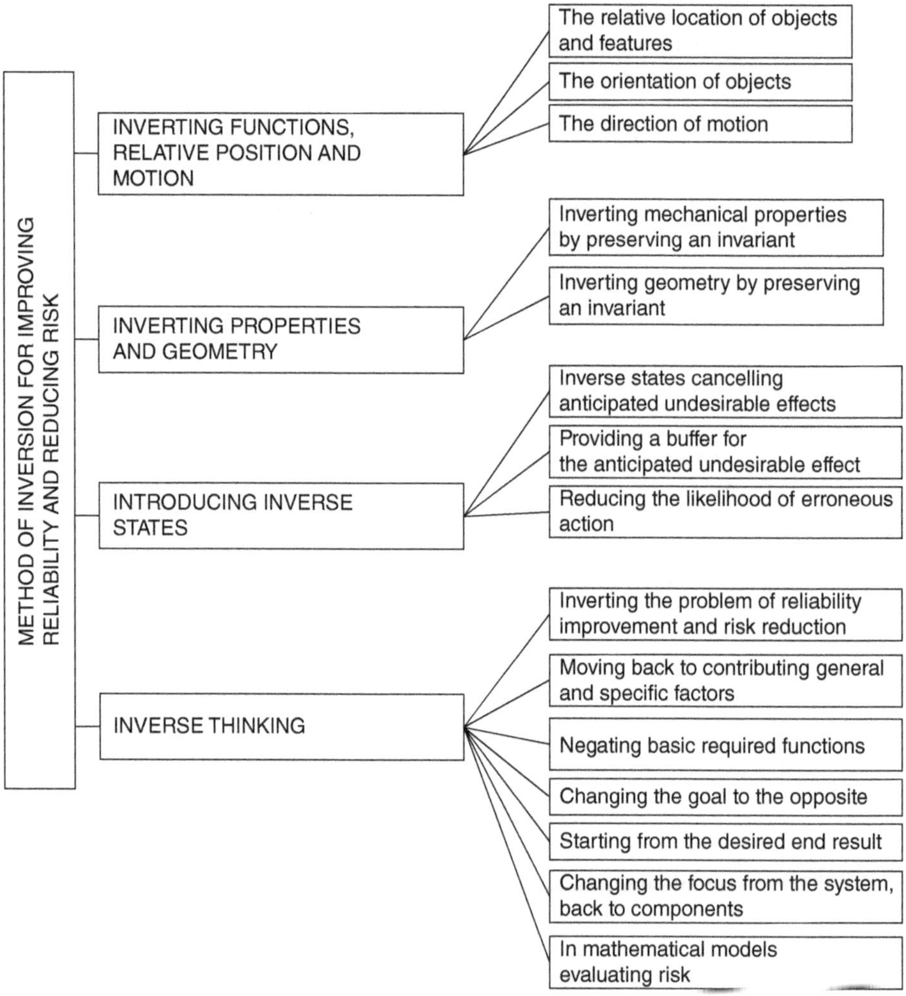

Figure 8.17 The method of inversion for improving reliability and reducing risk.

the shortest distance between two ships, for example between ships S_1 and S_2, the updating of the velocity of the second ship must be done regularly at the times $t_{11}, t_{12}, \ldots$, at which the first ship changes the magnitude or the direction of its velocity and also at the times $t_{21}, t_{22}, \ldots$, at which the second ship changes the magnitude or the direction of its velocity. During the updates, the updating velocity for the second ship is always equal and directly opposite to the velocity of the first ship. In this case, the modified travelled path of the second ship will no longer be a single straight-line segment but a series of connected line segments (Figure 8.16b).

The first ship in Figure 8.16a changed its velocity magnitude and direction at time t_{11} while the second ship changed its velocity magnitude and direction at time t_{21}. The modified path of the second ship is composed of the connected line segments S_2M_1, M_1M_2, and M_2E_2. The shortest distance between the ships is determined by determining the shortest distance between point S_1, which corresponds to the position of the first ship (which never changes during the repeated inversion of velocity) and the series of connected line segments $S_2M_1M_2E_2$. The shortest distance is obtained by determining sequentially the shortest distances to the individual segments and selecting the smallest distance.

The classification of the techniques for improving reliability and reducing risk based on the method of inversion, discussed in this chapter, is given in Figure 8.17. This classification introduces a structure in implementing the method of inversion and simplifies the process of searching for an appropriate solution. In addition, the structure provided by the classification makes the application of the method of inversion less dependent on inspiration and lateral thinking leaps. The classification is easy to use and does not require a significant investment of time. In addition, the classification acts as a useful checklist which helps avoid the omission of a relevant technique based on the method of inversion.

9

Reliability Improvement and Risk Reduction Through Self-Reinforcement

9.1 Self-Reinforcement Mechanisms

The present chapter contributes an important domain-independent reliability improvement and risk reduction method, referred to as 'the method of self-reinforcement'. A key feature of self-reinforcement is that increasing the external/internal forces intensifies the system's response against these forces. As a result, the driving net force towards precipitating failure is decreased. Self-reinforcement, therefore, increases the ability of the system to absorb overloading thereby increasing the resilience of the system. As it will be demonstrated later, in many cases, the self-reinforcement mechanisms achieve a remarkable reliability increase at no extra cost.

A substantial amount of research exists on improving reliability by 'reinforcement' of components and systems and virtually no research exists on 'self-reinforcement'. However, there is a very important difference between improving reliability by 'reinforcement' and by 'self-reinforcement'. An important feature of self-reinforcement which distinguishes it from mere 'reinforcement' is that increasing the external/internal force magnitudes increases the resistance against these forces. For the common reinforcement solutions, such an increase of the resistance is absent. *The reinforcement only provides compensation against an undesirable effect but not a proportional compensation.* Shot peening of the surface of automotive suspension springs, for example, is an example of reinforcement. It introduces compressive residual stresses in the surface of the spring coil and strengthens the resistance of a loaded spring against fatigue failure (Niku-Lari 1981; Bird and Saynor 1984). *Cold expansion*, used in aviation for creating compressive stresses at the surface of fastener holes (Fu et al. 2015), is also an example of reinforcement.

However, increasing the load on a shot-peened spring or the loading stress in the vicinity of a fastener hole, does not change the magnitude of the compressive residual stresses and does not increase the fatigue resistance of the components. Reinforcement from shot peening and cold expansion is present but not a self-reinforcement.

Mechanisms that are in fact only reinforcement mechanisms should not be confused with the 'self-reinforcement' mechanisms. The focus of this chapter is on self-reinforcement mechanisms for improving reliability and reducing risk.

Isolated solutions for improving the performance of components and systems making use of self-reinforcement do exist in the engineering design literature but they have never been linked with the method of self-reinforcement.

Methods for Reliability Improvement and Risk Reduction, First Edition. Michael Todinov.

Thus, a self-reinforcing effect from capturing rotational motion has been discussed for example in Pahl et al. (2007). Pahl et al. (2007) however, did not go beyond this isolated example and generalise the self-reinforcement method as a domain-independent method for improving the reliability of components and systems.

Self-locking devices, such as self-locking screws, self-locking grips, self-locking hooks, self-energising breaks, etc., have been known for a long time. Costache et al. (2016), for example, recently introduced self-locking grips for anchoring fibre-reinforced tendons. However, self-locking has not been recognised as a self-reinforcement method for improving reliability and reducing risk.

Self-balancing can also be a source of a self-reinforcing response. The 'roly-poly' toy which rights itself when pushed over can be given as an example. Recently, self-balancing inspired by the 'roly-poly' toy has been suggested by Zhang et al. (2013) for improving the side rolling stability of an e-bike. Again, self-balancing has never been recognised as a method for improving reliability by self-reinforcement.

Eliminating harmful factors and influences is the purpose of many inventions and Altshuller's TRIZ system (Altshuller 1984, 1996, 1999) captured a number of useful general design principles that could be used to eliminate harm and reduce risk. No discussion however, has been presented by Altshuller related to the method of self-reinforcement. No discussion regarding the method of self-reinforcement has been presented in more recent literature related to TRIZ (Terninko et al. 1998; Savransky 2000; Orloff 2006, 2012; Rantanen and Domb 2008; Gadd 2011).

In summary, despite the availability of occasional isolated self-reinforcement solutions used in engineering designs, these were not recognised as instances of the method of self-reinforcement and were not linked with this method. No analysis of the mechanisms through which self-reinforcement improves reliability and reduces risk or a classification of self-reinforcement mechanisms has ever been presented in the engineering design literature or the reliability and risk literature. Consequently, the present chapter attempts to fill this gap by providing an introduction to the mechanisms through which the method of self-reinforcement improves reliability and reduces risk.

Two principal ways of self-reinforcement were identified: *self-reinforcement by capturing a proportional compensating factor;* and *self-reinforcement by using feedback loops.*

It is shown that the self-energising, the fast growth, and fast transition provided by the positive feedback loops are important mechanisms for achieving self-reinforcement response.

The following basic mechanisms of self-reinforcement by capturing a proportional compensating factor have also been identified: (i) mechanisms transforming forces and pressure into a self-reinforcing response; (ii) mechanisms transforming motion into a self-reinforcing response; (iii) mechanisms achieving a self-reinforcing response by self-balancing and self-anchoring and (iv) mechanisms achieving self-reinforcement response by self-aligning.

The classification of the self-reinforcement mechanisms presented in this chapter (Figure 9.1) is based on a number of solutions, from various engineering fields, related to improving reliability and reducing risk, containing an element of self-reinforcement. Each of the available solutions was analysed for recurring reliability-improvement patterns and invariants. A certain level of abstraction was used to strip the available solutions from their

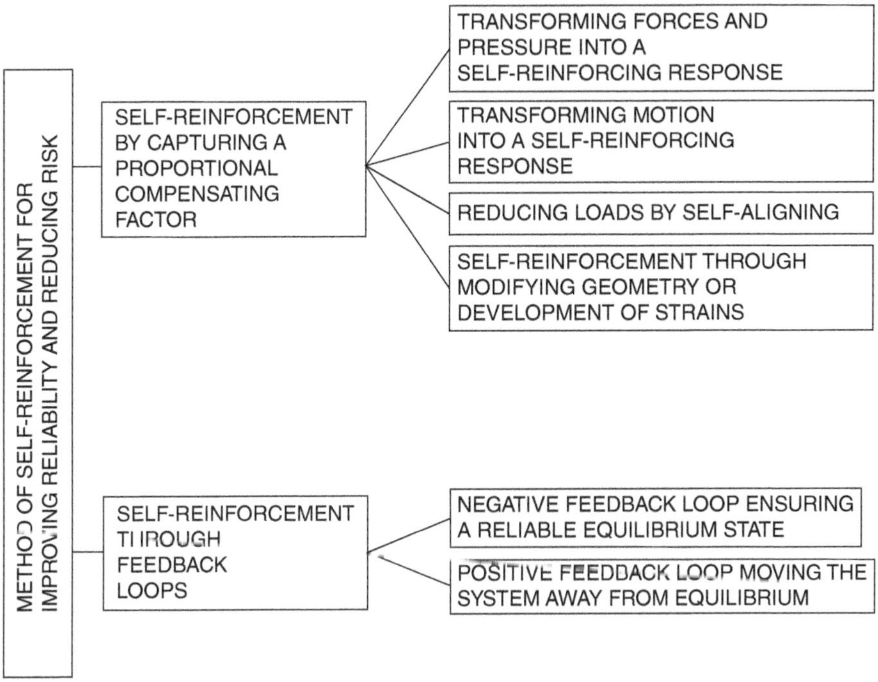

Figure 9.1 The method of self-reinforcement for improving reliability and reducing risk.

specific engineering context in order to uncover the underlying reliability-improvement patterns.

In what follows, the identified mechanisms of reliability improvement by self-reinforcement are discussed in detail.

9.2 Self-Reinforcement Relying on a Proportional Compensating Factor

Self-reinforcement by compensating proportional response is present in cases where a third factor is captured and used to provide a proportional, compensating, self-strengthening response. The proportional compensating response is channelled towards reducing the driving forces precipitating failure.

9.2.1 Transforming Forces and Pressure into a Self-Reinforcing Response

9.2.1.1 Capturing a Self-Reinforcing Proportional Response from Friction Forces

Friction forces can be captured and used as self-reinforcing counterforces. Common applications are self-locking screws and self-locking wedges. The higher the load in self-locking devices the higher is the magnitude of the excited friction forces and the larger is the reliability of the self-locking device. Self-reinforcement is also found in other self-locking

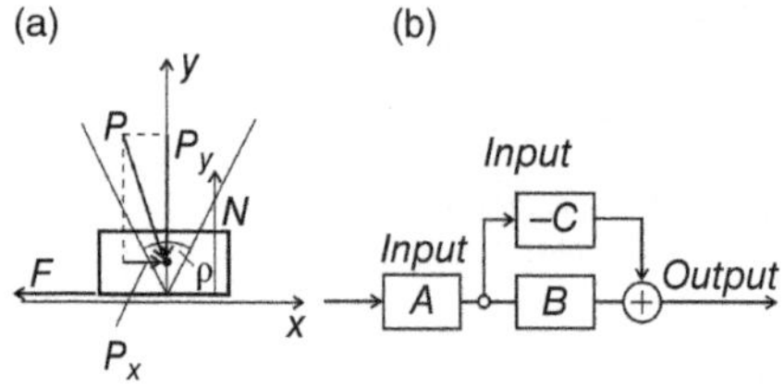

Figure 9.2 (a) A cone of friction; (b) self-reinforcement by capturing a compensating third factor whose magnitude is proportional to the undesirable outcome *B*.

devices: (i) self-locking grips in tensile testing machines; (ii) self-locking plate clamps; (iii) self-locking hooks; (iv) self-locking climbing equipment; and (v) self-locking marine cleats, etc.

Consider the block in Figure 9.2a loaded with the external force *P* and experiencing friction. Increasing the external force *P* increases proportionally the horizontal component P_x (undesirable outcome *B* in Figure 9.2b) working towards disturbing the equilibrium of the body (Figure 9.2a). It is a well-documented fact from statics (Meriam et al. 2016) that for the simple system in Figure 9.2a, no matter how large the loading force *P* is, if it lies within the friction cone defined by the friction angle ρ ($\mu = \tan \rho$, where μ is the static coefficient of friction), the body will remain in a state of equilibrium. Increasing the magnitude of the loading force *P* increases the component P_x pushing the component along the horizontal axis but also increases the normal force $N = P_y$ which excites the friction force *F*. The friction force *F* is proportional to the normal force *N*, $F = \mu N$ and can be interpreted as the third factor *C*, whose magnitude is increasing with increasing the magnitude of the undesirable component P_x (outcome *B* in Figure 9.2b) and is subtracted from the output (the component P_x). The two forces are combined and the friction force *F* fully compensates the undesirable force P_x. The body is in equilibrium. A self-reinforcement effect is present because with increasing the external load *P*, the resistance against moving along the horizontal axis *x* also increases.

The essence of the mechanism of self-reinforcement by capturing a compensating proportional response has been summarized with the abstract diagram in Figure 9.2b.

Factor *A* results in the undesirable outcome *B*. By appropriate modification (or directly), the effect *C* is captured to create a compensating response, whose magnitude is proportional to the magnitude of the undesirable outcome *B* but whose sign is opposite. Now, if the effect *C* is added to outcome *B*, the resultant undesirable outcome will be partially or fully cancelled. This is a self-reinforcement effect because with increasing the magnitude of the undesirable outcome *B*, the magnitude of the compensating effect *C* also increases. As a result, the driving force towards failure precipitation is reduced and reliability is improved. The self-reinforcement by capturing a compensating factor increases the capability of the system to absorb overloading shocks which also improves the resilience of the system.

9.2.1.2 Case Study: Transforming Friction Forces into a Proportional Response in the Design of a Friction Grip

Self-reinforcement by capturing a factor which proportionally compensates an undesirable outcome has been used in the design shown in Figure 9.3 of a friction grip.

The friction grip includes the strip 1 which is being pulled downwards with a force *P*, the block 2, cylinder 3, and wedge 4. The static friction coefficient between the cylinder 3 and the wedge 4 is μ_1; the friction coefficient between the cylinder 3 and the strip 1 is μ_2. The forces acting on the cylinder 3 are: the normal force N_1 and the friction force $T_1 = \mu_1 N_1$

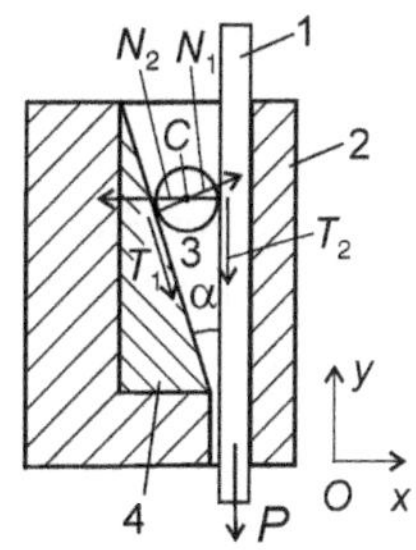

Figure 9.3 A self-reinforcing design of a friction grip based on exciting compensating friction forces.

from the wedge 4, the normal force N_2 and the friction force $T_2 = \mu_2 N_2$ from the strip 1. The cylinder 3 has a radius r and its weight has been ignored as being too small compared with the rest of the forces.

Under certain conditions, self-reinforcement will be present in the system because increasing the external load P will increase the friction forces T_2 and T_1 (factor C from Figure 9.2b) thereby retaining the strip and cylinder in equilibrium.

To determine the conditions under which there exists self-reinforcement, the equilibrium conditions for the cylinder 3 are considered. They include three equations: the sum of the projections of the forces along the x-axis must equal zero; the sum of the projections of the forces along the y-axis must equal zero and the sum of the moments of the forces with respect to the centre C of the cylinder must also equal zero:

$$N_1 \cos\alpha + \mu_1 N_1 \sin\alpha - N_2 = 0 \tag{9.1}$$

$$N_1 \sin\alpha - \mu_1 N_1 \cos\alpha - \mu_2 N_2 = 0 \tag{9.2}$$

$$\mu_1 N_1 r - \mu_2 N_2 r = 0 \tag{9.3}$$

In Eq. (9.3), r is the radius of cylinder 3.

From Eq. (9.3), $N_2 = \frac{\mu_1}{\mu_2} N_1$. Substituting $N_2 = \frac{\mu_1}{\mu_2} N_1$ in Eqs. (9.1) and (9.2) followed by dividing the equations by N_1 (assuming that $N_1 \gg 0$) results in:

$$\cos\alpha + \mu_1 \sin\alpha - \frac{\mu_1}{\mu_2} = 0 \tag{9.4}$$

$$\sin\alpha - \mu_1(1 + \cos\alpha) = 0 \tag{9.5}$$

From Eq. (9.5), $\mu_1 = \frac{\sin\alpha}{1+\cos\alpha} = \frac{2\sin(\alpha/2)\cos(\alpha/2)}{2\cos^2(\alpha/2)} = \tan(\alpha/2)$. From Eq. (9.4), $\mu_2 = \frac{\mu_1}{\cos\alpha+\mu_1 \sin\alpha}$.

Self-reinforcement (equilibrium of the cylinder 3) will only be present if, for the coefficients of friction, the following inequalities are simultaneously fulfilled:

$$\mu_1 \geq \tan(\alpha/2) \tag{9.6}$$

$$\mu_2 \geq \frac{\mu_1}{\cos\alpha + \mu_1 \sin\alpha} \tag{9.7}$$

If the two coefficients of friction are equal, Eq. (9.6) is obtained as a condition for self-reinforcement.

In this case study, the reliability improvement through self-reinforcement has been obtained at no extra cost.

9.2.1.3 Transforming Pressure into a Self-Reinforcing Response

This self-reinforcement mechanism is present *when external/internal forces or states are transformed/channelled towards inducing counter-forces which strengthen proportionally the system's response.*

Such a self-reinforcement application is given in Chapter 8, in the design of a cover for containers under pressure (Figure 8.2).

The design from Figure 8.2b channels the force due to the internal pressure into a self-reinforcing force which helps to form a reliable seal and prevent leakage. Increasing pressure increases the magnitude of the self-reinforcing force, strengthens the seal and increases the resistance to the increased pressure.

A similar effect is present for injection pistons with self-reinforcing seals. The elastic material of the seal is pressed against the walls of the cylinder. The higher the pressure, the tighter the seal formed between the piston and the cylinder. The reliability improvement is obtained at no extra cost.

9.2.1.4 Transforming Weight into a Self-Reinforcing Response

A powerful self-reinforcing effect, achieved by channelling weight as a self-reinforcement factor, has been discussed in Chapter 8 (Figure 8.4).

With increasing the weight of the component, the stress range decreases and the fatigue life increases. As a result, channelling the weight by inverting the location of the component and the supporting brackets results in a self-strengthened design characterised by an increased fatigue life and improved reliability. The larger the weight is, the larger is the reduction of the stress range, and the larger is the increase of the fatigue life.

9.2.1.5 Transforming Moments into a Self-Reinforcing Response

Self-reinforcement by capturing a factor providing a compensating proportional response can also be illustrated by an example related to capturing a counter-moment as a proportional compensating factor (Figure 9.4). Figure 9.4a features the overturning moment on a wall resisting the pressure from a large mass of liquid or granulated substance (water, concrete, sand, etc.). If an extension is firmly fixed to the lower end of the wall (Figure 9.4b), the distributed weight G of the mass of the liquid (granulated substance) creates an opposing

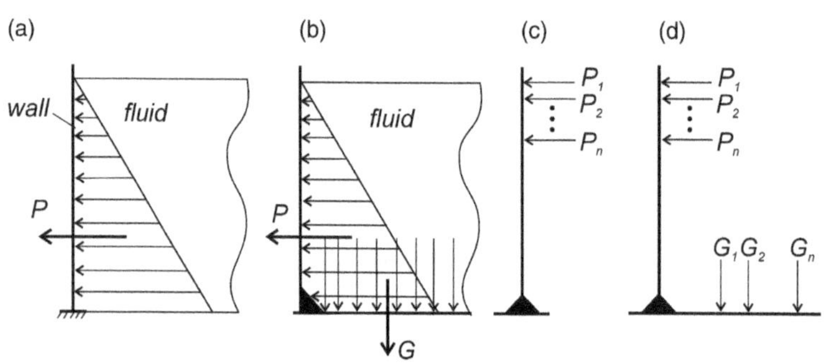

Figure 9.4 (a,b) Reducing the risk of overturning a wall supporting liquid or granulated substance by self-reinforcement. (c,d) Reducing the risk of overturning a crowd fence by self-reinforcement.

moment (factor C in Figure 9.2b) which counteracts the overturning moment of the force P (factor B in Figure 9.2b). The result is a reduced total overturning moment and reduced risk of overturning. Increasing the mass of the liquid (granulated substance) increases the overturning moment but also increases the opposing moment from the weight G.

The solution related to reducing the risk of overturning in Figure 9.4b can be replicated to reduce the risk of overturning of crowd fences. A fence without a self-reinforcement (Figure 9.4c) can be overturned relatively easily by the forces P_i created by people pushing the fence (Figure 9.4c). If the lower end of the fence on the crowd side is made wider (Figure 9.4d), people will have to stand on the fence while they push against it. As a result, the distributed weight of the people (the forces G_i in Figure 9.4d) standing on the lower part of the fence create an opposing moment that counteracts the overturning moment of the forces P_i from people pushing the fence. The more people push on the fence, the more weight forces will be available for counteracting the overturning moment. The weight of people is effectively channelled towards self-reinforcing. The capability of the system to resist and absorb overloading is increased which results in improved resilience. The reliability improvement in these examples is obtained at a low cost.

9.2.1.6 Self-Reinforcement by Self-Balancing

Unbalanced forces cause premature wear out, fatigue degradation and failure. As a rule, improving the level of balancing in a system improves the uniformity of the load distribution, reduces the magnitudes of the inertia forces, the loading stresses, and increases the reliability of the system.

Self-balancing improves the capability of the system to adapt to adversity and recover, which enhances the resilience of the system.

Self-reinforcement by self-balancing is also present in cases where negative factors are made to cancel one another.

Self-balancing is often achieved through symmetrical design which eliminates unwanted inertia forces and moments in rotating machinery. Static balancing, for example, guarantees that the mass centre of the rotating mass is on the rotation axis (Uicker et al. 2003). Increasing the speed of rotation increases the unbalanced inertia force but also increases the counterbalancing force. The result is a self-strengthening response.

The self-reinforcement by generating a similar in magnitude and opposite in direction counterforce, which subtracts from the force creating the undesirable effect, is at the heart of eliminating or reducing end-thrust forces. Force balancing can be done by a symmetrical design and used to minimise the axial forces on turbine shafts (Matthews 1998). For the design in Figure 9.5a, the axial force F from the turbine 1 needs to be counterbalanced by the thrust bearing force F' from the thrust bearing 2. For the symmetric design in Figure 9.5b, the increase of the axial force F from turbine 2, is accompanied by a similar increase of the equal and opposite force F' from turbine 1. The two forces counterbalance and the thrust bearing is no longer needed. The reliability of the assembly has been improved.

A sudden increase of the force F will cause an equal and opposite counterbalancing force F' counterbalancing the increase of the loading force F. As a result, the capability of

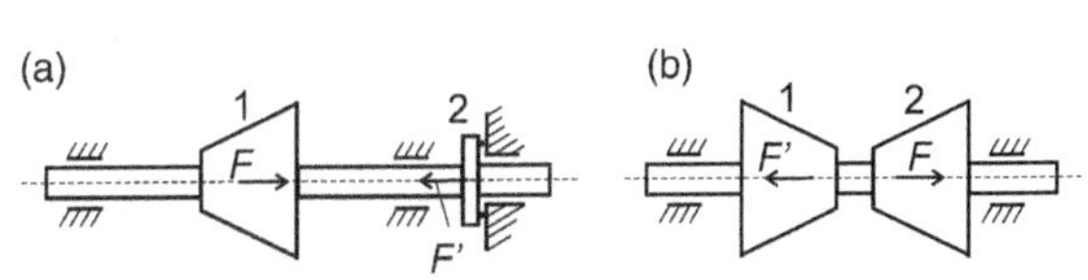

Figure 9.5 Self-balancing achieved through symmetrical design.

the system to absorb overloading shocks is improved thereby improving the resilience of the assembly.

The herringbone gears provide another example of self-reinforcement by counterbalancing. The axial forces in herringbone gear meshing are counterbalanced, which eliminates the need for thrust bearings. An increase in the transmitted torque increases simultaneously the magnitude of each axial force and because they act in opposite directions, the result is a very small resultant axial force.

Another example of self-balancing achieved through symmetrical design can be found in symmetrical epicyclic gear mechanisms where self-balancing of the radial forces acting on the central shaft is present.

Twisting wires to cancel their magnetic interference is also an example of self-reinforcement by self-balancing (Figure 9.6a). The flow of current through the wire results in an electromagnetic field around the wire which could generate noise in the neighbouring wires. Twisted wires carry equal and opposite currents whose electromagnetic fields cancel.

Finally, self-balancing can be implemented to reduce the magnitude of required activation forces if a very fast response is needed. The pipe in Figure 9.6b is under pressure p. If pressure needs to be released quickly, an actuation force F must be applied to the valve. The actuation force F must overcome not only the resistance of the spring but also the force C due to the pressure p (Figure 9.6b). In the self-reinforcing design in Figure 9.6c, a counterbalance force D is present. To release the pressure, the actuation force F must now overcome only the resistance of the spring and a small resultant force $C - D$. The required actuation force F for the design in Figure 9.6c is significantly smaller than that for the design in Figure 9.6b and the quick pressure release is more reliable. Even in the presence of a very large pressure p, the valve will still be released with a relatively small force F. Increasing pressure p increases the pressure force C but also proportionally increases the counterforce D. A self-reinforcing effect is present.

9.2.1.7 Self-Reinforcement by Self-Anchoring

Self-reinforcement by self-anchoring is present in cases where the loading forces excite counterforces which oppose the loading forces. This results in a reduced driving force for failure initiation. Self-anchoring bolts and screws are often used as a secure attachment to walls. Increasing the pulling force increases the anchoring effect and the pull-out resistance.

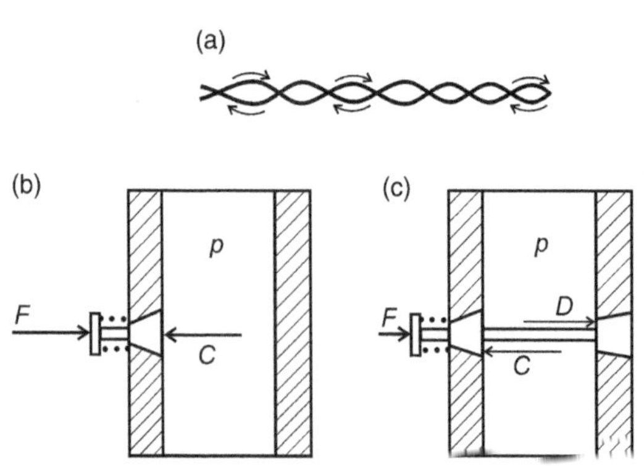

Figure 9.6 Other cases of self-reinforcement by self-balancing.

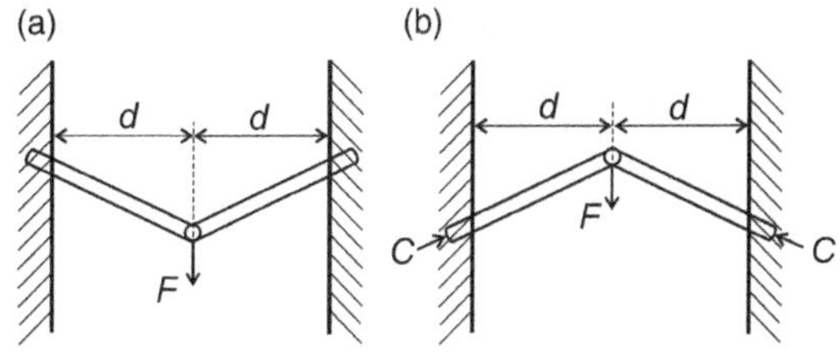

Figure 9.7 Self-reinforcement by self-anchoring.

Consider the cantilever double-beam structure in Figure 9.7a including two short links with a large cross section. The loading force F excites a moment $F \times d$ and friction forces which retain the beams in the wall.

If the excited friction forces are insufficient to retain the links, the links will be pulled out and the structure will collapse.

In Figure 9.7b, in addition to the friction forces, the loading force F also excites additional anchoring forces C in the supports. Increasing the magnitude of the loading force F increases the magnitude of the anchoring forces C. Because of the existing self-anchoring effect, the assembly in Figure 9.7b is more resistant against collapse compared with the assembly in Figure 9.7a. The short lengths of the links and the large cross sections exclude buckling as a failure mode.

Consider the device in Figure 9.8 designed for lifting steel drums 4 with mass m. The U-shaped brackets 1 connect the lifting ring 3 with the hooks 2. The device hooks over the end lips of the drum. From the equilibrium condition of the forces: $2 \times 2T \times \cos \beta = mg$, the tension force T acting in each rod of the U-shaped bracket can be determined: $T = \frac{mg}{4\cos \beta}$. The retaining force R with which the hook presses on the drum (Figure 9.8) can be approximated by $R = 2T \times \sin \beta = \frac{mg}{2} \tan \beta$. With increasing the weight of the drum mg, the retaining force R also increases. Self-reinforcement is present.

Increasing the angle β towards 90° results in a significant increase of tanβ and the retaining force R. In both cases, the reliability improvement by self-anchoring is obtained at no extra cost.

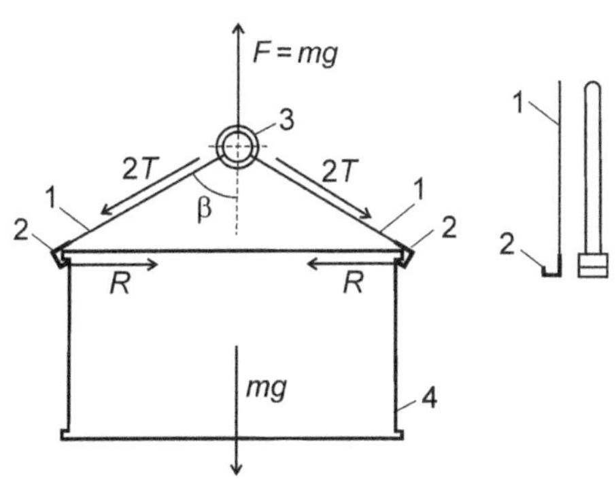

Figure 9.8 Hoisting a steel drum by self-reinforcement achieved through self-anchoring.

9.2.2 Transforming Motion into a Self-Reinforcing Response

This self-reinforcement mechanism is, for example, present in racing cars where wings are used for creating an aerodynamic downforce which enhances the grip with the racing track and prevents the loss of grip at high speeds while cornering. Increasing the speed of the car increases the inertia force acting on the car moving around a corner but also proportionally increases the downforce which strengthens the grip with the racing track. This is an example of self-reinforcement obtained by capturing and using motion to provide a compensating, self-strengthening response. The higher the speed, the higher the magnitude of the downforce, and the larger is the magnitude of the self-reinforcing response.

The next self-reinforcing effect from capturing rotational motion has been taken from an example discussed in Pahl et al. (2007) (Figure 9.9a,b).

The bending stresses at the base of a blade due to the tangential force P (undesirable outcome B, Figure 9.9a) can be reduced if the centrifugal force resulting from the rotation of the blade is channelled to create an opposing moment counteracting the moment created by the tangential force P. Channelling the centrifugal force to create an opposing moment (proportional compensating factor C) can be done by inclining the blades at an angle from the vertical (Figure 9.9b). For the design in Figure 9.9b, the centrifugal force F resulting from the rotation of the blade creates a proportional moment counteracting the bending moment from the tangential force P. This reduces the overall bending moment and the stresses at the base of the blade. The larger the tangential force P, the larger is the angular speed ω, the larger is the centrifugal force (proportional to ω^2), the larger is the counteracting bending moment created by the centrifugal force.

As a result, the capability of the assembly to absorb and withstand a sudden increase in the rotational speed is increased which improves the resilience of the assembly and reduces the risk of failure. The reliability improvement by self-reinforcement is obtained at no extra cost.

9.2.3 Self-Reinforcement by Self-Alignment

The mechanism of self-reinforcement by self-alignment works by using the external forces to alter the position/orientation of the object which results in reduced loading stresses. Self-alignment improves the capability of the system to adapt to adverse operational conditions which enhances its resilience.

An example of a self-alignment mechanism can be given with the self-aligning ball bearing whose outer rings have spherical raceways (Bergman et al. 2009). The self-aligning ball bearing allows for inevitable misalignments of the shaft leading to increased levels of the loading stresses. Without the benefit of the self-alignment, the life of the ball bearing will be reduced significantly.

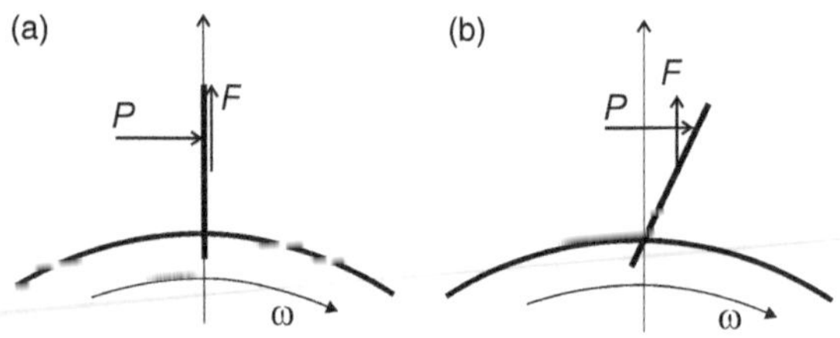

Figure 9.9 Reducing the stresses at the base of a blade attached to a rotor by capturing a proportional compensating moment.

9.2.3.1 Case Study: Self-Reinforcement by Self-Alignment of a Rectangular Panel Under Wind Pressure

Self-reinforcement can also be achieved if the loading force induces self-alignment which reduces the loading force. A similar application has been discussed in Orloff (2012). Consider a common case where the surface of a rectangular panel with side lengths L and M is perpendicular to the wind velocity (Figure 9.10a). In this case, the dynamic pressure of the wind $(1/2)\rho_a v_a^2$ is transformed into a force $F_w = C_d(1/2)\rho_a {v_a}^2 A_p$ acting on the panel, where C_d is the drag coefficient characterising the panel, $\rho_a = 1.2$ is the mass density of air, v_a is the air velocity, and $A_p = L \times M$ is the projected surface area of the panel perpendicular to the velocity of the wind. The wind force F_w can reach a large magnitude because it increases with the square of air velocity. Thus, wind with velocity v_a, perpendicular to the panel, whose mass centre G is at a height h above the ground creates the force $F_w = (1/2) \times C_d \times \rho_a \times v_a^2 \times A_p$ and the moment

$$M_w = F_w h = (1/2) \times C_d \times \rho_a \times v_a^2 \times A_p \times h \tag{9.8}$$

This bending moment could reach significant magnitude which results in significant bending stresses at the base of the support.

However, if the panel has been made to rotate about an axis (point O in Figure 9.10b) which is off the axis going through the mass centre G, the wind pressure will rotate the panel until the resisting moment counterbalances the moment rotating the panel. Suppose that this balancing occurs at an angle α. In this case, the projected area is $A_p = L \times M \times \cos\alpha$, which results in a bending moment M_w with a significantly smaller magnitude.

Increasing the velocity of the wind produces a self-reinforcing response. Indeed, increasing the air velocity v_a increases the magnitude of the angle α at which the resisting moment counterbalances the moment rotating the panel. Increasing the magnitude of the angle α reduces the projected area $A_p = L \times M \times \cos\alpha$ which results in a corresponding reduction of the bending moment M_w at the base of the support. In other words, increasing the wind velocity intensifies the system's response against the increased wind velocity. As a result, the capability of the assembly to absorb overload and adapt to adverse operational conditions is improved which enhances the resilience of the assembly. The reliability improvement has been obtained at no extra cost.

The principle behind this self-reinforcing design can even be used for everyday objects such as umbrellas. The umbrella is made flexible and changes its area (partially collapses) when turned against the wind. Increasing the strength of the wind causes reduction of the exposed surface and reduces the forces on the umbrella which reduces the risk of failure.

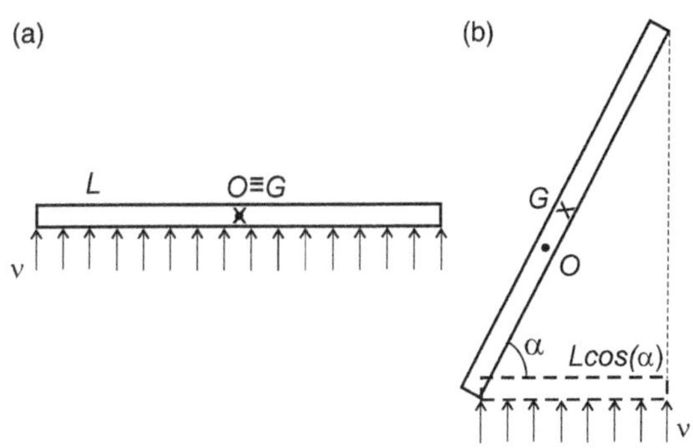

Figure 9.10 Self-reinforcement by self-alignment (reducing the stresses at the base of a panel with length L and height M).

9.2.4 Self-Reinforcement Through Modified Geometry and Strains

Modifying the geometry of a loaded component is often associated with a more uniform load distribution which decreases the stresses in the material and results in improved load-bearing capacity. Such are, for example, the nuts with special geometry that are associated with more even distribution of the load and higher load-carrying capacity of the thread (Matthews 1998). The higher the load, the more uniform is the load distribution along the thread. Changing the shape of a component to a symmetric shape often results in counterbalancing forces (Matthews 1998) which increase the load-bearing capacity.

In materials, a self-reinforcing effect is present during plastic deformation which is known as *strain hardening* (Dowling 1999). The resistance of the material to plastic deformation (the self-reinforcing effect) increases with increasing the deformation strains.

Another example of improving reliability by self-reinforcement can be found in the deformation self-strengthening of Hadfield steels discussed in Lindroos et al. (2015). Increasing the impact loading improves the resistance of the steel to abrasive wear. A similar self-reinforcing effect is present in cubic zirconia ceramic with dispersed tetragonal zirconia particles (Kelly and Francis-Rose 2002). For an existing crack in the ceramic material, increasing the crack tip opening load intensifies the stress field around the tip of the crack. The intensified stress field triggers the transformation of the embedded tetragonal zirconia particles into monoclinic zirconia which has a larger specific volume. The volume expansion associated with the phase transformation strains exerts a crack-closing force on the tip of the crack and resists the crack propagation. These are examples of self-reinforcement resulting from phase transformation strains.

9.3 Self-Reinforcement by Feedback Loops

9.3.1 Self-Reinforcement by Creating Negative Feedback Loops

This mechanism utilises negative feedback control and achieves self-reinforcement and reliability improvement through the process described next. The feedback loop is a circular chain of causes and effects. The negative feedback loop seeks a goal around which it stabilises the system. The negative feedback samples the output of the system and information about the deviation of the output from the goal is used to control the input in such a way that the deviation is diminished. If the output signal is above the goal, the negative feedback loop corrects the input signal in such a way that the output is decreased towards the goal. If the output signal is below the goal, the negative feedback loop corrects the input signal in such a way that the output is increased towards the goal (Figure 9.11a).

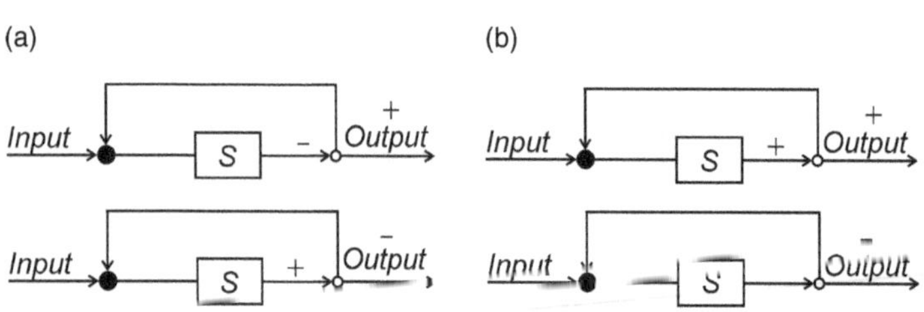

Figure 9.11 (a) Negative and (b) positive feedback loops.

The larger the deviation of the output from the goal, the larger is the input correction counteracting the deviation of the output.

Negative feedback loops are widely used in control systems to guarantee specified position, temperature, pressure, humidity, concentration, speed of rotation, etc.

The negative feedback loop is commonly found in the coupling of an induction motor and a machine. An increase in the angular velocity of the rotor from the point of stable operation causes the driving torque to drop below the resisting torque which decreases the angular velocity of the rotor. A decrease in the angular velocity of the rotor causes the driving torque from the induction motor to increase above the resisting torque. The result is an increase of the angular velocity of the rotor.

Commonly, in negative feedback control systems, the output is measured and fed back to an error detector at the input. A controller is then correcting the parameters of the system/process so that the deviations from the set output become as close to zero as possible. The availability of low-cost electrical devices and sensors makes it possible to provide more flexibility and regulate mechanical systems to a finer degree compared with all-mechanical systems. The negative feedback loop improves the capability of the system to recover if the balance is disturbed which results in improved system resilience.

In general, interpreting and acting upon a flow of information about the state of a system, in order to stabilise its output, is essentially a negative feedback control. Whenever the information flow indicates that a particular system parameter has drifted from the set value, an action is taken to return the system back towards the set value of the parameter. In this respect, measurements from sensors, together with actions following the interpretation of the measured values, is a common form of a negative feedback loop. For example, the measured speed provided by a radar speed sign helps motorists control their speed according to the permitted speed limit.

9.3.2 Positive Feedback Loops

While the application of negative feedback control loops is well-understood and well-documented, there is a profound lack of discussion in the reliability and risk literature on the application of positive feedback loops to improve reliability and reduce risk.

The positive feedback loop is also a circular chain of causes and effects. The mechanism of positive feedback loops consists of correcting the input in such a way that a further deviation of the output occurs in the same direction (Figure 9.11b). If the output is increasing, the positive feedback loop leads to a further increase of the output, and if the output is decreasing, the positive feedback loop leads to a further decrease of the output (Figure 9.11b). In this way, the mechanism of positive feedback loops works towards moving the system away from equilibrium. Feedback loops are presented by feedback loop diagrams (Figure 9.12). The links on the diagrams imply a direction of causation, and not simply a time sequence. Thus, the positive link from *A* to *B* in Figure 9.12a means 'when *A* increases then *B* increases' while the negative link from *B* to *C* means 'when *B* increases then C decreases'.

A positive feedback loop may not necessarily be based on a single stage control from the output directly to the input. A positive feedback loop could include a long chain of cause and effect relationships the final stage of which affects the input in such a way that a deviation of the output in one direction leads to further deviation of the output in the same direction. Thus, in the loop depicted in Figure 9.12a, increasing the input factor *A*, leads to

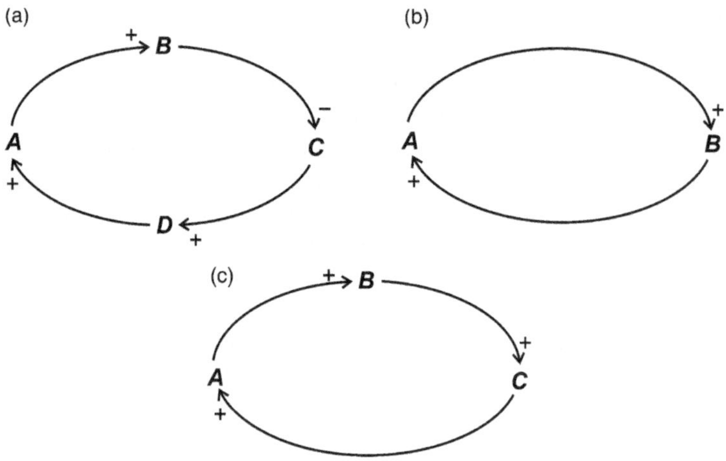

Figure 9.12 Positive feedback loop diagrams.

an increase of factor *B*. The increase of factor *B* leads to a decrease of factor *C* which causes an increase of factor *D*. The increase of factor *D* finally cases an increase of the input *A*. As a result, an increase of factor *A* leads to a further increase of factor *A*.

If an initial base level exists, a slight deviation from the base level tends to produce a further deviation due to the positive self-reinforcing feedback loop. In this sense, the positive feedback loop works towards increasing the gap between the current state of the system and its base-level state. If the initial deviation is positive with respect to the base level, the positive feedback loop initiates a further positive deviation, which is a process of growth. If the initial deviation is negative, the positive feedback loop initiates a further negative deviation which is a process of decay/decline.

Positive feedback loops work towards destabilising the system's output and moving it away from equilibrium. The positive feedback loops are important reasons behind prolonged droughts, the global warming, bank runs, etc. This is why, it is widely believed that positive feedback loops are always associated with destruction and collapse. However, positive feedback loops are also important drivers behind growth that can be successfully used for reliability improvement and risk reduction.

There are two principal pathways to reducing risk through positive feedback loops: (i) by eliminating or inhibiting self-reinforcing feedback loops with negative impact; and (ii) by promoting positive self-reinforcing feedback loops with positive impact.

9.3.3 Reducing Risk by Eliminating or Inhibiting Positive Feedback Loops with Negative Impact

Significant risk reduction can be achieved by discovering and eliminating positive feedback loops destabilising a system.

An example of a positive feedback loop with negative impact is an exothermal chemical reaction. The heat (cause *A*) released during the chemical reaction (Figure 9.12b), if not removed, causes the reaction to occur at a higher speed (effect *B*) which in turn leads to more released heat *A*, etc. This positive feedback loop could lead to a thermal runaway and chemical explosion. Such positive feedback loops are present in lithium-ion batteries used as sources of electrical power in various electronic devices and electric vehicles. The critical temperature at which the side reactions causing thermal runaway start is known as the thermal runaway onset temperature. The thermal runaway onset temperature can be reached during an accident causing a short circuit in some of the battery cells, during overheat or overcharge (Feng et al. 2017). Thus, eliminating the possibility for a runaway chemical reaction by redesigning the chemical process essentially reduces risk by removing a positive feedback loop with negative impact.

The negative impact from a positive feedback loop in lithium ion batteries due to a thermal runaway, for example, can be prevented if additives are used to switch off the current at high temperatures (Feng et al. 2017). This measure effectively introduces a negative feedback loop which stabilises the system.

The negative impact from the positive feedback loop in electrical batteries can be prevented by inhibiting the feedback loop by adequately cooling the battery cells so that a specified temperature of the battery cells is continuously maintained or by modifying the materials in the battery cell to improve their thermal stability. Reducing the likelihood of a thermal runaway can also be done by a properly designed battery management system that ensures a proper cooling of the battery cells.

It is important to note that before reaching the onset temperature at which the positive feedback loop and the thermal runaway reaction start, a negative feedback loop is in place. Thus, increasing the temperature of the cell leads to a larger thermal gradient which leads to a larger heat transfer and loss of heat. As a result, further increase of the temperature is halted. The system is in stable equilibrium. However, beyond a certain triggering temperature, the larger thermal gradient and heat transfer are not sufficient to stop the temperature increase of the cell which marks the start of the positive feedback loop.

Interconnected networks often exhibit positive feedback loops with negative impact. In interconnected power and telecommunication networks, the telecommunication network is powered from the power network while the power network is controlled by the telecommunication network.

The failure of a number of nodes in the power network due to overloading caused by a random failure (factor *A* in Figure 9.12b) causes failures of a number of nodes in the connected telecommunication network (factor *B* in Figure 9.12b). Failure of nodes in the telecommunication network cause further failures in the power network and this process continues until the collapse of both networks.

The positive feedback loop with negative impact created vulnerability that can be eliminated, for example, by decoupling the two networks in times of crisis. Decoupling destroys the positive feedback loops between the networks and helps bring the systems into a stable condition. Since the interdependency between systems is increasing, the importance of eliminating positive feedback loops with negative impact is also increasing.

Removing positive feedback loops with negative impact due to capacitive or inductive couplings is also an essential component of debugging circuits. Thus, removing positive feedback loops with negative impact from amplification circuits reduces unwanted spontaneous oscillations and greatly improves the stability of the amplifier.

A positive feedback loop in a loaded structure could cause a fast deterioration and collapse. The process continues as follows. An overload causes some of the load-carrying elements of the structure to fail. The load is then redistributed and carried by a smaller number of elements. The higher load per element causes the rate of deterioration of the elements to increase which causes more of the elements to fail. This, in turn, increases the load on the rest of the elements and the process continues until the entire structure collapses.

This process has a powerful analogue in organisations. A positive feedback loop may be triggered if a group of dissatisfied experts leaves an organisation because of excessive work load (factor *A* in Figure 9.12b). If no sufficient number of appointments are made because of ill-conceived management strategy, the duties of people who have left are distributed across the remaining people. The increased work load (factor *B*) on the remaining staff causes further dissatisfaction (factor *A*) and more experts leave. This self-reinforcing process may continue until the organisation is depleted of high quality staff and the service it provides deteriorates beyond repair.

In human behaviour, a frequent positive feedback loop with profound negative impact is created by the factors 'belief' and 'choice'. For example, belief in the incorrect Weibull model (Weibull 1951) for determining the probability of fracture initiated by flaws determined its choice as a model for many decades. The choice of the incorrect Weibull model then reinforced the belief in this model to the extent of ignoring experimental data sets clearly contradicting the strictly increasing Weibull function. This interaction of belief and choice led to a firmly entrenched false modelling paradigm, blind to the mounting contradicting evidence from experiments, computer simulations, and theoretical arguments (Todinov 2009b, 2010).

In the economy, the withdrawal of investment from a country, triggered by a political crisis, causes poverty, and poverty leads to a further withdrawal of investment.

9.3.3.1 Case Study: Growth of Damage Sustained by a Positive Feedback Loop with Negative Impact

In addition to providing the conditions for a rapid growth of a beneficial outcome, the positive feedback loops often work in the opposite direction and provide the conditions for a rapid growth of damage.

Suppose that at time t, the rate of increasing the damage y is positive and proportional to the existing damage y. Initially, at time $t=0$, the damage is equal to y_0. At time Δt, the damage y will increase to $y_1 = y_0 + \Delta y_0 > y_0$. At time $2\Delta t$, the already increased damage y_1 will give rise to an even larger increase Δy_1 ($\Delta y_1 > \Delta y_0$) and the damage becomes $y_2 = y_1 + \Delta y_1 > y_1$, etc. The growth of damage is sustained by a positive feedback loop because the more damage is produced, the more the damage is generated.

It needs to be pointed out that the growth of damage sustained by a positive feedback loop is not necessarily an exponential growth.

Assume that the rate of increase dy/dt of the damage y is proportional to the existing damage y raised to some power $1+\varepsilon > 0$, where $\varepsilon \neq 0$, k is a some positive constant ($k > 0$) and t is the time:

$$dy/dt = ky^{1+\varepsilon} \tag{9.9}$$

If the initial damage y is equal to y_0 (at $t=0$), separating the variables of Eq. (9.9) yields $dy/y^{1+\varepsilon} = kdt$ which, after integration and determining the integration constant, yields the growth law

$$y = \frac{1}{\left(\frac{1}{y_0^\varepsilon} - k\varepsilon t\right)^{1/\varepsilon}} \tag{9.10}$$

As can be seen, if $\varepsilon > 0$ the damage growth sustained by the positive feedback loop is extremely rapid and after a finite time $t \to \frac{1}{y_0^\varepsilon k\varepsilon}$, $y \to \infty$.

At the beginning, the damage growth is slow, which can give a wrong perception of safety. With time, what appeared to be a minor problem, very quickly transforms into a major problem. By the time the late stages of damage growth are reached, it may be too late to do anything to limit the extent of damage.

An example of damage growth by a positive feedback loop can be given with the growth of fatigue cracks. The fatigue crack growth is given by the Paris law (Paris et al., 1961; Paris and Erdogan, 1963)

$$da/dt = ka^{m/2} \tag{9.11}$$

where a is the current length of the fatigue crack, k is a constant depending on the geometry, the material, and the loading stress range, and m is a material constant (the Paris exponent). The initial size of the crack (at time $t=0$) is a_0.

For typical values of the Paris exponent, the constant m is greater than two ($m>2$). Consequently, setting $\varepsilon = m/2 - 1$ gives

$$da/dt = ka^{1+\varepsilon} \tag{9.12}$$

which is essentially Eq. (9.9). Its integration yields

$$a = \frac{1}{\left(\frac{1}{a_0^\varepsilon} - k\varepsilon t\right)^{1/\varepsilon}} \tag{9.13}$$

which describes the length of the crack with time. Initially, the fatigue crack growth sustained by the positive feedback loop is slow but it accelerates with increasing the length of the crack. If the crack length that triggers fast fracture is a_f, the rate of the fatigue crack growth increases as the crack length approaches this value. With increasing time, the rate of damage accumulation *da/dt* increases (Figure 9.13).

During damage growth sustained by a positive feedback loop, the frequency of the damage monitoring intervals should increase as damage progresses. The length of the damage inspection intervals Δt_i, in the initial stages of the damage accumulation, is significantly larger than the length Δt_f of the damage inspection intervals in the final stages of damage accumulation (Figure 9.13).

The growth of damage caused by a positive feedback loop can be reduced by inhibiting the positive feedback loop so that the damage accumulation is slowed down significantly.

Suppose that an implemented inhibition measure causes the rate of accumulation of damage to be proportional to the square root of the damage y. The constant ε in Eq. (9.9) is then $\varepsilon = -1/2$. Substituting $\varepsilon = -1/2$ in the solution (9.10) yields the polynomial growth law

$$y = \left(\sqrt{y_0} + \frac{kt}{2}\right)^2 \tag{9.14}$$

which describes the damage accumulation with time. The damage growth described by Eq. (9.14) corresponding to $\varepsilon = -1/2$ is significantly slower than the damage growth corresponding to $\varepsilon > 0$ given by Eq. (9.10).

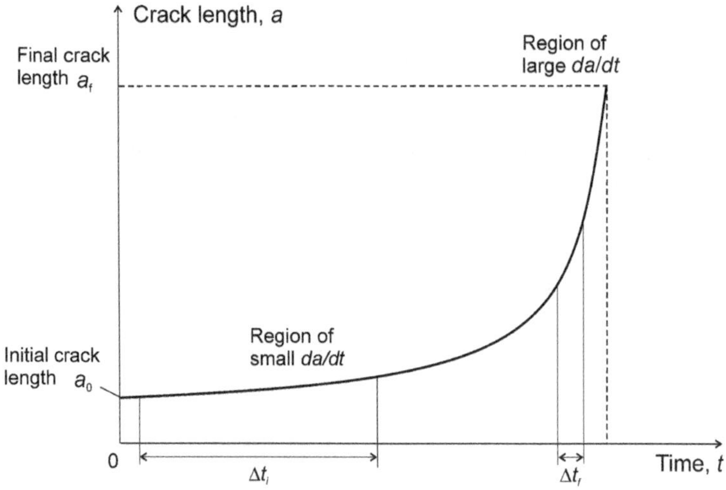

Figure 9.13 In a positive feedback loop, the rate of damage accumulation increases with time.

Even in the case where the constant ε is reduced to zero ($\varepsilon = 0$), the integration of Eq. (9.9) yields the exponential growth:

$$y = y_0 e^{kt} \tag{9.15}$$

which is slower than the damage growth given by Eq. (9.10), where $\varepsilon > 0$.

The exponential growth described by Eq. (9.15) is the growth of capital due to continuous compounding.

9.3.4 Self-Reinforcement by Creating Positive Feedback Loops with Positive Impact

A positive feedback loop with positive impact is a self-reinforcing loop which promotes and sustains growth.

A positive feedback loop with positive impact is present in the important case where an initial investment *A* in reliability improvement leads to a more superior product *B* which, in turn, leads to increased sales *C*, and the increased sales *C* lead to increased investment *A* in the reliability improvement. The dynamics of this process has been illustrated in Figure 9.12c. The result is a sustained growth towards a superior product characterised by increased reliability and reduced risk of failure.

In capturing the benefits from positive feedback loops with positive impact, accelerators can be used.

In business, positive feedback loops are powerful engines of growth. Business creates value, which attracts customers. Customers bring profits that are reinvested in the business, which leads to more value creation. Surveys can be used as accelerators of this positive feedback loop. Comments from customers in the surveys are used to tailor the content of the business to the customers' expectations, which leads to improved services, expanding the business, and further growth.

Consider the common case of a positive feedback loop where the rate of change of factor *x* is proportional to factor *y* and the rate of change of factor *y* is, in turn, proportional to factor x (Figure 9.12b). Formally, this relationship between the two factors can be written as the system of differential equations

$$\frac{dx}{dt} = k_1 y$$

$$\frac{dy}{dt} = k_2 x \tag{9.16}$$

where k_1 and k_2 are positive constants of proportionality, and t is the time. Suppose that the initial conditions $x(0) = 1$ and $y(0) = 1$ have been specified at time $t = 0$.

The system (9.16) can be solved easily, for example, by applying a Laplace transform. This leads to the solutions:

$$x(t) = \frac{k_1 + \sqrt{k_1 k_2}}{2\sqrt{k_1 k_2}} \exp\left(\sqrt{k_1 k_2}\, t\right) - \frac{k_1 - \sqrt{k_1 k_2}}{2\sqrt{k_1 k_2}} \exp\left(-\sqrt{k_1 k_2}\, t\right) \tag{9.17}$$

$$y(t) = \frac{k_2 + \sqrt{k_1 k_2}}{2\sqrt{k_1 k_2}} \exp\left(\sqrt{k_1 k_2}\, t\right) - \frac{k_2 - \sqrt{k_1 k_2}}{2\sqrt{k_1 k_2}} \exp\left(-\sqrt{k_1 k_2}\, t\right) \tag{9.18}$$

As can be verified from the solutions, with increasing time t, both factors $x(t)$ and $y(t)$, mutually reinforcing each other's growth tend to grow infinitely.

Positive feedback loops can also be used in bistable systems to guarantee a fast transition into a more reliable state. If the input is slightly higher than the equilibrium state, the positive feedback causes the output to move in the same direction, until the output reaches the nearest upper equilibrium state. If the input is slightly lower than the equilibrium state, the positive feedback causes the output to move fast in the same direction until it reaches the nearest lower equilibrium state. Once an equilibrium state has been reached, the system remains in that state. The larger the deviation of the output from the specified level, the larger is the input correction increasing the deviation of the output in the same direction. In electrical switches, for example, the positive feedback control minimises the length of arcing during switching thereby minimising the wear out and extending the life of the switches.

Positive feedback loops are also used in circuits which trigger a particular action if a signal crosses a predetermined threshold value. The Schmitt trigger circuit, for example, uses a positive feedback control to force the output in one of the two stable logic states.

9.3.4.1 Case Study: Positive Feedback Loop Providing Self-Reinforcement by Self-Energising

Positive feedback can also be used with success for improving the reliability of a system and reducing the risk of failure. The self-energising friction holder which supports the weight G of a sheet in Figure 9.14 is an example of a self-reinforcement through a positive feedback loop. Once the friction pads of the holder touch the sheet, a small normal force N appears and excites a proportional friction force $F_{fr} = \mu N$ where μ is the coefficient of friction between the friction pad and the sheet. The friction force $F_{fr} = \mu N$ is causing rotation of the link OP around the pin O. However, the geometric constraint causes the normal force N to increase which, in turn, increases the friction force $F_{fr} = \mu N$, and so on. This process of self-excitation continues until the friction forces $F_{fr} = \mu N$ acting on both sides of the sheet counterbalance the weight G of the sheet.

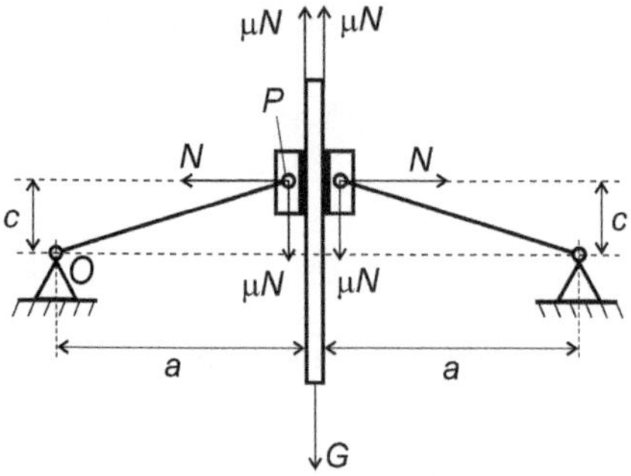

Figure 9.14 Positive feedback loop in a self-energising friction holder.

As result of the positive feedback loop, the friction forces $F_{fr} = \mu N$ quickly reach large magnitudes even though the friction pads were initially energised with a very small load. The result is a reliable support of the weight G of the sheet.

The smallest coefficient of friction μ^* necessary to achieve self-energising can be determined from the equilibrium condition of the link OP. From the equilibrium condition of the moments with respect to the pin O:

$$Nc - \mu Na = 0 \tag{9.19}$$

which is equivalent to $c - \mu a = 0$. Self-energising is present if $c \leq \mu a$. The minimum coefficient of friction for self-energising is given by $\mu^* = c/a$. Thus, for $c = 15$ mm and $a = 60$ mm, $\mu^* = 15/60 = 0.25$.

10

Improving Reliability and Reducing Risk by Minimising the Rate of Damage Accumulation and by a Substitution

10.1 Improving Reliability and Reducing Risk by Minimising the Rate of Damage Accumulation

10.1.1 Classification of Failures Caused by Accumulation of Damage

Failures caused by accumulation of damage with mechanisms such as fatigue, corrosion, wear, and creep have been known for a long while.

Dasgupta and Pecht (1991) distinguished 'overstress failure modes' (brittle fracture, ductile fracture, yield, buckling, etc.) and 'wearout failure modes' (fatigue, corrosion, stress-corrosion cracking, wear, creep, etc.). Overstress failures occur when load exceeds strength. If load is smaller than strength, the load has no permanent effect on the component. Conversely, wearout failures are characterised by damage which accumulates irreversibly and does not disappear when the load is removed. Examples are the fatigue failures and the failures caused by wear, creep, corrosion, and other degradation mechanisms. A common trait of degradation failure mechanisms is that failure is precipitated as soon as the accumulated damage reaches a critical threshold level (Blischke and Murthy 2000). Reducing the rate of damage accumulation and delaying the failure mode is therefore a powerful way of improving reliability and reducing risk.

Although wearout failures are very common, it is surprising that minimising the rate of damage accumulation has never been formulated as a domain-independent principle for risk reduction. Furthermore, no overview of general techniques by which the rate of damage accumulation can be minimised and reliability maximised has ever been reported.

To fill this knowledge gap, solutions related to improving reliability and reducing risk were collected, whose underlying basis is reducing the rate of damage accumulation. A certain level of abstraction was used to strip the available solutions from their specific engineering context and uncover the underlying mechanism for reducing the rate of damage accumulation. From the body of available solutions, various reliability improvement patterns and mechanisms emerged which were captured and distilled into categories, classes, and individual techniques. A classification has been presented in Figure 10.1. In what follows, an extended discussion is provided on a number of key methods for reducing the rate of damage accumulation.

Methods for Reliability Improvement and Risk Reduction, First Edition. Michael Todinov.

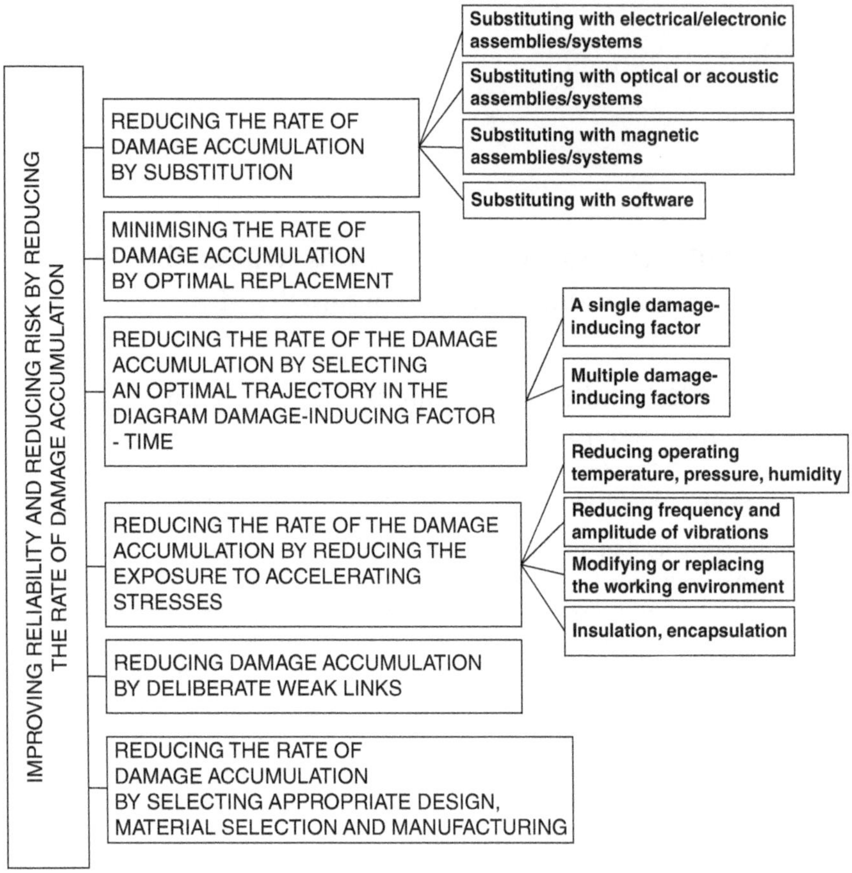

Figure 10.1 Classification of various domain-independent methods for reducing risk by reducing the rate of damage accumulation.

10.1.2 Minimising the Rate of Damage Accumulation by Optimal Replacement

Suppose that the probability of failure in the elementary time interval t, $t+\Delta t$, given that the component has survived time t is given by $h(t)\Delta t$, where $h(t)$ is referred to as the *hazard rate* (Figure 10.2a). The reliability, associated with a time interval $(0,t)$ is then given by the expression

$$R(t) = \exp\left(-\int_0^t h(v)dv\right) \tag{10.1}$$

where v is a dummy integration variable. The integral $H(t) = \int_0^t h(v)dv$ in Eq. (10.1) is also known as the *cumulative hazard rate*. Using the cumulative hazard rate, reliability can be presented as (Barlow and Proschan 1975):

$$R(t) = \exp(-H(t)) \tag{10.2}$$

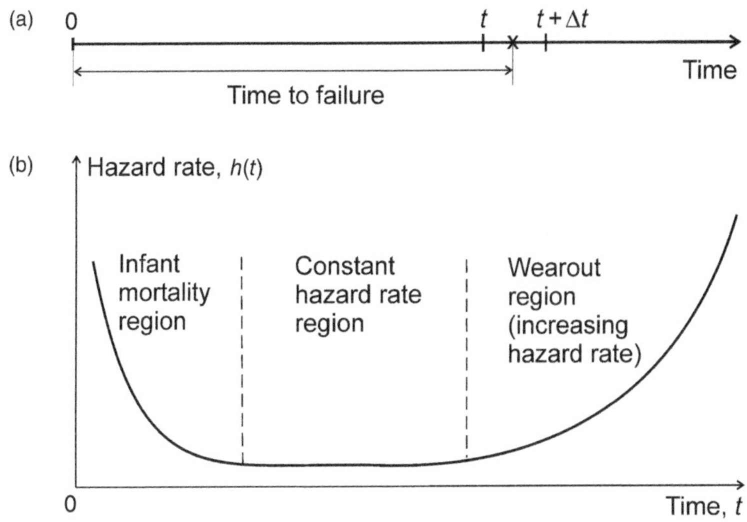

Figure 10.2 (a) Time to failure in the small time interval t, $t+\Delta t$; (b) reliability bathtub curve.

Reliability $R(t)$ can be increased by decreasing the hazard rate $h(t)$, which decreases the value of the cumulative hazard rate $H(t)$. Correspondingly, the cumulative distribution of the time to failure becomes

$$F(t) = 1 - \exp(-H(t)) \tag{10.3}$$

The accumulation of damage is characterised by the hazard rate $h(t)$ which normally increases with time (age). If the hazard rate practically does not depend on age, it remains constant, ($h(t) = \lambda = \text{constant}$) and the time to failure distribution (10.3) transforms into the negative exponential distribution $F(t) = 1 - \exp(-\lambda t)$. Indeed, from $h(t) = \lambda = \text{constant}$, the cumulative hazard rate becomes $H(t) = \int_0^t \lambda dv = \lambda t$.

The negative exponential distribution is the model of the time to failure in the constant hazard rate region (Figure 10.2b). Failures in this region are not due to age, wear-out or degradation; they are due to random causes. This is why preventive maintenance in this region has no effect on the hazard rate and the time to failure distribution.

The *wearout region* of the bathtub curve in Figure 10.2b is characterised by a hazard rate increasing with age, due to accumulated damage and degradation of properties caused by wear, erosion, corrosion, fatigue, creep, etc. Unlike preventive maintenance in the constant hazard rate region, *preventive maintenance in the wearout region significantly enhances the life of the system.*

This can be demonstrated by considering that for components experiencing a non-constant hazard rate $h(t)$, the integral $S = \int_0^a h(t)dt$ gives the expected number of failures in the finite time interval $(0,a)$ which is numerically equal to the area S beneath the hazard rate curve (Figure 10.3a). Reliability associated with the time interval $(0,a)$ is given by $R(a) = \exp(-S)$. Consequently, replacing a component in the wearout region at time t^* (Figure 10.3b) reduces the expected number of failures from S to S', where S' is the hatched area beneath the combined hazard rate before and after the component replacement, $S' < S$, (Figure 10.3b). Consequently, a component replacement results in an increase

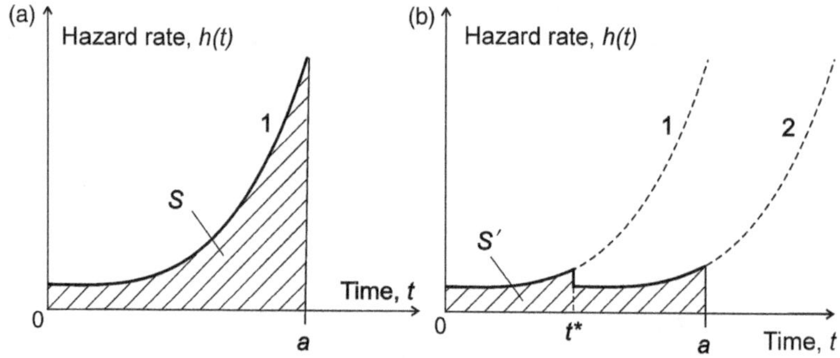

Figure 10.3 (a) The area beneath the hazard rate curve is numerically equal to the expected number of failures in the interval (0,a). (b) Decreasing the hazard rate in the wearout region by a component replacement reduces the expected number of failures in the interval (0,a).

of the probability of surviving the time interval (0,a) from $R(a) = \exp(-S)$ to $R'(a) = \exp(-S')$.

In the wearout region, reliability is enhanced significantly by preventive maintenance consisting of replacing old components. This effectively delays the wearout phase and increases reliability (Figure 10.3b).

Consider now a time interval (0,a) and a system consisting of a number of components logically arranged in series that undergo fast wear-out. A failure of any of the components constitutes a system failure. A certain number of spare components is kept for each of the working components.

A sudden component failure is often dangerous for the system because it may cause further secondary failures. Furthermore, the recovery of the system after an uncontrolled shutdown following a sudden failure is more difficult and associated with significant costs. Consequently, a question of significant interest is how to find the optimal replacement times which maximise the probability of surviving the operational time interval (0,a).

Consider a system that consists of two components logically arranged in series, with a single spare component available for each of the working components. Because each component can be replaced at any time during the time interval (0,a), a combination of replacement times for the components can be represented by a point in the square domain D with side a (Figure 10.4).

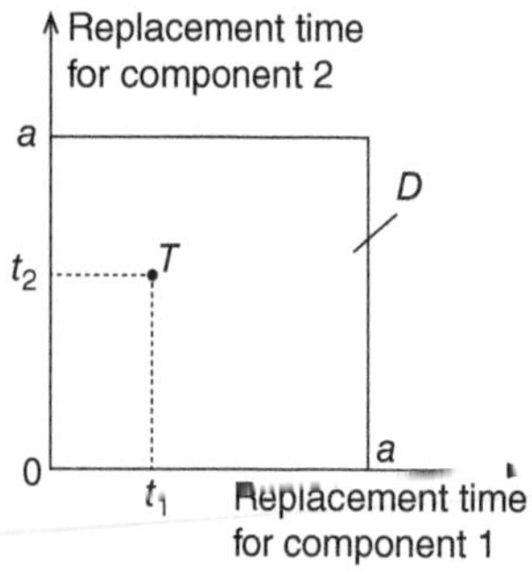

Figure 10.4 All possible combinations t_1, t_2 of replacement times for the two components building the system can be represented by the points from the square domain D.

For a specified point T from the domain D, which defines the replacement times for the components, the probability of a system failure can be determined by considering the following:

- The replacement times are always within the interval $(0,a)$.
- A system failure occurs if the time to failure of any of the two components is within the interval $(0,a)$.
- A system failure also occurs if, after a replacement of a failed component within the time interval $(0,a)$, the replacing component also fails within the interval $(0,a)$.

Consider a system with components logically arranged in series, which undergo intensive wearout, with hazard rate functions strictly increasing with time. Assume also that the derivative of the hazard rate functions characterising the separate components exists at any point in the time interval $(0,a)$.

Theorem 10.1 For an operational time interval with length 'a', if n spares are available for each component building the system, the optimal replacement intervals $a/(n+1)$ for each component minimise the rate of damage accumulation and maximise the probability that the system will survive the operational time interval.

Proof: The case related to $n = 1$ spares for each component, will be proved first.

Indeed, because of the strictly increasing hazard rate function $h(t)$ characterising each component, it can be shown by a differentiation that a replacement at $x=a/2$, at half of the operational interval $(0,a)$, minimises the expected number of failures $H(x)$ of the component during the time interval $(0,a)$. The total expected number of failures $H(x)$ within the time interval $(0,a)$, at a replacement time x, is given by (see Figure 10.3b)

$$H(x) = \int_0^x h(t)dt + \int_0^{a-x} h(t)dt, 0 \leq x \leq a \tag{10.4}$$

Differentiating $H(x)$ with respect to the unknown replacement time x gives

$$\frac{d}{dx}H(x) = h(x) - h(a-x)$$

The local extrema can be obtained by equating this expression to zero:

$$h(x) - h(a-x) = 0 \tag{10.5}$$

Because $h(x)$ is a strictly increasing function of x, $h(x)-h(a-x)=0$ only if $x=a/2$. This value is a local minimum because the second derivative at $x=a/2$, is positive:

$$\left.\frac{d^2}{dx^2}H(x)\right|_{x=a/2} = h'(a/2) + h'(a/2) > 0,$$

because $h(x)$ is a strictly increasing function and therefore $h'(a/2)>0$. At the local minimum, $x=a/2$, the value of the function $H(x)$ is $H(a/2) = 2\int_0^{a/2} h(t)dt$.

At the ends of the interval $0 \leq x \leq a$, $H(0) = H(a) = \int_0^a h(t)dt$. This value is larger than the value $2\int_0^{a/2} h(t)dt$ corresponding to the local minimum because $h(x)$ is a strictly increasing function of x. Furthermore, there are no points in the interval $0 \leq x \leq a$ where the first derivative is not defined. Consequently, the local minimum at $x=a/2$ is also a global minimum.

Consequently, at $x = a/2$ $H(x)$ has a global minimum which minimises the expected number of failures in the interval $(0,a)$ and maximises the probability of surviving the operational interval $(0,a)$. As a result, for $n = 1$, the statement of the theorem has been proved.

Now, assume that the statement is valid for any arbitrary $n - 1 \geq 1$. In other words, assume that for $n - 1$ spare parts available for each component, the optimal replacement times are at intervals a/n (inductive hypothesis).

Let the points 1, 2, …, n in Figure 10.5a define the optimal replacement times which deliver the smallest expected number of failures over the time interval $(0,a)$. Let x denote the length of the time interval from the start 0 of the operational interval $(0,a)$ to the time of the first replacement. Let y denote the length of the time interval from the last replacement time to the end of the operational interval $(0,a)$. If the smaller interval $(0, a - y)$ is considered (Figure 10.5b), the $n-1$ replacement times within this shorter interval must also provide an optimal replacement. Otherwise, the replacement times inside the shorter time interval $(0, a - y)$ could be rearranged to yield a smaller expected number of failures. This smaller expected number of failures, together with the expected number of failures over the length y of the last time interval, will yield a smaller overall expected number of failures over the interval $(0, a)$ which contradicts the assumption that the initial replacement times on the interval (0, a) are characterised by the smallest possible expected number of failures.

According to the inductive hypothesis, if the $n-1$ replacement times in the shorter time interval $(0, a - y)$ are the optimal replacement times, they must divide the interval into equal-length time sub-intervals. In other words, for the length of the sub-interval x:

$$x = \frac{a - y}{n} \tag{10.6}$$

must be fulfilled.

Now consider the smaller interval $(1,a)$ (Figure 10.5c), where the $n-1$ replacement times 2, 3, …, n within this shorter interval must also be optimal replacement times. Otherwise, the replacement times inside the shorter time interval $(1,a)$ could be rearranged to yield a smaller expected number of failures. According to the inductive hypothesis, if the replacement times in the shorter time interval $(1,a)$ are the optimal replacement times, they must divide the interval $(1,a)$ into equal-length time sub-intervals. In other words, for the length of the last sub-interval with length y:

$$y = \frac{a - x}{n} \tag{10.7}$$

must be fulfilled.

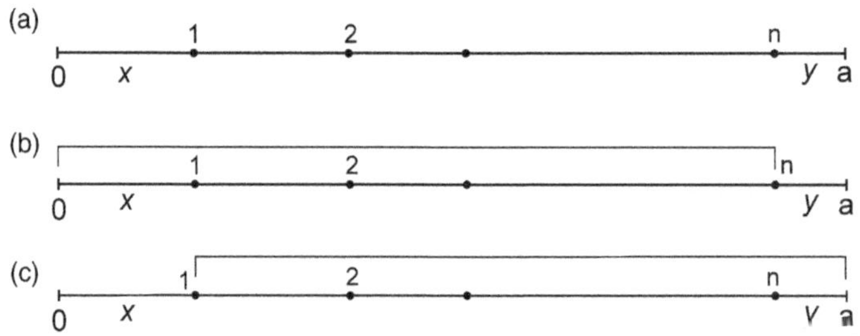

Figure 10.5 The optimal replacement times for a component with n spares should be at equal intervals $a/(n + 1)$.

Dividing the Eqs. (10.6) and (10.7) yields

$$\frac{x}{y} = \frac{a-y}{a-x}$$

which is equivalent to

$$(x-y)[x+y-a] = 0 \tag{10.8}$$

Since $x+y-a<0$, $x=y$ must be fulfilled. From $x = \frac{a-x}{n}$, $x=a/(n+1)$. Since x must be equal to all sub-intervals from 1 to $n-1$ and also equal to y, all sub-intervals defined by the replacement times 1,2, …, n must be equal to $a/(n+1)$. The theorem has been proved.

Example 10.1 ***Minimising the Rate of Damage Accumulation by Optimal Replacement***
Given an operation interval with length $a=2.5$ years. The system contains two components logically arranged in series with times to failure following the Weibull distributions (with increasing hazard rate):

$$F_1(t) = 1 - \exp[-(t/eta_1)^{m1}]; m_1 = 3.6;\; eta_1 = 3$$

and

$$F_2(t) = 1 - \exp[-(t/eta_2)^{m2}]; m_2 = 1.8;\; eta_2 = 2.7$$

A single spare is available for each component.

According to the proved theorem, the replacement time t for each of the components building the system is equal to half of the operation interval $t=a/2=2.5/2=1.25$.

This result has been verified by a simulation algorithm for determining the optimal replacement times, whose details have been omitted. The procedure for global optimisation yielded an optimal replacement time of 1.25 years for each component which confirmed the theoretical result. The probability of system failure within 2.5 years, corresponding to these replacement times, is 0.44.

10.1.3 Minimising the Rate of Damage Accumulation by Selecting the Optimal Variation of the Damage-Inducing Factors

10.1.3.1 A Case Related to a Single Damage-Inducing Factor

Suppose that the damage accumulation from a particular degradation mechanism is a function of time and a particular damage-inducing factor p. In the case of fatigue loading, for example, the damage-inducing factor can be the stress amplitude. In the case of excessive corrosion or wear, the damage-inducing factor can be 'temperature' or 'sliding speed', etc.

Suppose that damage is accumulated at M different intensity levels $p_1, \ldots, p_M$ of the damage-inducing factor p (Figure 10.6). At each intensity level p_i, the component is exposed to damage accumulation for time Δt_i. The times t_i, corresponding to constant intensity levels p_i of the damage-inducing factor p, mark the critical level of damage D_c precipitating failure. After the critical level of damage D_c the component is considered to have failed (Figure 10.6). It is also assumed that the sequence in which the various levels of the factor p are imposed does not affect the component's life.

Damage factorisation is present if, for a constant level p of the controlling factor, *the rate of damage accumulation dD/dt can be factorised as a function F(D) of the current accumulated damage 'D' and a function G(p) of the damage-inducing factor p*,

$$dD/dt = F(D)\,G(p) \tag{10.9}$$

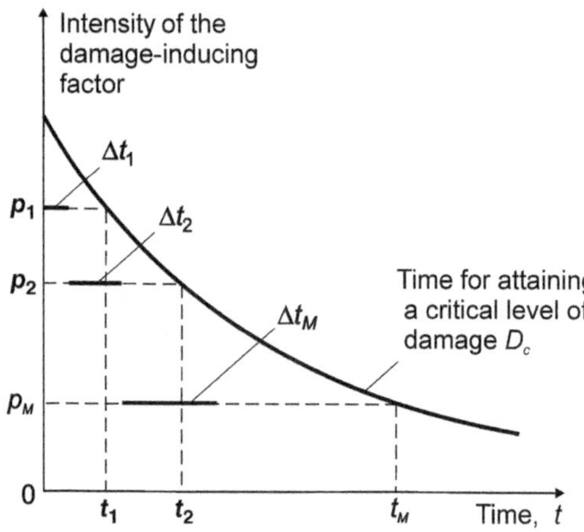

Figure 10.6 Exposure for times Δt_i at different intensity levels p_i of the damage-inducing factor p.

The critical level of damage D_c at different levels of the damage-inducing factor will be attained when the sum

$$\frac{\Delta t_1}{t_1} + \frac{\Delta t_2}{t_2} + \dots \tag{10.10}$$

becomes unity for some k:

$$\frac{\Delta t_1}{t_1} + \frac{\Delta t_2}{t_2} + \dots + \frac{\Delta t_k}{t_k} = 1 \tag{10.11}$$

The time t_c to attain the critical level of damage D_c is then equal to

$$t_c = \Delta t_1 + \Delta t_2 + \dots + \Delta t_k \tag{10.12}$$

Conversely, if the time for obtaining the critical level of damage D_c can be determined using the additivity rule (10.11), the factorisation (10.9) must necessarily hold. *In other words, the damage factorisation law* (10.9) *is a necessary and sufficient condition for the additivity rule* (10.11) (Todinov 2001). This also means that if the rate of damage accumulation cannot be factorised, the additivity rule (10.11) is not valid and must not be used.

Effectively, according to the additivity rule (10.11), the total time t_c required to attain a specified level of damage D_c is obtained by adding the absolute durations Δt_i (Eq. 10.12) spent at each intensity level i of the damage-inducing factor p, until the sum of the relative durations $\Delta t_i/t_i$ becomes unity. The fraction $\Delta t_i/t_i$ of accumulated damage at a particular intensity level p_i of the damage-inducing factor p is the ratio of the time Δt_i spent at level p_i and the total time t_i at level p_i needed to attain the level D_c of damage precipitating failure.

Ideally, to minimise the rate of damage accumulation and maximise reliability, the variation (trajectory) of the damage-inducing factor should be selected in such a way that *a minimum possible time is spent at the levels of the damage-inducing factor where the damage accumulating rate is high. Most of the time should be spent at the levels of the damage-inducing factor where the damage accumulating rate is low.*

Consider stainless steels, with chromium added for corrosion resistance. When such a steel is cooled from high temperature, a precipitation of chromium carbide at the grain boundaries occurs, resulting in chromium-depleted zones along the grain boundaries. Most commonly, this occurs during welding and the process is known as *sensitisation*. If such a weld is exposed to a corrosive environment, the chromium depleted zones along the grain boundaries are preferentially attacked and the fracture toughness of the weld is severely compromised.

Now consider the two cooling cycles of a chromium stainless steel weld in Figure 10.7a. The first cooling cycle '1' corresponds to a fast cooling of the weld and the second cooling cycle '2' corresponds to a normal cooling of the weld.

For the two cooling curves (trajectories) '1' and '2' in Figure 10.7a, conducted for the same length of time t^*, cooling curve '1' is preferable because less time is spent at temperatures where the rate of precipitation of chromium carbides is high. Therefore, curve '1' is associated with a smaller amount of chromium carbides along the grain boundaries (less damage accumulated along the grain boundaries).

An important application of the additivity rule is the evaluation of the amount of damage induced by fatigue cycles. In this case, the damage-inducing factor p is the loading stress range and a measure of the extent of damage is the length a of the fatigue crack. The additivity rule (10.11), also known as the *Palmgren–Miner rule*, has been proposed as an empirical rule in case of damage due to fatigue controlled by crack propagation (Miner 1945). The rule states that in a fatigue test at a constant stress amplitude $\Delta\sigma_i$, damage could be considered to accumulate linearly with the number of cycles. Accordingly, if at a stress amplitude $\Delta\sigma_1$, the component has n_1 cycles of life, which correspond to an amount of damage (crack length) a_c, after Δn_1 cycles at a stress amplitude $\Delta\sigma_1$, the amount of damage will be $\frac{\Delta n_1}{n_1} a_c$. After Δn_2 stress cycles spent at stress amplitude $\Delta\sigma_2$, characterised by a total life of n_2 cycles, the amount of damage will be $\frac{\Delta n_2}{n_2} a_c$ and so on. Failure occurs when, at a certain stress amplitude $\Delta\sigma_M$, the sum of the partial amounts of damage attains the amount a_c, i.e. when

$$\frac{\Delta n_1}{n_1} a_c + \frac{\Delta n_2}{n_2} a_c + \ldots + \frac{\Delta n_M}{n_M} a_c = a_c \tag{10.13}$$

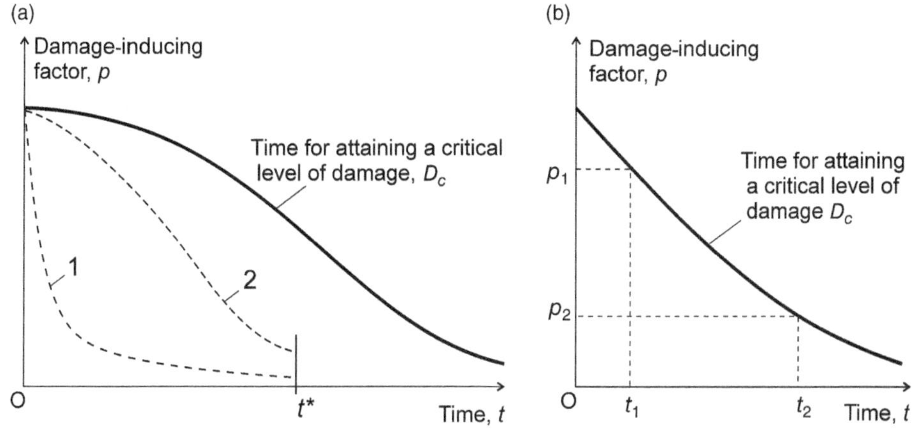

Figure 10.7 (a) Different trajectories result in different rates of damage accumulation. (b) Time to failure for different intensity levels of the damage-inducing factor.

is fulfilled. As a result, the analytical expression of the Palmgren–Miner rule becomes

$$\sum_{i=1}^{M} \Delta n_i / n_i = 1 \tag{10.14}$$

where n_i is the number of cycles needed to attain the specified amount of damage (crack length) a_c at a constant stress amplitude $\Delta\sigma_i$.

The Palmgren–Miner rule is central to estimating the fatigue life of components yet no comments are ever made as to whether it is compatible with the rate of damage accumulation law characterising the different stages of fatigue crack growth. The necessary and sufficient condition for the validity of the empirical Palmgren–Miner rule is the possibility to factorise the rate of damage accumulation *da/dn* as a function of the amount of accumulated damage a (the crack length) and as a function of the stress or strain amplitude (Δp):

$$da/dn = F(a)\,G(\Delta p) \tag{10.15}$$

The theoretical derivation of the Palmgren–Miner rule has been given in Todinov (2001). A widely used law for the rate of fatigue damage accumulation is the Paris power law (Paris et al. 1961; Paris and Erdogan 1963):

$$da(n)/dn = C\,\Delta K^m \tag{10.16}$$

where $\Delta K = Y\,\Delta\sigma\,\sqrt{\pi\,a}$ is the stress intensity factor range, C and m are material constants, a is the current crack size (damage), and Y is a geometry factor that can be presented as a function of the amount of damage a. Clearly, the Paris–Erdogan rate of damage accumulation law can be factorised as in (10.15) and therefore, it is compatible with the Palmgren–Miner additivity rule. In cases where this factorisation is impossible, the Palmgren–Miner rule does not hold. Such is, for example, the fatigue crack growth law

$$da/dn = B\Delta\gamma\, a^{\beta} - D \tag{10.17}$$

discussed in Miller (1993), which characterises physically small cracks. In Eq. (10.17), B and β are material constants, $\Delta\gamma$ is the applied shear strain range, a is the crack size, and D is a constant.

10.1.3.2 A Case Related to Multiple Damage-Inducing Factors

The rate of damage accumulation often depends on multiple damage-inducing factors. Such is the case for the rate *dQ/dt* of wear debris produced during sliding wear described by Archard's equation (Archard 1953):

$$dQ/dt = k\frac{Nv}{H} \tag{10.18}$$

where k is a constant, N is the magnitude of the normal load, v is the sliding speed, H is the hardness of the softest contact surface, and t is time.

This model states that both the sliding speed and the normal load affect the rate of damage accumulation. The dependence of the damage accumulation rate in the case of mutually independent damage-inducing factors, expressed by Eq. (10.18), is very simple. In this case, it is obvious that the rate of damage accumulation is minimised simply by minimising the magnitude of the normal load N and the sliding speed v and maximising the hardness H. In cases where the damage-inducing factors are mutually dependent (such as temperature and humidity), it is not at all clear what is the optimal variation of the damage-inducing factors yielding the smallest amount of accumulated damage.

In the case of multiple damage-inducing factors varying between a specified initial state and a final state, the optimal trajectory yielding the smallest amount of accumulated damage can be determined by dynamic programming (Bellman 1957). The algorithm will be presented for two damage-inducing factors (e.g. temperature and humidity) but it can be easily generalised for any number of damage-inducing factors.

Statement of the problem: Suppose that there are two damage-inducing factors p and q, whose initial values are p_0, q_0. At the end of the time interval $(0,t)$ the values are p_e, q_e. It is assumed, that at any set of values p^*, q^* for the damage-inducing factors, the time t^* for attaining a critical level of accumulated damage which precipitates failure, is specified and can always be evaluated (Figure 10.8a). Therefore, for a time step Δt, the increase ΔD_p of the damage by varying factor q and keeping factor p constant and the increase ΔD_q of the damage by varying factor p and keeping factor q constant can also be evaluated (Figure 10.8b).

It is required to determine the optimal trajectory in the space defined by the damage-inducing factors p and q (Figure 10.8b), such that the total amount of accumulated damage is minimised.

For example, the pair of damage-inducing factors p and q could be: *temperature–humidity, concentration–temperature, pressure–corrosion potential*, etc.

Dynamic algorithm for determining the optimal trajectory yielding the smallest total accumulated damage: The time interval $(0,t)$ during which the damage is accumulated is first divided into $2n$ steps with length $\Delta t = t/(2n)$ where n is a sufficiently large number to ensure the desired precision. Next, a square dynamic table (two-dimensional array with $n \times n$ entries) is built where n rows are allocated for the different levels of the damage-inducing factor q and n columns are allocated for the different levels of the damage-inducing factor p (Figure 10.8b). It is assumed that the real trajectory in the $p \times q$ phase space is approximated by a staircase-type curve in the $n \times n$ dynamic table, consisting of segments on which one of the damage-inducing factors is constant. This

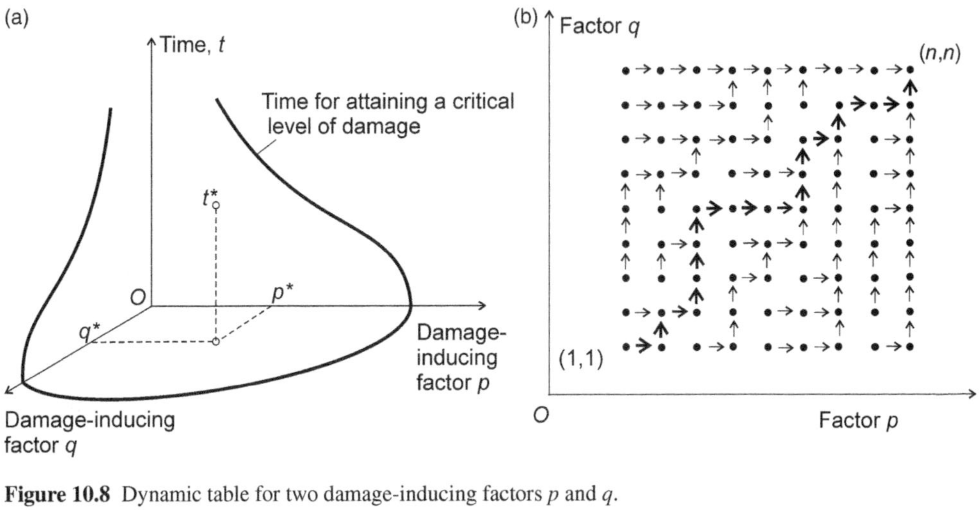

Figure 10.8 Dynamic table for two damage-inducing factors p and q.

approximation to the real trajectory can be made with any desired precision provided that a sufficiently large number ($2n$) of division points are taken on the time interval $(0,t)$.

The advantage of the dynamic programming algorithm consists of the fact that it finds solutions to sub-problems, stores them in the memory and describes the solution of each sub-problem in terms of already solved and previously stored solutions of other sub-problems. As a result, the sub-problems are solved only once, which makes the dynamic programming algorithm orders of magnitude more efficient than a brute-force method based on the enumeration of all possible trajectories on the square lattice of size $n \times n$. The computational time of a brute-force method based on scanning all possible trajectories increases exponentially with increasing the number $2n$ of the time intervals into which the interval $(0,t)$ is divided.

The dynamic algorithm for determining the optimal trajectory works as follows. The solutions of the sub-problems are kept in the array $x[][]$, which will be referred to as a 'dynamic table'. The size of the $x[][]$ array is $n \times n$. The information necessary to restore the optimal solution is kept in the array trac[][] which is also of size $n \times n$. To each entry of the array trac[][] corresponds either the value '1' or the value '2'. The value '1' stands for moving right ('1'), i.e. keeping factor q constant, and the value '2' stands for moving up, i.e. keeping factor p constant.

The last row $x[n][]$ of the array $x[][]$ corresponds to $q = q_e$ value of the factor q. The only possibility for all entries in the last row is to move 'right', i.e. to keep q constant, therefore the array values trac[n][] of the last row are all initialised with '1'.

The last column $x[][n]$ of the array $x[][]$ corresponds to $p = p_e$ value of the factor p. The only possibility for all entries in the last column is to move 'up', i.e. to keep the factor p constant. Therefore, the array values trac[][n] of the last column are all initialised with '2'.

The entries $x[i][j]$ of the array $x[][]$ are always initialised with the smallest amount of accumulated damage for moving from node (i,j) to node (n,n).

The entries of the dynamic table (array $x[][]$) are initialised by starting from entry $x[n][n]$ to which a value zero is assigned: $x[n][n] = 0$. If node $x[n][n-1]$ has been reached, the only direction in the next step is to move right hence, the entry $x[n][n-1]$ is initialised with the value $\Delta D_{(n,n-1),(n,n)}$ which is the smallest amount of accumulated damage from moving from node $(n, n-1)$ to node (n, n). By definition, $\Delta D_{(n,n-1),(n,n)}$ can always be evaluated and is a known quantity. If node $(n, n-2)$ has been reached, the only direction in the next step is again 'right' hence, node $(n, n-2)$ is initialised with the sum $\Delta D_{(n,n-2),(n,n-1)} + x[n][n-1]$ which corresponds to the sum of the smallest amount of damage in moving from node $(n, n-2)$ to node $(n, n-1)$ and the smallest amount of damage in moving from node $(n, n-1)$ to node (n, n). The entry $x[n][n-2]$ therefore contains the smallest amount of accumulated damage in moving from node $(n, n-2)$ to node (n, n).

In this way, values are assigned to all entries of the entire last row $x[n][]$. Next, in a similar fashion, values are assigned to all entries of the last column $x[][n]$ of the dynamic table by starting from the entry $x[n-1][n]$. If node $(n-1, n)$ has been reached, the only direction in the next step is upwards and the entry $x[n-1][n]$ is initialised with the value $\Delta D_{(n-1,n),(n,n)}$ which is the smallest amount of damage in moving from node $(n-1, n)$ to node (n, n).

By definition, $\Delta D_{(n-1,n),(n,n)}$ can always be evaluated and is a known quantity. If the trajectory is in the node $(n-2, n)$, the only direction in the next step is 'upwards' and the entry

$x[n-2][n]$ is initialised with the sum $\Delta D_{(n-2,n),(n-1,n)} + x[n-1][n]$ which corresponds to the sum of the smallest amount of damage in moving from node $(n-2, n)$ to node $(n-1, n)$ and the smallest amount of damage in moving from node $(n-1, n)$ to node (n, n). In this way, values are assigned to the elements of the entire last column $x[][n]$ of the dynamic table.

Next, the assignments in the dynamic table continue with node $(n-1, n-1)$. If node $(n-1, n-1)$ has been reached, there are now two possibilities for the next step: to move 'right' or to move 'up'. The decision depends on which quantity is smaller: $\Delta D_{(n-1,n-1),(n-1,n)} + x[n-1][n]$ or $\Delta D_{(n-1,n-1),(n,n-1)} + x[n][n-1]$. If $\Delta D_{(n-1,n-1),(n-1,n)} + x[n-1][n] < \Delta D_{(n-1,n-1),(n,n-1)} + x[n][n-1]$, the trajectory moves 'right'; if the converse is true, the trajectory moves 'up'. The entry $x[n-1][n-1]$ is then updated with the smaller of the two quantities to show the smallest amount of accumulated damage in moving from node $(n-1, n-1)$ to node (n, n). The entry trac$[n-1][n-1]$ of the track array is updated with '1' or '2' depending on whether the trajectory moves 'right' or 'up', respectively. The process of updating the entries of the $x[][]$ array and the trac[][] array continues with the remaining entries on the $n-1$ row starting from entry $x[n-2][n-1]$ and moving towards $x[n-2][1]$. Next, the row $x[n-3][]$ is updated and this process continues until all entries in the dynamic table are initialised. When the entry $x[1][1]$ is reached, by the way the dynamic table has been constructed, $x[1][1]$ will contain the smallest amount of total accumulated damage in moving from node (1, 1) to node (n, n).

Restoring the optimal trajectory is done by starting from the entry trac[1][1]. Initially, $i = 1$ and $j=1$. If the recorded value in trac$[i][j]$ is '1', a transition is made to the entry trac$[i][j+1]$. If the recorded value in trac$[i][j]$ is '2', the transition is to the entry trac$[i+1][j]$. The process is repeated until the element trac$[n][n]$ is finally reached. An exemplary optimal trajectory is shown in Figure 10.8b, with bold arrows.

If three damage-inducing factors are present, the dynamic table is a three-dimensional array ($x[][][]$), built in a similar fashion.

The running time of the dynamic algorithm for two damage-inducing factors is therefore $O(n^2)$; for three damage-inducing factors the running time is $O(n^3)$, and so on. Therefore, the dynamic optimisation algorithm runs in polynomial rather than exponential time, which is a big advantage to a brute-force optimisation algorithm.

To minimise the rate of damage accumulation and maximise the life of the component, the trajectory associated with the smallest amount of total accumulated damage must be followed.

10.1.3.3 Reducing the Rate of Damage Accumulation by Derating

Derating is a powerful tool available to the designer for reducing the rate of damage accumulation and the likelihood of failure. It follows directly from the conclusion made earlier: *to minimise the total accumulated damage, minimum possible time should be spent at the levels of the damage-inducing factor where the rate of damage accumulation is high.*

Derating to minimise the rate of damage accumulation is done by reducing the operating stresses below their rated levels. Life of many components and systems increases dramatically if the level of the damage-inducing factor is decreased because the rate of damage accumulation is decreased.

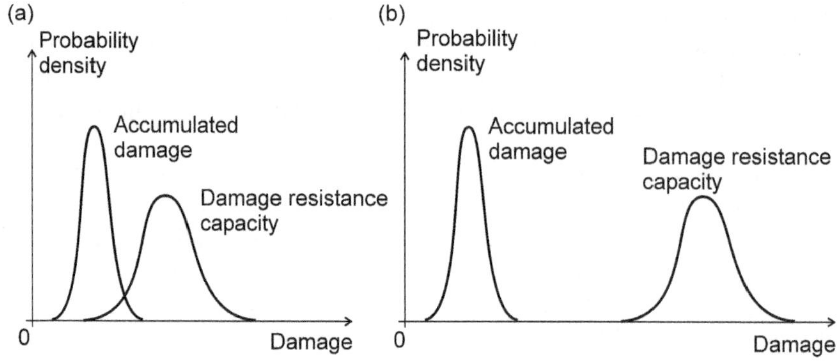

Figure 10.9 Stress and strength distribution: (a) before derating; and (b) after derating.

As can be seen from Figure 10.7b, reducing the intensity level of the damage-inducing factor from p_1 to p_2 enhances the component's life because of the increased time for the accumulated damage to attain the critical level D_c which precipitates failure. A typical application of this method is the reduction of the stress amplitude, which results in a significant decrease of the rate of fatigue damage accumulation.

Derating essentially 'overdesigns' components by separating the damage resistance capacity distribution from the distribution of the accumulated damage (Figure 10.9a) thereby reducing the interaction between the distribution tails (Figure 10.9b). The smaller the interaction of the distribution tails, the smaller the probability that the accumulated damage will exceed the damage resistance capacity, the smaller the probability of failure. In general, the greater the derating the greater the life of the component. However, derating is associated with inefficient use of the damage resistance capacity of components (Figure 10.9b).

Voltage and temperature are common derating stresses for electrical and electronic components. The life of a light bulb designed for 220 V, for example, can be enhanced enormously simply by operating it at a voltage below the rated level (e.g. at 110 V), which reduces significantly the rate of filament degradation. For mechanical components, common derating stresses reducing the rate of damage accumulation are the operating speed, stress amplitude, temperature, pressure, etc.

10.1.4 Reducing the Rate of Damage Accumulation by Deliberate Weaknesses

The essence of this mechanism consists of introducing deliberate weaknesses where damage accumulates instead of accumulating in the valuable (more expensive) part of the system. Instead of accumulating in the expensive component (with a large cost of failure or cost of replacement) the damage accumulates in a cheap component which is replaced when the accumulated damage reaches a critical level. Consequently, the deliberate weakness must be designed with a rate of damage accumulation higher than the rate of damage accumulation in the expensive component.

Examples featuring reducing the rate of damage accumulation by introducing deliberate weaknesses have been discussed in Section 3.3.3.

10.1.5 Reducing the Rate of Damage Accumulation by Reducing Exposure to Acceleration Stresses

10.1.5.1 Reducing Exposure to Acceleration Stresses by Reducing the Magnitude of the Acceleration Stresses

The environment is a major source of acceleration stresses which lead to increased rate of damage accumulation and a significant increase of the hazard rates of components. Examples of acceleration stresses are the temperature, humidity, radiation, vibration, pressure, concentration of particular ions, etc. (Nelson 2004). This list is only a sample of the possible acceleration stresses and can be extended significantly. A typical acceleration stress is the high temperature which increases the rate of damage accumulation and the hazard rates of the affected components. Humidity, corrosion, radiation, or vibrations also increase the rate of damage accumulation and the hazard rates of the affected components.

Failure to account for the negative impact of the acceleration stresses usually leads to optimistic reliability predictions – the actual reliability is smaller than the predicted.

Reducing the exposure to acceleration stresses by reducing the magnitude of the acceleration stresses can be illustrated with the most common acceleration stress – the temperature. In many cases, damage is accumulated at elevated temperatures and is caused by the diffusion of particular atoms (ions). As a result, the instantaneous rate of damage accumulation is proportional to the diffusion coefficient of these atoms (ions). Such is, for example, the process of decarburisation of steels which results in a reduced resistance to contact damage. The variation of the diffusion coefficient D with temperature, in solids, is given by the equation: $D = D_0 \exp(-E_D/RT)$, where T is the absolute temperature (K), E_D is the activation energy for diffusion (J mol^{-1}), R is the gas constant in $\text{J (K} \times \text{mol)}^{-1}$, and D_0 is a constant. From this equation, it can be seen that an increase in temperature results in a drastic increase of the coefficient of diffusion and the rate of damage accumulation. Conversely, reducing temperature of the environment drastically reduces the diffusion coefficient and damage accumulation. Limiting the exposure to elevated temperatures is of critical importance to reducing the damage accumulation. Detailed treatment related to the impact of the acceleration stresses on the life of complex systems has been presented elsewhere (Todinov 2011b).

10.1.5.2 Reducing Exposure to Acceleration Stresses by Modifying or Replacing the Working Environment

Reducing the exposure to acceleration stresses can also be attained by *modifying or substituting the working environment.*

Arc welding, shielded by an inert gas atmosphere such as argon or carbon dioxide, is an example of replacing the working environment. As a result, the accumulation of damage from the contact of the weld metal with oxygen is reduced and the reliability of welds is improved. This principle is used, for example, in MIG (metal inert gas) and TIG (tungsten inert gas) welding techniques. Another example of reducing the exposure to acceleration stresses is the encapsulation of integrated electronic circuits in resin in order to protect them from degradation caused by the environment.

Corrosion increases significantly the rate of damage accumulation. Material degradation due to corrosion (Ohring 1995) is often the root cause of failures entailing loss of life, damage to the environment, and big financial losses. Methods increasing the corrosion resistance include techniques such as *cathodic protection, thermochemical treatment of the surface and protective coatings*. Using *corrosion inhibitors* is an alternative way of reducing the rate of damage accumulation from corrosion. These are compounds that modify the corrosive environment thereby reducing the rate of damage from corrosion. A number of design measures to reduce the rate of accumulation of corrosion damage have been discussed in Mattson (1989).

10.1.6 Reducing the Rate of Damage Accumulation by Appropriate Materials Selection, Design, and Manufacturing

Modifying designs is often used to reduce the rate of damage accumulation. Thus, a design modification resulting in a reduced stress range reduces significantly the accumulation of fatigue damage (Todinov 2016a). Zones in components, characterised by large stress gradients within a small volume, are known as 'stress raisers'. Commonly, these are discontinuities in the geometry and material such as *fillets, notches, holes, threads, steps, grooves, keyways, rough surface finishes, quenching cracks*, and *inclusions*.

The stress raisers are reliability-critical design features because they intensify the rate of fatigue damage accumulation. As a result, they make it easy for a crack to initiate and propagate. Consequently, the appropriate design and manufacturing of the stress raisers should be guaranteed. The rate of damage accumulation at stress raisers can be reduced by avoiding sharp notches and corners, keyways, holes, abrupt changes in cross sections, badly machined fillet radii and grooves.

Poor design of the flow paths of fluids containing abrasive material, promotes rapid *erosion* that can be minimised by proper material selection and design. Structural design features promoting rapid erosion should be avoided. The recently introduced domain-independent risk reduction methods, *separation, segmentation*, and *inversion* (Todinov 2015), can be used successfully to produce designs associated with a small rate of damage accumulation. Thus, the separation of properties of gears achieved by induction heat treatment results in a hard surface and tough core and yields components with superior contact strength, fatigue resistance, and shock resistance. The segmentation of components contains the spread of damage and prevents the damage accumulation from reaching dangerous levels. Thus, segmenting a pipeline, glass panel, solid rod, etc. confines the spread of a crack within a single segment and limits the amount of accumulated damage.

Introducing inverse states counteracts loading stresses and reduces the intensity of damage accumulation. *Cold expansion*, for example used in aviation for creating compressive residual stresses at the surface of fastener holes, is an example of reducing the rate of fatigue damage accumulation by introducing an inverse state. This is done by passing a tapered mandrel through the hole. The compressive residual stress field created in the vicinity of the hole counters the tensile loading stresses during operation and impedes the formation and propagation of fatigue cracks at the edge of the hole.

Material selection is central to design and guarantees the combination of properties needed for the required function. Material selection and selection of appropriate microstructures plays a very important role in reducing the rate of damage accumulation.

Consider the common case of pipes transferring corrosive production fluids or pipes working in a corrosive environment. Selecting steel instead of appropriate corrosion-resistant polymer, alloy, or composite often results in costly failures due to high corrosion rates. Conversely, selecting polymer instead of metal or composite for components subjected to intensive cyclic loading often leads to high rates of fatigue damage accumulation and a short fatigue life. In a high-temperature environment, selecting metals instead of ceramics often leads to high rates of damage accumulation and premature failures.

The susceptibility to cavitation damage can be reduced by using cavitation-resistant materials, welded overlay of metals, sprayed metal coatings, or elastomeric coatings.

Selecting clean materials, with a significantly reduced content of critical flaws which serve as places for fatigue crack initiation, increases significantly the fatigue life of components. Most of the loading cycles are expended on the early stages of fatigue crack propagation when the crack is small. During the late stages of fatigue crack propagation, a relatively small number of cycles is sufficient to extend the crack until failure. The flaw size can be decreased by better material processing and better inspection for flaws.

Controlling the structure of materials during manufacturing, brings unique properties which limit the rate of damage accumulation and the risk of failure in safety-critical applications. Altering the steel microstructure, for example, by appropriate heat treatment, often results in a significantly improved fatigue resistance.

The rate of damage accumulation from erosion, for example, can be reduced significantly by appropriate heat treatment increasing the surface hardness. The rate of fatigue damage accumulation is reduced significantly by strengthening the components surface by gas carburising, gas nitriding, or by the deposition of hard coatings.

Eliminating low-strength surfaces by machining also significantly reduces the rate of fatigue damage accumulation. This is due to eliminating soft decarburised surfaces, surface discontinuities, folds and pores, and by eliminating coarse microstructure characterised by a low toughness.

10.2 Improving Reliability and Reducing Risk by Substitution with Assemblies Working on Different Physical Principles

A device or assembly performing a particular function can be built on different physical principles. Thus, a device measuring length can be built on mechanical, electrical, electromechanical, magnetic, optical, or acoustic physical principles. It can also incorporate a software component. Consider a device/assembly performing a particular function which is substituted with a device/assembly built on different physical principles but performing the same function. The purpose behind this reliability improvement method is *to eliminate or mitigate failure modes resulting from intensive damage accumulation by a substitution/replacement with an assembly/system working on a different physical principle.*

The failure modes characterising the substituting device are normally associated with less intensive damage accumulation but the eliminated failure modes are not necessarily associated with damage accumulation only.

The argument about which of several competing assemblies performing the same function is associated with less dangerous failure modes or failure modes with a smaller rate of damage accumulation (the mechanical, magnetic, electrical, acoustic, or optical assembly), *depends largely on the specific application, conditions, and requirements.*

In substituting mechanical assemblies with electrical, magnetic and optical assemblies and software, the failure modes of the substituting assembly need to be assessed carefully.

If a substitution of a mechanical assembly introduces new, more dangerous failure modes, such a substitution cannot be justified. Thus, if a mechanical system is substituted by a combination of electronics and software, all failure modes of the electronic part and the software part should be considered carefully. If the electronic part is subjected to high temperatures and vibrations or the software part has been developed quickly, by inexperienced developers using inappropriate algorithms, insufficiently tested, the substituting assembly could exhibit more dangerous failure modes compared with the original mechanical system it replaces. Bugs in the software part, for example, could cause failures that cannot be predicted.

Thus, solid-state relays accumulate less damage compared with mechanical relays in applications characterised by a frequent switching in high vibration, dusty, and humid environments. However, the mechanical relays accumulate less damage in cases of frequent current surges, voltage spikes, and currents whose magnitudes vary widely from very small to very large.

Mechanical and optical assemblies have an advantage over electronic assemblies in environments characterised by a high temperature and high levels of nuclear irradiation where the reliability of electronic circuits will be compromised.

Commonly, the substitution of mechanical assemblies with electrical, magnetic or optical assemblies and software eliminates intensive wear, fatigue, intensive corrosion and material degradation, jamming, misaligning etc., which are major factors causing failures of mechanical equipment. Event though electronic components also suffer wear, fatigue, and overstress failure modes which are typical for mechanical components, normally, the percentage of these failure modes is significantly smaller compared with the mechanical components. In addition, the quality of manufacturing of electronic components today is very high: typically of the order of 10 faulty components per million for components such as integrated circuits (O'Connor 2002).

The advantage from a substitution of mechanical assemblies is confirmed by a clear evolutionary trend in many systems which started as mechanical systems and were gradually replaced by electronic, optical, or magnetic systems. The evolution of the mechanical control of the air–fuel mixture and the ignition timing of car engines to an *engine control module* (ECM) is a common example. The evolution of mechanical watches to quartz watches, an electromechanical mouse to an optical mouse, and washing machines and air conditioning units to intelligent washing machines and air conditioning units are other common examples.

Many complex mechanisms, which occupy substantial space in books on theory of mechanisms, can be successfully replaced by a combination of simple mechanisms coupled with servomotors and software. In this way, the complexity needed to guarantee the required kinematics is transferred from the mechanism to the software. The result is significantly simplified mechanical assemblies, with smaller inertia forces, fewer possibilities for jamming, wear, and misalignment.

More often, the substitution of complex mechanical assemblies with electrical, magnetic, optical or acoustic assemblies and software reduces the complexity of design, the number of moving parts, and increases precision. In addition, replacing mechanical assemblies with electrical, optical or acoustic assemblies often improves maintainability which results in reduced downtime and increased availability of the device. Thus, replacing a mechanical measuring system with a magnetic or optical measuring system often eliminates the need for calibration and lubrication which are necessary for conducting an accurate measurement. For example, measuring the thickness of the tube in continuous tube production by using radiography is superior to measuring the thickness by using mechanical devices. The radiographic measurement is continuous and does not require stopping the production line. In addition, it is more precise than measurement conducted by using a mechanical device.

10.2.1 Increasing Reliability by a Substitution with Magnetic Assemblies

A good example of substituting a mechanical assembly with a magnetic assembly is provided by the magnetic stirrer. Magnetic stirrers rely on a rotating magnetic field and a stir bar to stir chemically active liquids in hermetically closed vessels. Magnetic stirrers eliminate rotating seals needed for conventional stirring and the failure modes associated with them.

Another example of replacing a mechanical assembly with a magnetic assembly is provided by the magnetic worm drive (featured in the US patent US3814962; Baermann 1971) whose worm gear is made of permanent magnet material. The teeth of the worm gear and the worm wheel are also magnetised so that the like poles on the wheel and on the worm gear face one another. Magnetic repulsion transmits force from the rotating worm gear to the worm wheel which causes the rotation of the worm wheel.

The advantage of the magnetic worm drive compared with the conventional mechanical worm drive is the frictionless transfer of torque which eliminates contact stresses and wear. Wear is a major failure mode characterising mechanical worm drives. The need for lubrication is also eliminated, together with its failure modes (wrong oil used, oil degradation, oil contamination, low quantity of oil, too high oil temperature, too high or too low oil pressure, too high or too low oil viscosity, blockage of the oil galleries, abrasive particles in the oil). The absence of lubrication simplifies the system and increases its reliability. Furthermore, the clearance between the teeth of the worm gear and the worm wheel eliminates failure modes caused by misalignment which enhances the life of the bearings. The clearance also eliminates the spread of vibrations through the worm wheel which reduces wear and further enhances the reliability of the assembly. Finally, the absence of lubrication simplifies maintenance and improves availability.

The substitution with magnetic devices should be considered after considering all failure modes associated with the magnetic devices: health issues associated with electromagnetic radiation, excessive heating due to hysteresis losses, undesirable magnetisation, undesirable interference with the work of electronic devices, etc. Otherwise, eliminating a particular set of failure modes may cause other, equally undesirable failure modes.

10.2.2 Increasing Reliability by a Substitution with Electrical Systems

An example related to increasing reliability by substituting a mechanical component with an electrical component is the mechanical push button switch. The reliability of a switch is measured by the number of actuations that can be done by the switch before an unacceptable deviation from the required performance is obtained due to damage accumulation. The damage accumulation is due to the presence of mechanical contact which promotes: (i) mechanical deterioration caused by fatigue and wear; (ii) contact erosion due to spattering and fusing of contact material caused by arcing; and (iii) collection of dirt and corrosion products which prevent a good contact. Consequently, eliminating the mechanical contact has the potential to eliminate these failure modes.

Eliminating the mechanical contact could, for example, be achieved by substituting a mechanical push button switch with a switch whose operation is based on the Hall effect. This substitution increases the durability of the switch from tens of thousands to tens of millions of actuations. The Hall effect is the potential difference appearing on a conductor carrying electrical current and placed in a magnetic field. The potential difference is due to the asymmetric distribution of charge caused by the Lorenz force which is experienced by charges moving in a magnetic field.

Furthermore, the substitution with electrical systems often provides a better protection against some environmental stresses. Thus, an electrical module can be protected from corrosion by encapsulation in corrosion-resistant resin, which is not possible for a mechanical assembly.

The substitution with electrical and electronic devices also should be considered after considering all failure modes associated with the substituting devices: risk of overheating and fire, increased tendency to failure in the cases of increased temperature, vibrations, and humidity; increased tendency to failure in the case of damaged insulation, increased dependency on the failure-free operation of the power source, etc. Otherwise, eliminating a particular set of failure modes may cause other, equally undesirable failure modes.

10.2.3 Increasing Reliability by a Substitution with Optical Assemblies

Another example of improving reliability by eliminating mechanical contact is with the replacement of the contact measurement of the temperature of metal surfaces with optical (contactless) measurement by using infrared thermometers (pyrometers) (Childs 2001). The advantages of eliminating the mechanical contact in temperature measurements are numerous:

- By using infrared technology, measurements can be made at temperatures greater than 1300 °C. At these temperatures, even if contact thermometers were available, they would have a very limited life while the reliability of the optical (infrared) thermometers is unaffected.
- By using infrared technology, temperature measurements on hazardous surfaces can be made (for example, on high-voltage surfaces) for which contact temperature measurements are highly problematic and unreliable.
- No interference is present during optical measurements (no energy is lost from the surface during the measurement) compared with contact measurements. Because of the lack of distortion of the measured temperature, the measurement is more accurate compared with a contact measurement.
- By eliminating the mechanical contact, reliable measurements can be made at a very high speed.
- By eliminating the mechanical contact, no damage is done to the surface whose temperature is measured, which cannot be said about measurements done by welding thermocouples on the metal surface.

Fibre optic technology is another area where the rate of damage accumulation is greatly reduced by replacing electrical lines for data transmission with fibre optic lines. Compared with electrical data transmission lines, fibre optic lines are better suited for harsh environments characterised by high humidity, high pressure, electromagnetic interference, high voltage, and extreme temperature variations. Fibre optic transmission lines are not sensitive to electromagnetic interference and accumulate significantly less damage compared with electrical lines. For this reason, fibre optic sensors are better suited for a continuous condition monitoring in composite materials experiencing strain from loading or thermal expansion.

Replacing mechanical strain gauges with optical strain gauges improves the reliability of measurement because it eliminates variability due to the preparation of the surface, necessary to operate the mechanical strain gauges. The variability due to the preparation of the surface leads to poor repeatability and poor reliability of the measured strain. The optical strain gauges do not require physical contact, therefore the variability associated with the different properties of the physical contact are significantly reduced.

The substitution with optical assemblies should also be considered after considering all failure modes associated with the substituting optical devices: increased tendency to failure in contaminating environment; instability in dusty and humid environment, etc. Otherwise, eliminating a particular set of failure modes may cause other, more undesirable failure modes.

10.2.4 Increasing Reliability and Reducing Risk by a Substitution with Software

Software components guarantee flexibility and do not exhibit deterioration which is a major contributing factor to unreliability. Furthermore, replicating the software does not result in manufacturing variability of the software component.

The substitution of mechanical assemblies with electrical and software components also permits the introduction of sensing capabilities. These make the systems able to reset their goals autonomously and better adapt under changing external environment which significantly enhances the resilience of the systems. This also enhances the functionality of mechanical systems and enables them to meet a broad spectrum of user requirements. For example, some modern air conditioning units are capable of sensing both temperature and humidity and adapt their function to the environment through fuzzy logic reasoning.

Software units processing signals from sensors and operating actuators, often substitute control systems working on purely mechanical principles. The advantages are (i) greater flexibility; (ii) better diagnostics of the problem; (iii) better adaptation to changing environmental and operating conditions; and (iv) more precise and adequate control. These manifest into an increase of the overall reliability. An example of such a solution is the programmable ECM in modern cars which controls the valve timing, ignition timing, transient fuelling, air/fuel ratio, the optimal amount of fuel injected in the engine at different combinations of engine speed and throttle position, water temperature correction when the engine is cold, etc. Before ECMs these parameters were controlled by mechanical or pneumatic units.

Autonomous vehicles are a good example of a substitution with electrical and software components capable of sensing their environment and navigating without human input. This helps avoid a number of driving failure modes associated with lack of sleep, lack of concentration, information overload, insufficient reaction speed, etc., typical for human drivers. Software is the major component in the control system of the autonomous car, capable of interpreting the information coming from sensors, identifying obstacles, lanes, vehicles, pedestrians, taking appropriate navigation routes and executing necessary actions for driving and collision avoidance. Machine vision, artificial neural networks (ANNs) and deep neural networks (DNNs) are some of the components of the software for visual object recognition.

Software components can serve as a basis of intelligent equipment that provides guidance and warnings for the user about its correct operation. This helps to avoid dangerous actions and prevent injuries and other failure modes. Software components can also deliver effective troubleshooting which reduces risk by reducing the consequences in the case of failure.

Here, it needs to be pointed out that all safety-critical software components need to be carefully developed and extensively verified and tested. Often, software components are developed by inexperienced developers using inappropriate design, data structures and algorithms. In some cases, the software components are developed quickly, without a formal verification or testing. In these cases, the substituting assembly, incorporating software with bugs, could exhibit more dangerous failure modes than the mechanical or electromechanical system it replaces. In addition, one of the main advantages of the software components, lack of variability after copying into other systems, transforms into common cause failures due to the common software bugs.

11

Improving Reliability by Comparative Models, Permutations, and by Reducing the Time/Space Exposure

11.1 A Comparative Method for Improving System Reliability

Calculating the absolute reliability built in a product is often a very difficult task. Here are some of the reasons:

- In many cases, reliability–critical data (failure frequencies, strength distribution of the flaws, repair times) are not available for the components.
- The physical processes and physical mechanisms underlying the failure modes remain unknown or are associated with large uncertainties.
- The complex influence and uncertainty associated with the environment, the operational loads, and the duty cycles.
- The variability associated with reliability–critical design parameters (e.g. the state of manufactured surfaces, components tolerances, unbalanced forces, internal environment, duty cycles, etc.).
- The non-robustness of the reliability prediction models.

Key reliability-controlling parameters are associated with uncertainty which does not allow the absolute reliability level to be revealed. Major sources of uncertainty are the natural variation of material properties and the uncertainty associated with their measurement, the uncertainty in determining the times to failure, estimating load magnitudes, etc. Furthermore, even if this information was available, for common, widely used reliability models, even relatively small errors in the reliability parameters lead to large errors in the model predictions which renders reliability predictions of questionable value. Consider, for example, a simple assembly which includes a large number of identical components, logically arranged in series, from the same batch. The reliability of the assembly is estimated by raising the estimated reliability of a single component to a power equal to the number of the components in the assembly. As demonstrated in Chapter 1, a small error in estimating the reliability of the component would result in an unacceptably large error (uncertainty) in the estimated reliability of the assembly, which renders the reliability prediction meaningless.

One possible way to avoid the problems with the reliability data and with the amplification of uncertainties associated with the component reliabilities is to build and analyse a comparative reliability model and on its basis to select the most reliable system. This is a powerful strategy for reliability improvement and risk reduction that can be implemented in the absence of any numerical data (Todinov 2009a).

Methods for Reliability Improvement and Risk Reduction, First Edition. Michael Todinov.

11.1.1 Comparative Method for Improving System Reliability Based on Proving an Inequality

A comparative method based on proving an inequality can be used to rank the reliabilities of competing systems and to select the most reliable system. For two competing systems 'a' and 'b' this method can be summarised by the following steps:

- Build the reliability networks of the competing systems from the functional diagrams.
- Determine the system reliabilities of the competing systems R_a and R_b by constructing analytical expressions.
- Subtract the system reliabilities and prove either the inequality $R_a - R_b > 0$ or $R_a - R_b < 0$ regarding the system reliabilities.
- Select the system with the superior reliability.

This method works particularly well with systems including the same number of components. In some cases, for competing systems with the same reliability network topology, the only available information is the ranking of components in terms of their reliability, without it being possible to attach numerical values to their reliabilities. Such is the case where old, new, and medium-age mechanical/electrical components of the same type are used in the same system. Because of inevitable component wear-out and deterioration of mechanical and electrical components, it is sensible to assume that the new components (type A) are more reliable than the medium-age components (type B) and the medium-age components are more reliable than the old-age components (type C). In this case, subtracting the system reliabilities and using the information related to ranking the component reliabilities often yields an inequality that can be proved easily and help to identify the more reliable system.

Here is an example featuring two systems '1' and '2' built with the same types of components: a type-A component and identical type-B_1 (new) component and type-B_2 (old) component. The reliability networks of the competing systems are given in Figure 11.1. If a type-B_1 component is more reliable than a type-B_2 component, building a reliability model of the systems and using algebraic inequalities helps to identify quickly the more reliable system.

Indeed, let a, b_1, and b_2 denote the reliabilities of components A, B_1, and B_2. The reliability of system '1' is

$$R_1 = 1 - (1 - ab_1)(1 - b_2) = b_2 + ab_1 - ab_1b_2 \tag{11.1}$$

The reliability of system '2' is

$$R_2 = 1 - (1 - ab_2)(1 - b_1) = b_1 + ab_2 - ab_1b_2 \tag{11.2}$$

If the inequality $R_2 - R_1 > 0$ or the inequality $R_2 - R_1 < 0$ can be proved, this would mean that one of the systems possesses superior reliability. Subtracting the system reliabilities yields:

$$R_2 - R_1 = b_1 + ab_2 - b_2 - ab_1 = b_1 - b_2 - a(b_1 - b_2) = (b_1 - b_2)(1 - a) \tag{11.3}$$

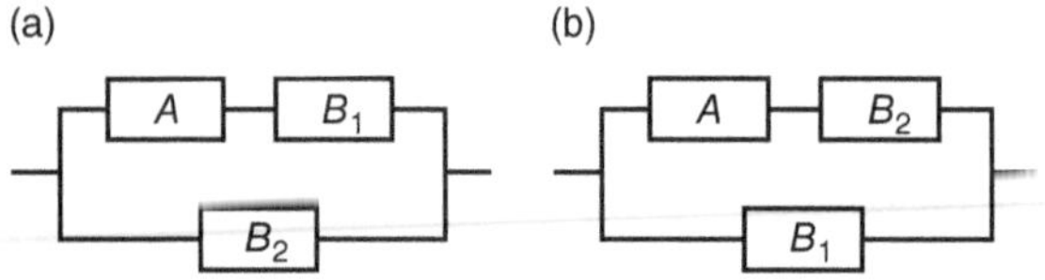

Figure 11.1 Comparing the reliabilities of competing systems: (a) system '1'; (b) system '2'.

The ranking of the component reliabilities yields $b_1 > b_2$ and $1 - a > 0\,(a < 1)$; from which it follows that $R_2 - R_1 > 0$. Therefore, the second system is the more reliable system.

11.1.2 The Method of Biased Coins for Proving System Reliability Inequalities

In proving the inequalities $R_a - R_b > 0$ or $R_a - R_b < 0$, a method referred to as *the method of biased coins* will be introduced.

The method is based on the following idea. Suppose that there are n independent events $A_1, A_2, \ldots, A_n$, not necessarily mutually exclusive. For a particular event A_m from the set of n events, it is known that if A_m does not occur, then at least one of the remaining events has occurred. It is also known that there are at least two events which can be simultaneously present. In other words, there are events A_i and A_j for which $P(A_i \cap A_j) \neq 0$.

Then, it can be shown that for the probabilities of the events, we have

$$P(A_1) + P(A_2) + \ldots + P(A_m) + \ldots + P(A_n) > 1 \tag{11.4}$$

Suppose that $0 < r_1 < 1, 0 < r_2 < 1, 0 < r_3 < 1, f_1 = 1 - r_1, f_2 = 1 - r_2, f_3 = 1 - r_3$ are positive real numbers between 0 and 1. Relationship (11.4) and the biased coins technique can then be illustrated by proving the inequality:

$$r_1 r_2 + r_2 r_3 + r_3 r_1 + f_1 f_2 + f_2 f_3 + f_3 f_1 > 1 \tag{11.5}$$

Suppose that three biased coins are tossed independently. The probabilities of landing 'head' for the first, second, and third coins are r_1, r_2, and r_3, respectively. The probabilities of landing 'tail' for the separate coins are therefore $f_1 = 1 - r_1, f_2 = 1 - r_2, f_3 = 1 - r_3$.

There are six random events in this example: event A_1: first and second coins landing 'head', $P(A_1) = r_1 r_2$; event A_2: second and third coins landing 'head', $P(A_2) = r_2 r_3$; event A_3: third and first coins landing 'head', $P(A_3) = r_3 r_1$; event A_4: first and second coins landing 'tail', $P(A_4) = f_1 f_2$; event A_5: second and third coins landing 'tail', $P(A_5) = f_2 f_3$ and event A_6: third and first coins landing 'tail', $P(A_6) = f_3 f_1$.

Consider event A_6. If event A_6 does not occur (if the third and the first coins do not both land 'tail'), then the outcomes for these two coins can only be HH, HT, TH (where H is 'head' and T is 'tail'). Whatever the outcome from the toss of the second coin (H or T) at least one of the rest of the events $A_1, \ldots, A_5$ must occur. At the same time, events A_6 and A_5 can be simultaneously present P(A_6 and A_5)<>0. As a result, relationship (11.4) holds and so does inequality (11.5).

The application of the method of biased coins in selecting a system with superior reliability will be illustrated by a classical example of selecting between two systems built on n distinct types of independently working components, where for one of the systems the redundancy is at a system level (Figure 11.2a) and for the other system, the redundancy is at a component level (Figure 11.2b).

The reliabilities of the distinct types of components in the reliability networks from Figure 11.2a and Figure 11.2b are: $0 < r_1 < 1; 0 < r_2 < 1; \ldots; 0 < r_n < 1$. The probabilities of failure of the n distinct types of components are $f_1 = 1 - r_1, f_2 = 1 - r_2, \ldots, f_n = 1 - r_n$. All components work and fail independently from one another. The reliability of the system in Figure 11.2a is:

$$R_a = 1 - (1 - r_1 r_2 r_3 \ldots r_n)^2 \tag{11.6}$$

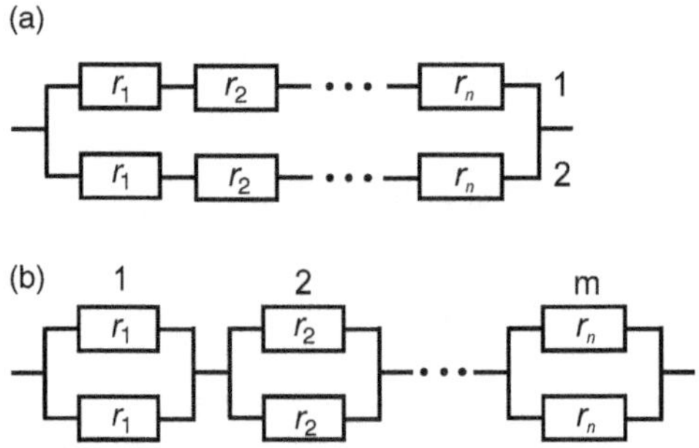

Figure 11.2 A system with redundancy at (a) system level and (b) component level.

while the reliability of the system in Figure 11.2b is

$$R_b = [1 - f_1^2] \times [1 - f_2^2] \times \ldots \times [1 - f_n^2] \tag{11.7}$$

If the inequality $R_b - R_a > 0$ can be proved, this would mean that the reliability of the second system is superior. Proving the inequality $R_b - R_a > 0$ is equivalent to proving

$$(1 - r_1 r_2 r_3 \ldots r_n)^2 + (1 - f_1^2)(1 - f_2^2) \ldots (1 - f_n^2) > 1 \tag{11.8}$$

Because $f_i = 1 - r_i$, $0 < r_i < 1$, and $0 < f_i < 1$, the problem has a clear analogue with tossing n unfair coins, independently. A single biased coin corresponds to each distinct component type. Thus, for the system in Figure 11.2a only n distinct biased coins are present because there are only n distinct types of components. Note that the failure/working states of each component obtained with probabilities r_i and f_i is analogous to flipping a biased coin which has a probability r_i to land 'head' and a probability f_i to land 'tail'. To obtain the state of each component for the network in Figure 11.2a, the n distinct biased coins are tossed twice. This gives the states of $2n$ independently working components. To obtain the state of each component for the network in Figure 11.2b, each of the biased coins is tossed twice: the first toss determines the state of the component in the upper branch; the second toss determines the state of the corresponding component (the same type of component) in the second branch. Note also that because the components work independently and the tosses of each biased coin are statistically independent from one another, the two sequential tosses of n distinct biased coins, needed to obtain the state of the components in the network from Figure 11.2a, are equivalent to n pairs of two tosses with each biased coin, needed to obtain the state of the network in Figure 11.2b. Therefore, it can be safely assumed that the state of the components from the network in Figure 11.2b are also obtained after two sequential tosses of all n biased coins.

If, after tossing the ith coin, the result is a 'head' with probability r_i and a 'tail' with probability f_i, the expression $(1 - r_1 r_2 r_3 \ldots r_n)$ represents the probability that after a single toss of all n biased coins, at least one 'tail' will be obtained. Accordingly, the expression $(1 - r_1 r_2 r_3 \ldots r_n)^2$ represents the probability of event A_1 that after two tosses of all n biased coins, at least one 'tail' will be obtained in each toss. Accordingly, the probability of event A_1 is given by

$$P(A_1) = (1 - r_1 r_2 r_3 \ldots r_n)^2 \tag{11.9}$$

The expression f_1^2 expresses the probability that after two tosses of all n biased coins, the first coin has landed 'tail' in both tosses (probability of the simultaneous occurrence of two independent events). The expression $(1 - f_1^2)$ expresses the probability that after two

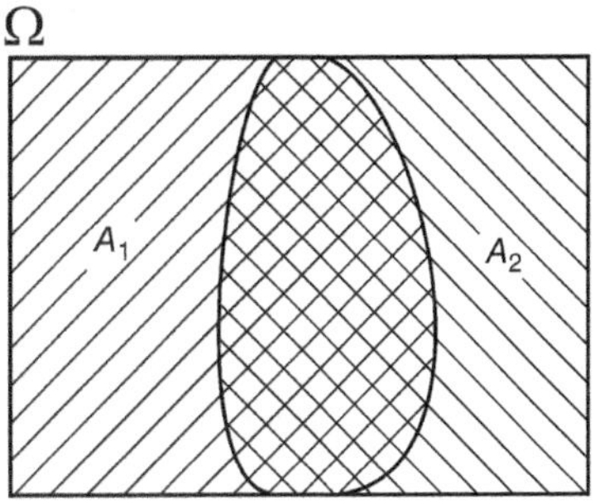

Figure 11.3 Venn diagrams for the probabilities of events A_1 and A_2, composing the left-hand part of inequality (11.8).

tosses of all n biased coins the first coin has landed 'head' at least once. Accordingly, the expression $(1-f_1^2)(1-f_2^2)\dots(1-f_n^2)$ represents the probability of event A_2 that after two tosses of all n biased coins each coin has been a 'head' at least once. Accordingly, the probability of event A_2 is given by

$$P(A_2) = (1-f_1^2)(1-f_2^2)\dots(1-f_n^2) \tag{11.10}$$

Now note that at least one of the events A_1 or A_2 must occur after tossing twice all n biased coins.

Indeed, suppose that A_2 has not occurred. This means that there is a biased coin which has never been a 'head' during the two tosses. This, however, implies event A_1: that at least one 'tail' will be obtained in each toss of all the n biased coins. Suppose now that A_1 has not occurred. This means that there was a toss for which all coins landed 'head'. This, however, implies event A_2: that after two tosses of all n biased coins, each coin has been a 'head' at least once.

Furthermore, events A_1 and A_2 can certainly occur simultaneously. This is the case where, for example, in the first toss the first $n-1$ coins landed 'tail' and the last coin landed 'head' while in the second toss, the first $n-1$ coins landed 'head' and the last coin landed 'tail'.

Consequently, $P(A_1 \cap A_2) > 0$. As a result, the conditions for Inequality (11.4) are fulfilled and $P(A_1) + P(A_2) = (1 - r_1 r_2 r_3 \dots r_n)^2 + (1-f_1^2)(1-f_2^2)\dots(1-f_n^2) > 1$. The Venn diagram of the events A_1 and A_2 looks like the one shown in Figure 11.3, where the probability of the certain event is $P(\Omega) = 1$. The physical property "superior reliability of redundancy at a component level compared with a redundancy at a system level" has been derived from pure mathematical reasoning. The property must be true because, if not, the underlying mathematical inequality (11.8) must be be wrong, which is impossible.

11.1.2.1 Case Study: Comparative Method for Improving System Reliability by the Method of Biased Coins

Consider the reliability networks in Figure 11.4. The two competing reliability networks are built on n distinct types of components, which correspond to n distinct biased coins.

The reliabilities of the corresponding distinct types of components in the networks from Figure 11.4a,b are unknown: $0 < r_1 < 1$; $0 < r_2 < 1$; …; $0 < r_n < 1$. The probabilities of failure of the components are $f_1 = 1 - r_1, f_2 = 1 - r_2, \dots, f_n = 1 - r_n$. All components work and fail independently from one another. The reliability of the system in Figure 11.4a is:

$$R_a = (1 - f_1 f_2 f_3 \cdots f_n)^m \tag{11.11}$$

while the reliability of the system in Figure 11.4b is

$$R_b = 1 - (1 - r_1^m)(1 - r_2^m)\dots(1 - r_n^m) \tag{11.12}$$

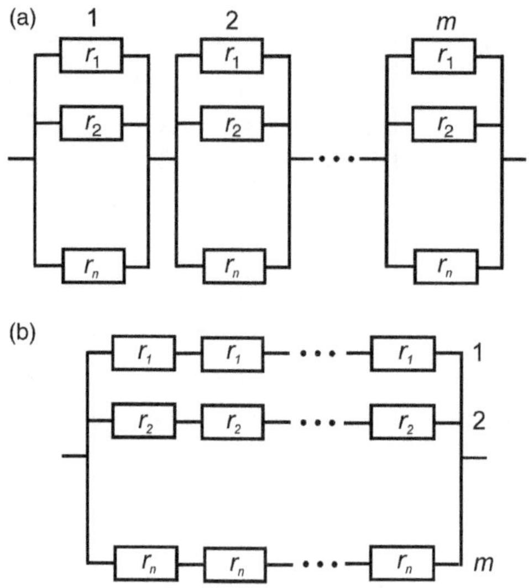

Figure 11.4 Reliability networks of two competing systems containing the same types of components.

If the inequality $R_a - R_b > 0$ or $R_a - R_b < 0$ can be proved, this would mean that the reliability of one of the systems is superior. Proving the inequality $R_a - R_b > 0$ is equivalent to proving

$$(1 - f_1 f_2 f_3 \dots f_n)^m + (1 - r_1^m)(1 - r_2^m)\dots(1 - r_n^m) > 1$$

Because $f_i = 1 - r_i$ and both $0 < r_i < 1, 0 < f_i < 1$, if after tossing the ith coin, the result is a 'head' with probability r_i and a 'tail' with probability f_i, the expression $(1 - f_1 f_2 f_3 \dots f_n)$ represents the probability of event A_1 that after a single toss of all n coins, at least one 'head' will be obtained. Accordingly,

$$P(A_1) = (1 - f_1 f_2 f_3 \dots f_n)^m \tag{11.13}$$

represents the probability of event A_1 that after m tosses of all n coins, at least one 'head' will be obtained in each toss.

The expression r_1^m expresses the probability that after m tosses of all n coins, the first coin has landed 'head' in every single toss (probability of the simultaneous occurrence of m independent events). The expression $(1 - r_1^m)$ expresses the probability that after m tosses of all n coins, the first coin has landed 'tail' at least once. Accordingly,

$$P(A_2) = (1 - r_1^m)(1 - r_2^m)\dots(1 - r_n^m) \tag{11.14}$$

represents the probability of event A_2 that after m tosses of all n coins, each coin has been a 'tail' at least once.

Suppose that event A_2 has not occurred. This means that there is a coin which has never been a 'tail' in all m tosses. This implies event A_1: that at least one 'head' will be obtained in each toss. Now suppose that event A_1 has not occurred. This means that there was a toss in which all coins landed 'tail'. This implies event A_2: after m tosses of all n coins, each coin has been a 'tail' at least once.

Furthermore, events A_1 and A_2 can certainly occur simultaneously. This is indeed the case if in the first toss the first coin is a 'tail' and the rest are all 'head', in the second toss, the second coin is a 'tail' and the rest are all 'head', and so on.

Therefore, $P(A_1 \cap A_2) > 0$. As a result, the condition (11.4) is fulfilled and $P(A_1) + P(A_2) = (1 - f_1 f_2 f_3 \ldots f_n)^m + (1 - r_1^m)(1 - r_2^m) \ldots (1 - r_n^m) > 1$. This proves that the first system has a superior reliability to the second system. The Venn diagram of the events A_1 and A_2 looks like the one shown in Figure 11.3. This is an example related to a physical property (series-parallel systems have a superior system reliability compared with parallel-series systems) derived from pure mathematical reasoning. The physical property must be true because, if not, the underlying mathematical inequality must be be wrong, which is impossible.

11.1.3 A Comparative Method Based on Computer Simulation for Production Networks

An important application of this approach is comparing quickly the performance of competing network topologies and selecting the topology with the best performance. The comparative method for assessing the performance of competing network topologies by using computer simulation can be summarised by the following steps:

- Assume common flow capacities, failure frequencies, and repair times for the corresponding components/edges of the compared networks.
- Determine the performance of the networks by using a relevant software tool.
- Select the topology with superior performance.

Production availability is an important indicator of the performance of production systems (Ebeling 1997). It is defined as the ratio of the total amount of production delivered by the system for one year, in the presence of failures, to the total amount of production which can be potentially delivered in the absence of failures (Todinov 2007, 2013a). Even a small percentage decrease in the production availability (1–2%) translates into big financial losses over the entire period of operation. Maximising the production availability is already an essential part of the design of new production systems. The production availability of an oil and gas production network, for example, is its most important characteristic, which determines the profitability of the installation. The availability of electrical distribution networks, for example, determines the quality of electricity supply to customers.

Consider, for example, the competing production networks in Figure 11.5a,b where, for the sake of simplicity, all edges have the same flow rate capacity of 40 flow units per day, hazard (failure) rate of four expected failures per year and downtime for repair 10 days. Edges (3,8) and (4,9) from the network in Figure 11.5a and edges (2,8) and (4,10) from the network in Figure 11.5b are redundant. Without a supporting comparative model, it is not clear which network topology has superior availability. Applying the method for determining production availability, described in Chapter 8, yields production availability of $\psi_a = 70\%$ for the network in Figure 11.5a and $\psi_b = 75\%$ for the network topology in

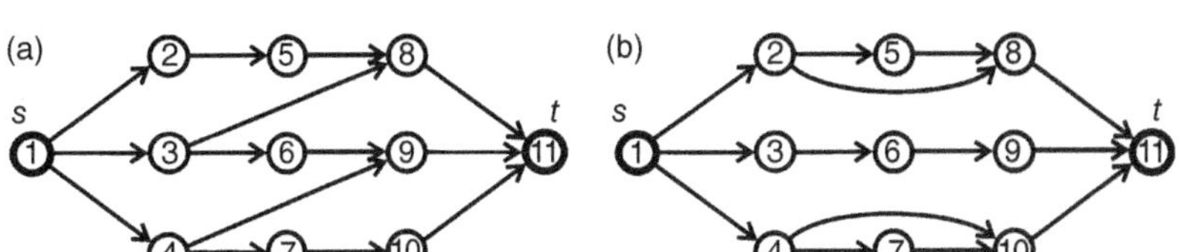

Figure 11.5 Two competing networks with different topology.

Figure 11.5b. Despite the seemingly insignificant differences in the competing topologies, the impact on the production availability is significant. No real reliability data were necessary to deduce which system topology possesses superior production availability.

11.2 Improving Reliability and Reducing Risk by Permutations of Interchangeable Components and Processes

Often the only available information is the ranking of components in terms of their reliability, without it being possible to attach any value to their failure frequencies. Such is the case where old and new components of the same type are used in the same system. Because of inevitable component wear-out and deterioration, it is usually sensible to assume that the new components are more reliable than the old components.

Consider the system in Figure 11.6a which transports cooling liquid from three sources s1, s2, and s3 to the chemical reactor *t*.

The cooling system consists of identical pipeline sections (the arrows in Figure 11.6a). Each pipeline section is coupled with a pump for transporting the cooling fluid through the section. Suppose that the pipeline sections and the pumps are old (sections 'a' in Figure 11.6a) and prone to failure due to corrosion, fatigue, wear, deteriorated seals, etc. The cooling system fulfils its mission if at least one cooling line delivers cooling fluid to the chemical reactor. Suppose for the sake of simplicity that all pipeline sections are in the same state of deterioration and each section is characterised by the same reliability 0.4, associated with one year of operation. Because of the deteriorated sections, the cooling system will benefit from risk-reduction consisting of replacing deteriorated pipeline sections with new sections (sections 'b' in Figure 11.6b). Consequently, the replacement of any of the nine pipeline sections is a possible risk-reduction option. Now suppose that the available budget is sufficient for purchasing and replacing exactly three pipeline sections 'b'. Each new pipeline section is characterised by a reliability 0.9 for one year of operation.

Because the pipeline sections work independently from one another and because all of them are identical (Figure 11.6a), it seems that any three pipeline sections can be replaced with new ones (Figure 11.6b), with the same effect.

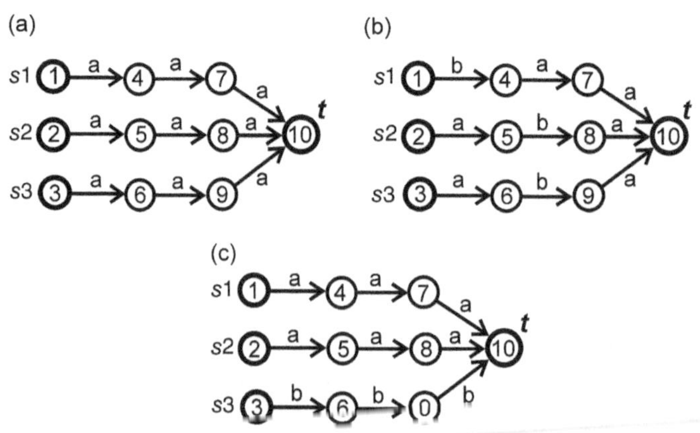

Figure 11.6 A safety-critical cooling system consisting of three parallel branches.

This conclusion, however, is incorrect. The total removed risk of system failure is highest if the available budget is spent on replacing three pipeline sections forming an entire cooling branch (Figure 11.6c), as opposed to replacing three randomly selected sections inside the system (Figure 11.6b).

Indeed, the reliability of the parallel–series arrangement in Figure 11.6b is:

$$R_b = 1 - (1 - 0.4^2 \times 0.9)^3 = 0.37 \tag{11.15}$$

This is the probability that there will be at least one branch carrying cooling fluid through working components from a source to the sink *t*.

The reliability of the parallel–series arrangement in Figure 11.6c is significantly higher:

$$R_c = 1 - (1 - 0.4^3)^2 \times (1 - 0.9^3) = 0.76 \tag{11.16}$$

The variant presented in Figure 11.6c is an example of *a well-ordered parallel–series system*. A well-ordered parallel–series arrangement is obtained if the available components are used first to build the branch with the highest possible reliability; next, the remaining components are used to build the branch with the second-highest possible reliability, and so on, until the entire parallel–series arrangement is built.

If there are three types of components with different age (new, medium age, and old), the maximum reliability is achieved if all new components are arranged in a single branch, the medium-age components in another branch, and all old components are grouped in a separate branch (Figure 11.7).

Because of the absence of a framework of domain-independent methods for reducing risk, the opportunity to increase reliability by a simple permutation of the same type of components in the system, at no extra cost, remained hidden to researchers and practicing engineers reducing risk in specific domains, particularly for parallel–series arrangements, which are very common. Indeed, almost any safety-critical system based on *n* detectors working in parallel for detecting the release of toxic gas, increased pressure, increased temperature, etc., is a parallel–series system. The system detects the critical event if at least one of the detectors working in parallel detects the critical event. The parts composing each detector are normally logically arranged in series (the detector fails if any of its parts fails).

The result stated for the maximum reliability of well-ordered parallel–series systems has been verified by a computer simulation. The computer simulation consisted of specifying the reliabilities of the interchangeable components in the branches and calculating the reliability of the well-ordered system. Next, a 'random scrambling' of the interchangeable components in the branches is initiated, by generating random indices of components from different branches and swapping their reliability values. The swapping guarantees that any resultant system includes exactly the same set of components as the initial system. After each 'random scrambling', the reliability of the scrambled system was calculated and compared with the reliability of the well-ordered system. If the reliability of the well-ordered system was greater than or equal to the reliability of the scrambled system,

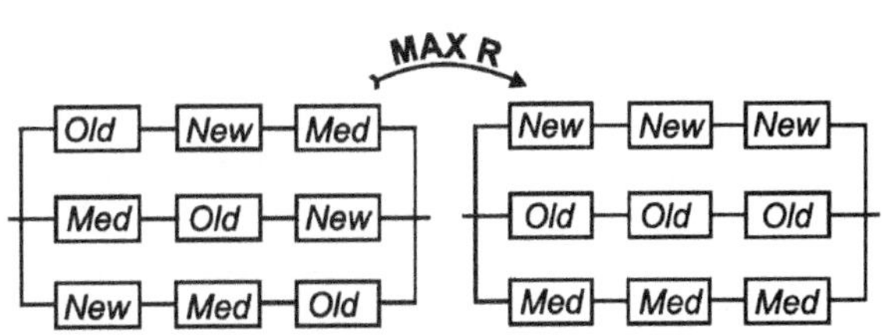

Figure 11.7 Minimising the risk of failure of a parallel–series system by permutation of interchangeable components.

the content of a counter was increased. At the end, the probability that the well-ordered system has reliability not smaller than the reliability of the scrambled system was calculated. In all of the conducted simulations, this probability was always equal to one, which confirms empirically that well-ordered systems are indeed characterised by the largest reliability/availability.

These results can be summarised by stating a domain-independent risk-reduction principle: *The well-ordered parallel–series system is characterised by the smallest possible risk of failure*.

This principle will be proved by contradiction and making use of the extreme principle. Suppose that there is a system which is not well-ordered, and which possesses the highest possible reliability. The branches in a parallel–series system can always be rearranged in such a way that for any two branches i, j for which $i<j$, the branch with index i is equally reliable or more reliable than branch j ($R_i \geq R_j$). If the system is not a well-ordered system, there will be two branches a and b ($a<b$) with reliabilities $R_a \geq R_b$, where there will be at least one component in branch b with a larger reliability than the reliability of the analogous interchangeable component in branch a. Suppose that $R_a = a_1 a_2 \times \ldots \times a_{na}$ and $R_b = b_1 b_2 \times \ldots \times b_{nb}$ are the reliabilities of branches a and b and na, nb are the number of components in branches a and b, respectively. Without loss of generality, suppose that the two analogous interchangeable components mentioned earlier are the last components in branches a and b ($a_{na} < b_{nb}$).

The reliability of the initial system can be presented as

$$R_{sys1} = 1 - (1 - a_1 a_2 \times \ldots \times a_{na})(1 - b_1 b_2 \times \ldots \times b_{nb}) \times [1 - R_{rest}] \tag{11.17}$$

where R_{rest} is the reliability of the rest of the parallel–series arrangement.

After swapping components a_{na} and b_{nb}, the reliability of the resultant system becomes

$$R_{sys2} = 1 - (1 - a_1 a_2 \times \ldots \times a_{na-1} b_{nb})(1 - b_1 b_2 \times \ldots \times b_{nb-1} a_{na}) \times [1 - R_{rest}] \tag{11.18}$$

Subtracting (11.18) from (11.17) yields:

$$R_{sys1} - R_{sys2} = (a_{na} - b_{nb})(a_1 a_2 \times \ldots \times a_{na-1} - b_1 b_2 \times \ldots \times b_{nb-1}) \times [1 - R_{rest}] \tag{11.19}$$

Because $R_a = a_1 a_2 \times \ldots \times a_{na} \geq R_b = b_1 b_2 \times \ldots \times b_{nb}$ by the way the branches have been arranged in descending order according to their reliability ($R_a \geq R_b$), and because $a_{na} < b_{nb}$ (by assumption), the inequality

$$a_1 a_2 \times \ldots \times a_{na-1} > b_1 b_2 \times \ldots \times b_{nb-1} \tag{11.20}$$

holds, which means that in Eq. (11.19) $a_1 a_2 \times \ldots \times a_{na-1} - b_1 b_2 \times \ldots \times b_{nb-1} > 0$.

Since $1 - R_{rest} > 0$, the right-hand side of Eq. (11.19) is negative, which means that the resultant system (after the swap of the two components of the same type) has a higher reliability. This contradicts the assumption that the initial system (which is not well ordered) possesses the highest possible reliability. It was demonstrated that the reliability of a system which is not well-ordered, can be improved by swapping components between parallel branches. Because a parallel–series system can either be a well ordered or not well-ordered system, the well ordered system has the highest reliability. The risk-reduction principle has been proved.

Figure 11.8 Three groups of people working towards achieving the same goal.

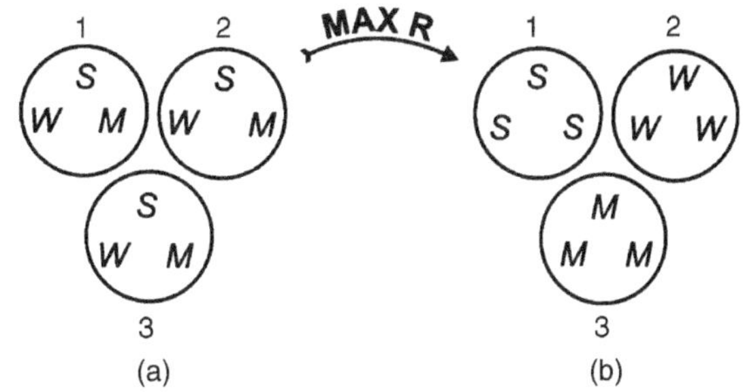

This principle provides an opportunity to remove the maximum amount of system risk *by concentrating the available budget on renewing single parallel branches as opposed to randomly replacing aged components in the system.*

This result also provides the valuable opportunity to improve the reliability of common systems with parallel–series logical arrangement of their components *without the knowledge of their reliabilities and without any investment*. Unlike traditional approaches, which invariably require resources to achieve reliability improvement and risk reduction, a system risk reduction can also be achieved by appropriate permutation of the available interchangeable components in the parallel branches.

Components of similar level of deterioration (reliability levels) should be placed in the same parallel branch (see the example in Figure 11.7).

The risk-reduction principle based on permutation of interchangeable components has wide applications reaching far beyond its initial engineering context.

Consider a common example of three groups of people (teams 1, 2, and 3), each of which includes three independently working team members. The teams work in parallel towards achieving the same goal (Figure 11.8a). The goal is achieved if at least one of the teams succeeds in achieving the goal. Within each team, the task of achieving the goal is divided into sub-tasks among the team members. Each person in a team must accomplish their sub-task successfully in order for the team to achieve the goal. The level of training of each team member is from one of the following categories: strong (S), weak (W), and medium (M). A person with a strong level of training has a better chance of accomplishing a task successfully compared with a person with medium training or weak training. A person with medium training has a better chance of accomplishing the task successfully compared with a person with weak training.

Separating the people in groups with a similar level of training (Figure 11.8b) yields the highest chance of achieving the goal. Note that the risk of not achieving the goal has been reduced at no extra cost.

11.3 Improving Reliability and Availability by Appropriate Placement of the Condition Monitoring Equipment

Monitoring provides an early warning of adverse network changes and is critical to the efficient design and operation of repairable networks and the availability of commodity supply.

Consider the system (Figure 11.9) which transports cooling liquid from three sources s1, s2, and s3 to the chemical reactor *t*. The cooling system consists of identical pipeline

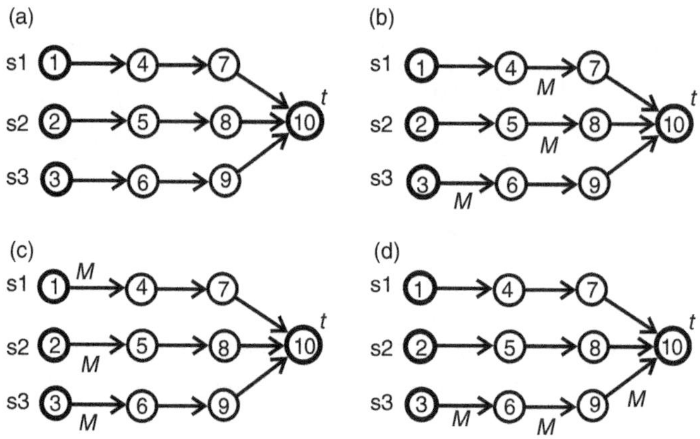

Figure 11.9 Monitoring the branches of a repairable flow network.

sections (the arrows in Figure 11.9). Each pipeline section is coupled with a pump for transporting the cooling fluid through the section. The cooling system fulfils its mission if at least one cooling line delivers cooling fluid to the chemical reactor.

Suppose that the pipeline sections and the pumps are in the same state of deterioration and each section is characterised by the same reliability 0.4, associated with one year of operation (the sections in Figure 11.9).

Introducing condition monitoring on the sections improves their reliability by providing early warning about problems: lack of proper lubrication of the pumps, deterioration of the seals and leaks, blockages of the filters, increased current in the coils of the electromotors, etc. If a problem is registered, a maintenance action is initiated, and failure is averted. Suppose that, as a result of the monitoring, the reliability of a monitored section increases from 0.4 to 0.9. Suppose also that due to budget constraints, it is not possible to purchase monitoring devices for all of the sections. The available budget is sufficient for monitoring exactly three pipeline sections. The critical question is how to choose which sections to monitor so that the reliability of the system is maximised.

For a large system, the number of possible distinct configurations of the monitoring equipment locations is huge (Figure 11.9b,c). Consequently, a brute-force testing of the reliability of the network at each possible configuration of the monitoring locations is simply not a feasible solution. Despite the recent intensive research on networks and system reliability, no viable solution has ever been presented for the long-standing problem of maximising the reliability of a complex system by determining the most appropriate places monitoring. The argument presented in Section 11.2 provides a solution for the network in Figure 11.9. The maximum reliability is obtained when all of the sections from a single branch are monitored (Figure 11.9d, the branch from s3).

The presence of monitoring on some of the sections of the network in Figure 11.9d also provides an early warning about the imminent failure of the section so that a replacment section can be ordered in advance. As a result, the act of monitoring reduces the downtime for repair for components. The improved reliability and reduced downtime improves the availability of the components.

Another important question that could be posed for the repairable flow network in Figure 11.9 is about the components that need to be monitored so that the availability of the system is maximised. Simulations conducted by using a discrete-event simulator for the production availability of repairable flow networks described in Todinov (2013a) show that the maximum production availability is obtained when all of the sections from a single branch are monitored (Figure 11.9d, the branch corresponding to s3).

The optimal set of monitored components provides the right balance between cost, risk, and performance. Optimal locations of the monitoring equipment will translate into low downtimes, high network availability and smaller maintenance resources, smaller amount of consumed fossil fuels, and less carbon emissions. The smaller maintenance resources also mean low operational costs and high profits.

Identifying the optimal places for network monitoring is also the key to maximising the potential of the existing infrastructure at a minimal cost. High network reliability, availability, and throughput capacity translate into a high operational availability and safety and high quality of service to existing and future customers.

11.4 Improving Reliability and Reducing Risk by Reducing Time/Space Exposure

11.4.1 Reducing the Time of Exposure

A typical example of limiting the risk of failure by reducing the time of exposure is reducing the length of operation in order to reduce the probability of encountering an overstress load. Indeed, if the critical overstress load follows a homogeneous Poisson process with density ρ_{cr} and the length of the time interval is a, the probability of encountering an overstress load during the time interval $(0,a)$ is $p_f = 1 - \exp(-\rho_{cr}\, a)$. This probability can be reduced significantly by reducing the length of the time interval a. If the process or action is conducted within a very small time interval ($a \approx 0$), the probability of encountering an overstress load also tends to zero.

$$\lim_{a\to 0}[p_f = 1 - \exp(-\rho_{cr}\, a)] = 0$$

For a particular type of road accident following a homogeneous Poisson process with density ρ, along a road with length L_1, the probability of no road accident, associated with the length L_1, is $p_1 = \exp(-\rho L_1)$. If the road length is decreased by a factor of m ($L_2 = L_1/m$), the probability of no accident is $p_2 = \exp(-\rho L_1/m)$. Taking logarithms of these probabilities gives

$$\frac{\ln p_1}{\ln p_2} = m \tag{11.21}$$

from which $p_2 = (p_1)^{1/m}$. From the last expression, if for example, the probability of no road accident, associated with the length L_1, is $p_1 = 0.41$, decreasing the length of the road four times increases the probability of no road accidents of the specified type to $0.41^{1/4} = 0.8$. Unlike short journeys, long journeys are likely to be affected by delays caused by road accidents. Consequently, for long journeys, a delay caused by an accident should be accounted for in the estimated overall time of the journey.

Suppose that a random load, characterised by a cumulative distribution function $F_L(x)$, is applied a number of times during a finite time interval with length t and the times of load application follow a homogeneous Poisson process with intensity ρ. Suppose that the strength is characterised by a probability density distribution $f_S(x)$. It is also assumed that the load and strength are statistically independent random variables. The probability of no failure (the reliability) associated with the finite time interval $(0,t)$ can be calculated from the overstress reliability integral derived in Todinov (2004b)

$$R(t) = \int_{Smin}^{Smax} \exp\{-\rho\, t\, [1 - F_L(x)]\}\, f_S(x)\, dx \tag{11.22}$$

The term $\exp\{-\rho\, t[1 - F_L(x)]\}$ in the *overstress reliability integral* gives the probability that none of the random loads in the time interval $0,t$ will exceed strength with magnitude x. With reducing the time of exposure t, the probability $\exp\{-\rho\, t[1 - F_L(x)]\}$ that the load will not exceed strength increases significantly which leads to a significant increase in the reliability $R(t)$.

Another example of limiting risk by reducing the time of exposure is present in cases where the accumulated damage is proportional to the time of exposure (e.g. corrosion, erosion, wear, etc.). In this case, reducing the time of exposure prevents damage from reaching a critical level. Thus, reducing the amount of time spent in a hazardous area is an important measure limiting the damage to health. In some cases, the extent of the damage (e.g. in the case of carbon monoxide poisoning or radiation damage) is strongly correlated with the amount of time spent in the hazardous area.

Often, reducing the time of exposure simply does not permit the negative effect to develop. A typical example is increasing the speed by which a rotating shaft goes through its natural frequencies which does not allow developing large resonance amplitudes. Another example is reducing the time of cutting plastic materials by increasing the speed of cutting faster than the plastic deformation can spread, which eliminates the deformation of the processed plastic material.

11.4.2 Reducing the Space of Exposure

A typical example of limiting the risk of failure by reducing the space of exposure is reducing the length of a piece of wire in order to reduce the probability that a critical defect will be present. Indeed, if the critical flaws in the wire follow a homogeneous Poisson process with density λ_{cr} and the length of the wire is L, the probability of encountering a critical flaw along a length $(0,L)$ is $p_f = 1 - \exp(-\lambda_{cr} L)$. This probability can be reduced significantly by reducing the length L of the wire.

If the piece of wire has a very small length L, the probability of having a critical flaw on the length L is also very small:

$$\lim_{L \to 0} [1 - \exp(-\lambda_{cr} L)] = 0$$

Another example is limiting the risk of an error in a long chain of the same type of calculations. Assuming that the errors follow a Poisson distribution, if λ is the number of errors per unit number of calculations, the probability of an error associated with the total number of calculations N is given by $p_f = 1 - \exp(-\lambda N)$. Reducing the number of calculations N, dramatically reduces the probability p_f of a calculation error.

11.4.2.1 Case Study: Reducing the Risk of Failure of Wires by Simultaneously Reducing the Cost

Wires with uniform circular cross section contain critical flaws whose locations along the loaded length of wire follow a homogeneous Poisson process. During operation, it has been observed that given that overloading to a specified limit is present, if a critical flaw is present on the loaded length, the wire fails. All wires without a critical flaw survive the overloading. The records of past failures show that 70% of the wires fail during the overloading and all failures have been caused by a critical flaw on the length of the wire.

The question of interest is by how much the probability of failure of the wires will change if the overloaded length of the wires is reduced by 80%.

The probability of survival of an overloaded piece of wire is given by

$$p_0 = e^{-\lambda_c L}$$

where L is the overloaded length of the wire and λ_c is the lineal number density of the critical flaws. According to the available data, $p_0 = 1 - 0.7 = 0.3$. Reducing the tested length by 80% gives the probability of survival:

$$p_1 = e^{-\lambda_c L \times 0.2} = (e^{-\lambda_c L})^{0.2} = p_0^{0.2} = 0.3^{0.2} = 0.79.$$

The probability of failure of the wire with reduced length will be $1 - 0.79 = 0.21$ (21%). Therefore, the probability of failure of the wires with reduced length will be reduced by 49%. A simple reduction of the loaded length reduced significantly the probability of failure. The probability of failure has been reduced at no extra cost. In fact, because of the reduction in wire material, the probability of failure of the wire has been reduced by simultaneously reducing the cost.

For experts in specific domains, it is not at all obvious that the probability of failure of a loaded component can be reduced to such an extent by a simple reduction of the loaded length.

11.4.2.2 Case Study: Evaluating the Risk of Failure of Components with Complex Shape

Suppose that V is the volume of a component with complex shape subjected to a complex loading. Let λ be the number density of the flaws which follow a homogeneous Poisson process in the stressed volume V. The probability that in the stressed volume V, there will be no critical flaws capable of causing failure is given by $\exp[-\lambda_{cr} L]$. The number density of critical flaws is $\lambda_{cr} = \lambda \times F_c$, where F_c is the probability that a single flaw will be critical, given that it resides in the stressed volume V. Consequently, the probability that the stressed component will not contain a critical flaw becomes $\exp[-\lambda V \times F_c]$ and the probability of failure of the component will be (Todinov 2006a)

$$p_f = 1 - \exp[-\lambda V \times F_c] \tag{11.23}$$

The conditional probability F_c of failure of the stressed component given that a single flaw with random size resides in the volume V can be determined easily from a finite elements solution.

Suppose that the stressed volume has been discretised into finite elements, for each of which, the maximum tensile stress is known. Suppose that the size distribution of the flaws is given by the cumulative distribution $F(d)$. The function $F(d)$ gives the probability that the size of a random flaw will not exceed a particular value d. The probability F_c that a single flaw will be critical, given that it resides in the stressed volume V, can be determined by the following algorithm:

Algorithm 11.1

$F_c = 0;$

For each finite element i **do steps 1–5**

1. Determine the maximum tensile stress σ in the ith finite element;
2. Determine the critical flaw diameter $d_{\sigma i}$ beyond which failure will be initiated by the maximal tensile stress σi in the ith finite element;
3. Determine the probability $1 - F(d_{\sigma i})$ that a flaw with random size will cause failure if it resides in finite element i;
4. Determine the probability $p_{c,\,i}$ that a flaw residing in the volume V will actually reside in the ith finite element and will cause failure:

$$p_{c,i} = (v_i/V) \times [1 - F(d_{\sigma i})]$$

5. Accumulate the probability $p_{c,\,i}$ into the total probability F_c: $F_c = F_c + p_{c,\,i}$

The probability $p_{c,\,i} = (v_i/V) \times [1 - F(d_{\sigma i})]$ that a flaw with random size (given that it is in the stressed volume V) will reside in the ith finite element and will cause failure is given by the product of the probability v_i/V that the flow will reside in the volume v_i of the ith finite element and the probability $1 - F(d_{\sigma i})$ that its diameter will be larger than the critical flaw diameter $d_{\sigma i}$ causing failure under the maximum stress σ characterising the ith finite element. Because a single flaw cannot be in the volume of more than one finite element, the probabilities $p_{c,\,i}$ are added as probabilities of mutually exclusive events.

At the end, the probability of failure of the component with complex shape and volume V is given by Eq. (11.23).

If the volume tends to zero ($V \approx 0$), the probability of having a critical flaw in the volume V also tends to zero:

$$\lim_{V \to 0} [1 - \exp(-\lambda\, V \times F_c)] = 0 \tag{11.24}$$

Now assume that the volume V has been discretised into n elements with the same volume $v = V/n$. Adding all elementary probabilities $p_{ci} = (v/V) \times [1 - F(d_{\sigma i})]$ gives:

$$F_c = \sum_{i=1}^{n} \{(v/V) \times [1 - F(d_{\sigma i})]\} = \frac{1}{n} \sum_{i=1}^{n} [1 - F(d_{\sigma i})] \tag{11.25}$$

for the conditional probability that a flaw with random size will be critical, given that it resides in the stressed volume V. In words, the conditional probability of failure given that the flaw resides in the volume V is equal to the average of the conditional probabilities of failure given that the flaw resides sequentially in each of the elementary volumes into which the volume V has been divided. Substituting F_c in Eq. (11.23) gives the probability of failure of the component with complex shape.

12

Reducing Risk by Determining the Exact Upper Bound of Uncertainty

12.1 Uncertainty Associated with Properties from Multiple Sources

Most of the failures, especially the ones occurring early in life, are quality failures due to the presence of substandard items that find their way into the end products. Variability associated critical design parameters is a major source of substandard items and unreliability. Material properties such as yield stress, fracture toughness, modulus of elasticity, density, resistivity, etc., often appear as design parameters. Production variability of material properties and the specified tolerances in the manufactured product often promote early-life failures.

The idea is to provide a bound for the worst possible variation of a key property, in components pooled from multiple sources, and comply the design with this worst possible bound.

Suppose that random entities are delivered from M different sources in certain proportions $p_1, p_2, \ldots, p_M$, $\sum_{k=1}^{M} p_i = 1$ (Figure 12.1). A particular key property X from each source k is characterised by a distribution $F_k(x)$ with mean μ_k and variance $V_k = \sigma_k^2$, where σ_k is the standard deviation. Quantifying the uncertainty related to the properties associated with pooled entities from different sources is of significant practical interest. Such is, for example, the probability $P(X \le x)$ that the pooled property will not exceed a specified value x. The pooled property X formed by sampling the distinct sources follows a *distribution mixture* $P(X \le x)$. The quantification of the variance σ^2 associated with the pooled properties is also of significant interest.

Suppose that products are delivered from M sources ($i = 1, M$) in certain proportions $p_1, p_2, \ldots, p_M$, $\sum_{i=1}^{M} p_i = 1$. The distributions of the property X characterising the individual sources are $F_i(x)$, $i = 1, 2, \ldots, M$, correspondingly. Thus, the probability $F(x) \equiv P(X \le x)$ of the event B *that the property* X *will not be greater than a specified value* x can be presented as a union of the following mutually exclusive and exhaustive events: $A_1 \cap B$: *the first source is sampled (event* A_1*) and the property is not greater than* x (the probability of this compound event is $p_1 F_1(x)$); $A_2 \cap B$: *the second source is sampled (event* A_2*) and the property is not greater than* x (the probability of this compound event is $p_2 F_2(x)$); …; $A_M \cap B$: *the Mth source is sampled (event* A_M*) and the property is not greater than* x (the probability of this compound event is $p_M F_M(x)$). According to the total probability theorem, the probability

Methods for Reliability Improvement and Risk Reduction, First Edition. Michael Todinov.

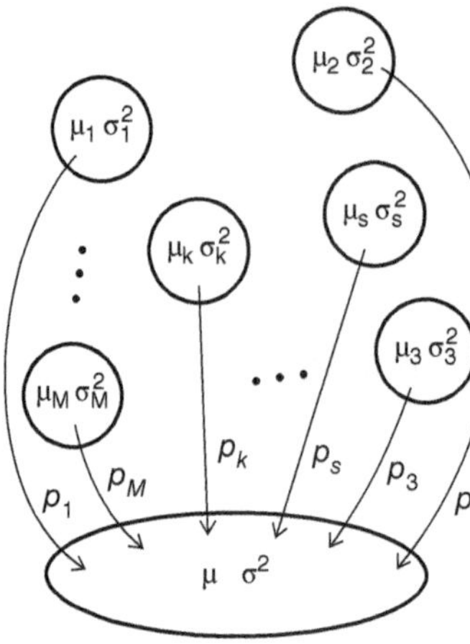

Figure 12.1 Variation of properties from multiple sources.

that the property X will not be greater than a specified value x is given by

$$F(x) \equiv P(X \leq x) = \sum_{k=1}^{M} p_k F_k(x) \tag{12.1}$$

which is the cumulative distribution of the pooled property from all sources. $F(x)$ is a mixture of the probability distribution functions $F_k(x)$ characterising the individual sources, scaled by the probabilities p_k, $k = 1, M$ with which they are sampled. After differentiating Eq. (12.1), a relationship between the probability densities is obtained:

$$f(x) = \sum_{k=1}^{M} p_k f_k(x) \tag{12.2}$$

where $f(x)$ is the probability density function characterising the pooled property and $f_k(x)$, $(k = 1, \ldots, M)$ are the probability density functions characterising the individual sources. Multiplying both sides of Eq. (12.2) by x and integrating:

$$\int_{-\infty}^{+\infty} x f(x)\, dx = \sum_{k=1}^{M} p_k \int_{-\infty}^{+\infty} x f_k(x)\, dx,$$

gives

$$\mu = \sum_{k=1}^{M} p_k \mu_k \tag{12.3}$$

for the mean value μ of a property from M different sources, characterised by means μ_k (Teicher 1963; Everitt and Hand 1981; Titterington et al. 1985).

Distribution mixtures are a useful tool in describing uncertainty associated with material properties. It is a well-established fact that a significant part of the scatter in mechanical properties of various materials is attributable to the inhomogeneity of the microstructure (Todinov 2002a, 2002b, 2003). The influence of the microstructural heterogeneities is particularly strong for the fracture properties compared with other properties (e.g. the material's elastic modulus) for which a smoothing effect appears. The reason is that the fracture criteria are particularly sensitive to microstructural heterogeneities leading to local zones of weaker fracture resistance. As a result, the analysis of the fracture properties of inhomogeneous microstructures benefits significantly from using distribution mixtures (Todinov 2002a, 2003).

Later, it will be demonstrated that distribution mixtures are also necessary in describing the uncertainty associated with the distribution of repair times.

Distribution mixtures can also be applied for assessing the robustness of manufacturing processes and engineering processes (Todinov 2009c). Robustness is an important property of manufacturing processes and engineering products and can be generally defined as 'the capability of a process or a product to cope with variability with minimal loss of functionality'. This is in line with the Taguchi 'on-target engineering philosophy' (Fowlkes and Creveling 1995) and with the fundamental components of quality defined by Juran and Gryna (1988): (i) the product features; and (ii) the product's conformance to those features. Product or process conformance to a required target means that quality is improved when the maximum variation of the output from the target is minimised.

Conformance to a customer-defined target also means that quality improves when variation in performance is minimised (Fowlkes and Creveling 1995).

This approach is also in line with the worst-case design philosophy (Pierre 1986) and the worst-case philosophy of the classical decision theory (Wald 1950).

The process capability index is defined as (Montgomery et al. 2001):

$$C_p = \frac{USL - LSL}{6\sigma} \tag{12.4}$$

where *USL* and *LSL* are the upper and lower specification limits, respectively, and σ^2 is the variance of the process.

A large process capability index means that fewer defective or non-conforming units will be produced. A process with a large capability index is a robust process. This means that the process mean can shift off-target and the percentage of faulty items can still remain very low.

A conservative estimate of the process capability index for properties pooled from multiple sources is important in order to assess the quality of the process. Such an estimate can be obtained if the exact upper bound estimate $\sigma_{\max}$ ($\sigma \leq \sigma_{\max}$) of the standard deviation σ of the process is used. In estimating the exact upper bound, usually, the mixing proportions p_i of the distribution mixture are unknown.

As a result, in quantifying uncertainty related to pooled properties from multiple sources, two fundamental problems emerge:

1. Quantifying uncertainty related to the pooled properties if the mixing proportions $p_1, p_2, \ldots, p_M$, $\sum_{k=1}^{M} p_i = 1$ characterising the separate sources are known.
2. Quantifying uncertainty related to the pooled properties if the mixing proportions $p_1, p_2, \ldots, p_M$, $\sum_{k=1}^{M} p_i = 1$ characterising the separate sources are unknown.

12.2 Quantifying Uncertainty in the Case of Known Mixing Proportions

Distribution mixtures are a necessary tool in describing uncertainty associated with the distribution of repair times. Indeed, consider the distribution of the repair times T, in a communication network composed of M different types of components (e.g. switches, routers, repeaters, servers, cables, printers, etc.). Each component type is characterised by a specific time to repair distribution $F_i(t)$ which gives the probability that the time to repair T of the component will be smaller than a specified value t: $F_i(t) = P(T \leq t)$. The failure rate of the type-i components in the network will be denoted by λ_i.

Suppose that a component failure has occurred. A question of fundamental importance is about the probability that the repair time of the component will not exceed a specified time t: $P(T \leq t)$.

In this example, the distribution of the repair time given that a failure of a component has occurred is a *distribution mixture*.

Indeed, the repair time T of a failed device can be smaller than a specified time t in m distinct, mutually exclusive ways. The repair time T can be smaller than t if the device belongs to the first type of components and its time to repair is smaller than t. The repair time T can also be smaller than t if the device belongs to the second type of components and its time to repair is smaller than t, and so on. The probability p_i that given failure, the failed component will be of type i is given by

$$p_i = \frac{\lambda_i}{\sum_{i=1}^{M} \lambda_i} \tag{12.5}$$

where λ_i is the failure rate of type-i components ($i = 1, 2, \ldots, M$). The probability that the failed component will be of type i and its repair time T will be smaller than t is then given by the product $p_i F_i(t)$, where p_i is specified by Eq. (12.5). Applying the total probability theorem yields the equation

$$F(t) = p_1 F_1(t) + p_2 F_2(t) + \ldots + p_m F_m(t) \tag{12.6}$$

for the distribution of the time to repair, *which, contrary to what is often suggested in reliability literature, is neither exponential, nor normal, nor log-normal, nor any other known standard distribution*. It is a distribution mixture.

Consider now a system built with three specific components: (i) power block; (ii) control module; and (iii) electromechanical device. The failure frequencies of the power block, control module, and electromechanical device are $\lambda_1 = 2$ year^{-1}, $\lambda_2 = 4$ year^{-1}, and $\lambda_3 = 1.5$ year^{-1}, respectively. The time to repair of the different types of device follow negative exponential distributions $F_i(t) = 1 - \exp(-t/\mu_i)$ with means $\mu_1 = 6.4$ days, $\mu_2 = 2.4$ days, and $\mu_3 = 11$ days.

Following Eq. (12.5), the probability p_i that given failure, the failed component will be of type i ($i = 1,2,3$) is given by

$$p_1 = \frac{\lambda_1}{\lambda_1 + \lambda_2 + \lambda_3} = \frac{2}{2 + 4 + 1.5} = 0.267$$

$$p_2 = \frac{\lambda_2}{\lambda_1 + \lambda_2 + \lambda_3} = \frac{4}{2 + 4 + 1.5} = 0.533$$

$$p_3 = \frac{\lambda_3}{\lambda_1 + \lambda_2 + \lambda_3} = \frac{1.5}{2 + 4 + 1.5} = 0.2$$

where λ_i is the failure rate of type-i components.

From Eq. (12.6), the probability $F(t) \equiv P(T \le t)$ that the repair time T of a failed component (in days) will be smaller than a specified value t (in days) is given by:

$$F(t) = 0.267 \times [1 - \exp(-t/6.4)] + 0.533 \times [1 - \exp(-t/2.4)] + 0.2 \times [1 - \exp(-t/11)]$$

The mean repair time is given by

$$\mu = p_1\mu_1 + p_2\mu_2 + p_3\mu_3 = 0.267 \times 6.4 + 0.533 \times 2.4 + 0.2 \times 11 = 5.2 \text{ days.}$$

12.2.1 Variance of a Property from Multiple Sources in the Case Where the Mixing Proportions are Known

In the case where the mixing proportions $p_1, p_2, \ldots, p_M$ are known, the variance V of the mixture distribution (12.1) for continuous probability density functions $f_k(x)$ characterising the existing sources can be derived as follows:

$$V = \int (x-\mu)^2 f(x)\,dx = \int (x - \mu_k + \mu_k - \mu)^2 \sum_{k=1}^{M} p_k f_k(x)\,dx$$

$$= \sum_{k=1}^{M} p_k \left(\int (x-\mu_k)^2 f_k(x)\,dx + \int 2(x-\mu_k)(\mu_k-\mu) f_k(x)\,dx + \int (\mu_k-\mu)^2 f_k(x)\,dx \right)$$

Because the middle integral in the expansion is zero, $\int 2(x-\mu_k)(\mu_k-\mu)f_k(x)\,dx = 0$, the expression for the variance becomes (Todinov, 2002a):

$$V = \sum_{k=1}^{M} p_k\,[V_k + (\mu_k - \mu)^2] \tag{12.7}$$

where V_k, $k = 1, M$ are the variances characterising the M individual distributions.

Consider an application example where the variance determined from Eq. (12.7) has been applied for modelling uncertainty associated with a quality control procedure.

12.2.1.1 Case Study: Estimating the Uncertainty in Setting Positioning Distance

An assembly procedure involves positioning identical components at a particular distance. During the operation, the quality requirement is that the standard deviation of the positioning distance of assembled components should not exceed 0.9 mm. Three robots position identical components. The first robot handles 20% of the components and is characterised by a mean positioning distance 900 mm and standard deviation 0.5 mm. The second robot handles 50% of the components and is characterised by a mean positioning distance 902 mm and standard deviation 0.4 mm, and the third robot handles 30% of the components and is characterised by a mean positioning distance 901 mm and standard deviation 0.7 mm.

A key question here is whether the quality requirement will be satisfied by the existing production equipment.

In this example, the positioning distance of the assembled components follows a distribution mixture. Since the probabilities with which a component will be handled by a particular robot are: $p_1 = 0.2$, $p_2 = 0.5$, and $p_3 = 0.3$, respectively, the mean μ of the positioning distance is $\mu = p_1\mu_1 + p_2\mu_2 + p_3\mu_3 = 0.2 \times 900 + 0.5 \times 902 + 0.3 \times 901 = 901.3$ mm. From Eq. (12.7), the variance is

$$V = \sum_{i=1}^{3} p_i[V_i + (\mu_i - \mu)^2] = 0.2 \times 0.5^2 + 0.5 \times 0.4^2 + 0.3 \times 0.7^2$$
$$+ 0.2 \times (900 - 901.3)^2 + 0.5 \times (902 - 901.3)^2$$
$$+ 0.3 \times (901 - 901.3)^2 = 0.887$$

from which, the standard deviation of the positioning distance is

$$\sigma = \sqrt{V} = \sqrt{0.887} \approx 0.94 \text{ mm.}$$

Clearly, the quality procedure will not be satisfied because the standard deviation of the positioning distance is greater than the required standard deviation of 0.9 mm.

In a common general case, suppose that a particular key property (tolerance, strength, weight, etc.) is obtained from pooling n batches, each containing m_i components. In each batch, the key property follows a particular distribution with mean μ_i and standard deviation σ_i, $i = 1, \ldots, n$. If all batches have been pooled together in a single, large batch, a question of significant interest is the distribution of the key property characterising the components in the large batch.

The distribution of the key property in the single large batch is a mixture of n distributions, where $p_i = \frac{m_i}{\sum_{k=1}^{n} m_k}$ is the probability of sampling the ith batch. The mean of the distribution of the key property in the pooled batch is

$$\mu = \sum_{i=1}^{n} p_i\mu_i$$

The standard deviation of the distribution is $\sigma = \sqrt{V}$, where

$$V = \sum_{i=1}^{n} p_i[\sigma^2 + (\mu_k - \mu)^2]$$

is the variance of the distribution mixture.

Although Eq. (12.7) has a simple form, the grand mean μ of the distribution mixture given by Eq. (12.7) is a function of the means μ_k of the individual distributions. An expression for the variance can also be derived as a function only of the pairwise distances between the means μ_k of the individual distributions (Todinov 2002a):

$$V = \sum_{k=1}^{M} p_k V_k + \sum_{i<j} p_i p_j (\mu_i - \mu_j)^2 \tag{12.8}$$

The expansion of $\sum_{i<j} p_i p_j (\mu_i - \mu_j)^2$ has $M(M-1)/2$ number of terms, equal to the number of different pairs (combinations) of distinct indices among M indices. For $M = 2$ individual distributions (sources), Eq. (12.8) becomes

$$V = p\,V_1 + (1-p)\,V_2 + p\,(1-p)\,(\mu_1 - \mu_2)^2 \tag{12.9}$$

For three sources $M = 3$, Eq. (12.8) becomes

$$V = p_1 V_1 + p_2 V_2 + p_3 V_3 + p_1 p_2(\mu_1 - \mu_2)^2 + p_2 p_3(\mu_2 - \mu_3)^2 + p_1 p_3(\mu_1 - \mu_3)^2 \quad (12.10)$$

Consider now the following application example.

Small samples are taken randomly from a three-component inhomogeneous structure (components A, B, and C). The probabilities p_1, p_2, and p_3 of sampling the structural constituents A, B, and C are equal to their volume fractions ξ_A, ξ_B, and ξ_C ($p_1 = \xi_A$; $p_2 = \xi_B$; $p_3 = \xi_C$). Suppose that the three structural constituents A, B, and C have volume fractions $\xi_A = 0.55$; $\xi_B = 0.35$; and $\xi_C = 0.1$. The mean yield strength of the separate constituents are: $\mu_A = 800\,\text{MPa}$, $\mu_B = 600\,\text{MPa}$, and $\mu_C = 900\,\text{MPa}$, and the standard deviations are: $\sigma_A = 20\,\text{MPa}$, $\sigma_B = 25\,\text{MPa}$, and $\sigma_C = 10\,\text{MPa}$, correspondingly.

The question of interest is the variance of the strength from random sampling of the inhomogeneous structure.

According to Eq. (12.3), the mean yield strength of the samples from all microstructural zones is

$$\mu = p_A \mu_A + p_B \mu_B + p_C \mu_C = 0.55 \times 800 + 0.35 \times 600 + 0.10 \times 900$$
$$= 740\,\text{MPa}$$

Considering that the variances V_k ($k = 1, 2, 3$) in Eq. (12.10) are the squares of the standard deviations σ_A, σ_B, and σ_C characterising the yield strength of the separate microstructural zones, the standard deviation of the yield strength from sampling all microstructural zones becomes

$$\sigma = [p_A \sigma_A^2 + p_B \sigma_B^2 + p_C \sigma_C^2 + p_A p_B(\mu_A - \mu_B)^2 + p_B p_C(\mu_B - \mu_C)^2$$
$$+ p_A p_C(\mu_A - \mu_C)^2]^{1/2} \quad (12.11)$$

After substituting the numerical values in Eq. (12.11), the value $\sigma \approx 108.85\,\text{MPa}$ is obtained for the standard deviation characterising the yield strength from sampling all microstructural constituents. The value $\sigma \approx 108.8\,\text{MPa}$ is significantly larger than the standard deviations $\sigma_A = 20\,\text{MPa}$, $\sigma_B = 25\,\text{MPa}$, and $\sigma_C = 10\,\text{MPa}$, characterising the yield strength from sampling the individual structural constituents.

The reason for the large variance becomes clear upon examining Eq. (12.8). In Eq. (12.8), the variance of the distribution mixture has been decomposed into two major components. The first component $\sum_{k=1}^{M} p_k V_k$ in Eq. (12.8) characterises only the variation of properties within the separate sources (individual distributions). The second component $\sum_{i<j} p_i p_j (\mu_i - \mu_j)^2$ of Eq. (12.8) characterises the variation of properties between the separate sources (individual distributions). Assuming that all individual distributions have the same mean μ ($\mu_i = \mu_j = \mu$), the terms $p_i p_j(\mu_i - \mu_j)^2$ in Eq. (12.8) become zero and the total variance becomes $V = \sum_{k=1}^{M} p_k V_k$. In other words, in this case, the total variation of the property is entirely a within-sources variation.

Now assume that all sources (individual distributions) are characterised by very small variances $V_k \approx 0$. In this case, the within-sources variance can be neglected: $\sum_{k=1}^{M} p_k V_k \approx 0$ and the total variance becomes $V = \sum_{i<j} p_i p_j (\mu_i - \mu_j)^2$. In other words, the total variation of the property is entirely a 'between-sources variation'.

12.3 A Tight Upper Bound for the Uncertainty in the Case of Unknown Mixing Proportions

12.3.1 Variance Upper Bound Theorem

If the mixing proportions p_k are unknown, the variance V in Eq. (12.8) cannot be evaluated. Depending on the actual mixing proportions p_k, the variance V may vary from the smallest variance V_k, characterising one of the sources, up to the largest possible variance obtained from sampling a particular combination of sources with appropriate probabilities p_i. A central question is to establish the exact upper bound for the variance of properties from multiple sources, irrespective of the mixing proportions p_k with which the sources are sampled. This exact upper bound can be obtained by using the numerical Algorithm 12.1 which is based on the next important result:

Variance upper bound theorem The exact upper bound of the variance of properties from sampling multiple sources is obtained from sampling not more than two sources.

The proof of the variance upper bound theorem is given in Todinov (2003) and will not be reproduced here.

In short, the global maximum of the right-hand side of Eq. (12.8) is attained either from sampling a single source/individual distribution, in which case one of the sampling probabilities p_i is unity and the rest are zero ($p_i = 1; p_{j \neq i} = 0$), or from sampling only two individual distributions k and m among all individual distributions composing the mixture distribution. In this case, $p_k \neq 0$ and $p_m \neq 0$; the rest of the p_i are zero ($p_i = 0$) for $i \neq k$ and $i \neq m$. If $V_{\max,k,m}$ denotes the local maximum of the variance from sampling sources (individual distributions) k and m ($k \neq m$), the global maximum $V_{\max}$ of the right-hand side of Eq. (12.8) can be found from $V_{\max} = \max\{V_1, V_2 \ldots V_M, V_{\max,k,m}\}$ where $k = 2, M$ and $m = 1, k-1$. Since there are $M \times (M-1)/2$ number of terms $V_{\max,k,m}$, the global maximum is determined after $M + M \times (M-1)/2 = M(M+1)/2$ checks. As can be verified from the algorithm presented in the next section, the maximum of the variance can be determined by two nested loops. The control variable i of the external loop takes on values from 2 to M (the number of sources) while the control variable j of the internal loop takes on values from 1 to $i-1$.

Mathematically, the upper bound variance theorem can be expressed as

$$V_{\max} = p_{\max} V_k + (1 - p_{\max}) V_s + p_{\max}(1 - p_{\max})(\mu_k - \mu_s)^2 \tag{12.12}$$

where k and s are the indices of the sources for which the exact upper bound of the variance is attained and $0 \leq p_{\max} \leq 1$ and $1 - p_{\max}$ are the mixing proportions with which the two sources are sampled. If $p_{\max} = 1$, the upper bound of the variance is obtained from sampling a single source (the kth source) only.

For the variance of properties V from sampling with unknown mixing proportions p_k, given by Eq. (12.8), the following key inequality holds

$$V \le V_{max}$$

where V_{max} is the variance upper bound obtained from Eq. (12.12). This inequality provides a tight upper bound for the variance because for a particular combination of mixing proportions p_k, an equality can be attained.

12.3.2 An Algorithm for Determining the Exact Upper Bound of the Variance of Properties from Multiple Sources

For a large number M of sources, determining the upper bound variance by finding directly the global maximum of Expression (12.8) regarding the probabilities p_k is a difficult task which can be greatly simplified by using the upper bound variance theorem. The algorithm is based on the upper bound variance theorem and consists of checking the variances of all individual sources and the variances from sampling all possible pairs of sources. As a result, finding the upper bound variance of the properties from M sources involves only $M(M+1)/2$ checks, which can be done by a computer.

According to the derivation of the upper bound variance presented in Todinov (2003), the upper bound of the variance either coincides with the variance of an individual distribution building the mixture or is obtained for some pair of distributions i,j sampled with mixing proportions $p_{k\_max} = 0.5 + \frac{V[i]-V[j]}{2(\mu[i]-\mu[j])^2}$ and $1 - p_{k\_max}$. The maximum variance obtained from sampling sources i and j is (Todinov 2002a)

$$V = V[i]/2 + V[j]/2 + \frac{(V[i]-V[j])^2}{4\,(\mu[i]-\mu[j])^2} + \left(\frac{\mu[i]-\mu[j]}{2}\right)^2 \tag{12.13}$$

The algorithm is given in pseudo-code, where the statements in braces {*op1*; *op2*; *op3*; …;} are executed as a single block. The variable max contains the largest variance at the end of the calculations; the constant M is the number of components (sources) composing the mixture distribution. Variables k_max and m_max contain the indices of the sources sampling from which yields the largest variance. If the maximum variance is attained from sampling a single source, k_max and m_max will both contain the index of this source.

Algorithm 12.1

```
max=V[1]; k_max=1; m_max=1;
for i from 2 to M do
{
if (max<V[i]) then do {
                        max=V[i]; k_max=i; m_max=i;
                        pk_max=1; pm_max=1;
                      }
for j=1 to i do
 {
 if |V[i] - V[j]| < (μ[i] - μ[j])^2 then do
```

```
{
candidate_max = V[i]/2 + V[j]/2 + (V[i] - V[j])^2 / (4 (μ[i] - μ[j])^2) + ((μ[i] - μ[j]) / 2)^2;
if (max < candidate_max) then do
        {
            max=candidate_max;
            k_max=i;
            m_max=j;
            pk_max = 0.5 + (V[i] - V[j]) / (2 (μ[i] - μ[j])^2);
            pm_max=1-pk_max;
        }
    }
  }
}
```

Variables pk_max and pm_max contain the probabilities of sampling the two sources which yield the largest variance. As can be verified, the statements in the internal loop are executed only if the condition $|V[i] - V[j]| < (\mu[i] - \mu[j])^2$ is fulfilled, which indicates a local maximum.

12.3.3 Determining the Source Whose Removal Results in the Largest Decrease of the Exact Variance Upper Bound

The inequality based on the variance upper bound theorem has a very important application: *it can be used for bounding the maximum possible variation of a property pooled from multiple sources.*

This application is particularly useful in cases where the mixing proportions from the n sources of variation are unknown and m sources of variation can be removed ($m < n$). A decision needs to be made about which m sources of variation out of the n available sources should be removed so that the maximum reduction of the maximum possible variation is achieved. In the case of unknown mixing proportions p_i ($i = 1, 2, \ldots, n$), the decision should be based on the sources of variation whose removal yields the smallest variance upper bound $V_{ub,\min}$ among the remaining $n - m$ sources of variation. This will ensure that irrespective of the mixing proportions p_i the variance V from the remaining $n - m$ sources of variation will not exceed the obtained variance upper bound $V_{ub,\min}$: $V \le V_{ub,\min}$.

Suppose that a single source of variation can be removed ($m = 1$) and a decision needs to be made about which source of variation should be removed so that the maximum reduction of the maximum possible variation is achieved or the smallest variance upper bound $V_{ub,\min}$ is obtained. This will ensure that irrespective of the mixing proportions p_i, the variance V from the remaining $n - 1$ sources of variation will not exceed the obtained variance upper bound $V_{ub,\min}$: $V \le V_{ub,\min}$, for any possible values of the mixing proportions.

The algorithm consists of finding the source (pair of sources) yielding the largest variance. Removing one of these sources results in the smallest exact variance upper bound $V_{ub,\min}$. If sampling from a single source yields the largest variance, removing this source yields the largest decrease in the exact variance upper bound and the smallest variance upper bound $V_{ub,\min}$. The algorithm will be illustrated by the following numerical example.

Suppose that the rates of degradation for a key property characterising items pooled from five processing units are characterised by individual distributions with variances $V_1 = 208$, $V_2 = 240$, $V_3 = 108$, $V_4 = 102$, $V_5 = 90$ and means $\mu_1 = 39$, $\mu_2 = 43$, $\mu_3 = 45$, $\mu_4 = 56$, and $\mu_5 = 65$, correspondingly.

The question of interest is the removal of which processing unit yields the largest decrease in the exact variance upper bound of the degradation rate.

Algorithm 12.1 yields that the maximum of the variance of the rate of degradation from all processing units is $V_{\max} \approx 323.15$, attained from sampling the fifth processing unit with probability $p_{\max} \approx 0.41$ and the first processing unit with probability $1 - p_{\max} \approx 0.59$. Removing the fifth processing unit yields the largest reduction of the variance upper bound of the degradation rate or the smallest variance upper bound $V_{ub,\min}$. Indeed, the calculations show that after removing the fifth processing unit, the exact variance upper bound $V_{\max 1} = V_{ub,\min} \approx 241.4$ of the degradation rate is attained from sampling the fourth processing unit with probability $p_{\max} \approx 0.09$ and the second processing unit with probability $1 - p_{\max} \approx 0.91$. The removal of the fifth processing unit yields a value of the variance upper bound $V_{ub,\min} \approx 241.4$ which cannot be improved (decreased) by the removal of any other processing unit instead.

In this example, the upper bound variance theorem has effectively been used for conducting *uncertainty importance analysis – identifying the source of variation whose removal yields the smallest value of the exact variance upper bound.*

If more m out of n sources of variation need to be removed, the described procedure is repeated m times.

12.4 Applications of the Variance Upper Bound

12.4.1 Using the Variance Upper Bound for Increasing the Robustness of Products and Processes

The variance upper bound theorem can be used as a basis for developing a worst-case design aimed at improving the robustness of processes, operations, and products originating from different sources. For such processes or products, the mean of the distribution mixture is not critical and can be easily adjusted to a specified target value. It is the deviation from the mean that leads to undesirable performance. Such is, for example, the positioning of identical components at a certain distance by different positioning devices. Each device is characterised by a mean positioning distance and a variance. While the mean positioning distance can be altered easily, no such control is normally available for the variance of the positioning distance. Reducing the variance of a process usually requires fundamental technological changes which need a substantial investment.

A conservative estimate of the process capability index for properties from multiple sources can be obtained by using an upper bound variance estimate σ_{max} produced by the Algorithm 12.1:

$$C_p^* = \frac{USL - LSL}{6\sigma_{\max}} \quad (12.14)$$

Determining a non-parametric and conservative estimate of the process capability index helps to stabilise the variation of the process within the control limits and reduce the number of faults in the end product. The non-parametric capability index can serve as a basis for ranking, comparing, and selecting competing manufacturing processes.

If no source of variation can be removed, the pair of distributions which yield the worst possible variation of the property can be identified. The next step involves verification whether the design can accommodate this worst-case variation of the property. In short, *the process of creating a more robust product based on the variance upper bound theorem is a process of accommodating the worst-case variation of a key property from its target value without loss of functionality.*

If sources of variation can be removed, by removing the source resulting in the most significant decrease of the exact variance upper bound, the overall variability of the property will be reduced and the capability index of the manufacturing process increased.

This is a way of obtaining a robust manufacturing process, whose variation is under control, irrespective of the actual mixing proportions from the manufacturing centres.

12.4.2 Using the Variance Upper Bound for Increasing the Robustness of Electronic Devices

Components building electronic circuits are characterised by properties such as resistance, capacitance, inductance, etc. Because of imprecision, during manufacturing, the actual magnitudes of these properties deviate from their stated nominal values.

Suppose that these components are part of safety-critical systems containing sensors measuring temperature, pressure, concentration, etc. in two different zones. Suppose that a difference exceeding a particular threshold of the sensors readings triggers an alarm or a shutdown system. Large deviations in the properties of the components building the measurement circuit are undesirable, because they lead to a deteriorated performance of the safety-critical devices.

Suppose that the components building the measurement circuit are manufactured by different centres/suppliers. Each centre/supplier is characterised by its individual distribution of the corresponding property. Usually, the variation of the property (resistivity, capacitance, inductance, etc.) associated with the common pool of manufactured components, for specified mixing proportions from the suppliers, is not the maximum possible variation that can occur. There exists a particular combination of sources and mixing proportions that yields the largest (worst-case) variation. The variance upper bound theorem makes it possible to calculate this worst-case variation. This is illustrated by the next case study.

12.4.2.1 Case Study: Calculating the Worst-Case Variation by the Variance Upper Bound Theorem

Suppose that electronic components are delivered from four suppliers. The mean resistances [Ω] characterising the individual suppliers are $\mu_R = \{500, 504, 510, 516\}$. The variances $V_R = \sigma_R^2$, equal to the squared standard deviations of the resistances characterising the individual suppliers are $V_R = \{102, 141, 166, 85\}$.

If, in the batch, the shares from the different suppliers are unknown, a calculation of the exact upper bound of the variance by using Algorithm 12.1 yields the value $V_{max} \approx 169$, attained from sampling two suppliers: the first supplier with a mixing proportion $p = 0.18$ and the third supplier with a mixing proportion $q = 1 - p = 0.82$.

For any other combination of suppliers and shares, a smaller variance will be produced. Indeed if, for example, the shares of the suppliers are: $p_R = \{0.15, 0.65, 0.15, 0.05\}$, where $\sum_{i=1}^{4} p_{Ri} = 1$, for the variance of the supplied components, Eq. (12.8) yields only $V \approx 150$.

The designer must make sure that the electronic circuit will operate satisfactorily under the worst possible combination of suppliers and mixing proportions, yielding the maximum possible variation of the resistance.

The maximum possible variance of the resistance is attained at its exact upper bound, determined by using the variance upper bound theorem.

12.4.3 Using the Variance Upper Bound Theorem for Delivering Conservative Designs

A conservative estimate of the likelihood of a faulty assembly can be obtained by sampling from the distributions of the reliability-critical properties which are associated with the exact upper bound of the variance.

The worst-case variation of the pooled reliability-critical property is determined by using the variance upper bound theorem. This is effectively an act of sampling from a mixture distribution containing at most two individual distributions. Suppose that sampling from the distribution mixture

$$F(x) = pF_k(x) + qF_s(x) \tag{12.15}$$

where p and $q = 1 - p$ are the mixing proportions, yields the largest variance of the property. $F_k(x)$, $F_s(x)$ are the two individual distributions in the mixture whose sampling yields the largest variance. Sampling the distribution mixture (12.15) involves two steps: (i) a random selection of an individual distribution to be sampled; and (ii) a random sampling from the selected individual distribution.

A random selection of the individual distribution $F_k(x)$, with probability p, and a random selection of the individual distribution $F_s(x)$, with probability $1-p$, can be done by simulating a uniformly distributed random variable u in the interval (0,1) by a standard procedure whose software implementation can be found in standard texts on Monte Carlo simulation (Ross 1997; Todinov 2016a). After generating the random variable u, a check is performed. If $u \leq p$ is fulfilled, the distribution $F_k(x)$ is selected; if $u > p$, the distribution $F_s(x)$ is selected. After selecting one of the distributions, a random sampling from the selected distribution can be performed by using standard methods (Ross 1997; Todinov 2016a).

After obtaining a sample from the worst-case combination of sources, a test is performed to check whether the sampled output parameter defines a faulty assembly. If a faulty assembly is present, a fault counter is incremented. At the end of the Monte Carlo trials, the probability of a faulty assembly is obtained as a ratio of the value stored in the fault counter and the total number of simulation trials.

12.4.3.1 Case Study: Identifying the Distributions Associated with the Worst-Case Variation During Virtual Testing

This approach will be illustrated by the example considered in Section 12.4.2.1. Suppose, that the resistance of the components from the individual suppliers of electronic components follows normal distributions with means 500Ω, 504Ω, 510Ω, 516Ω and squared standard deviations (variances) $V_R = \sigma_R^2 = \{102, 141, 166, 85\}$. According to the example solved in Section 12.4.2.1, the mixture distribution $F(x)$ yielding the exact upper bound of the variance, is obtained from sampling the first and the third supplier:

$$F(x) = pF_1(x) + qF_3(x) \tag{12.16}$$

where $p = 0.18$, $q = 1 - p = 0.82$, $F_1(x)$, and $F_3(x)$ are the individual Gaussian distributions characterising the first and the third supplier.

The distribution $F(x)$ of the resistance is a distribution mixture with mean

$$\mu = p\mu_1 + q\mu_3 = 0.18 \times 500 + 0.82 \times 510 = 508.2 \tag{12.17}$$

where $\mu_1 = 500$, $\mu_3 = 510$, and variance

$$\sigma^2 = p\sigma_1^2 + q\sigma_3^2 + pq(\mu_1 - \mu_3)^2 = 169.2 \tag{12.18}$$

where $\sigma_1^2 = 102$ and $\sigma_3^2 = 166$.

Suppose now that two electronic components are assembled in two symmetrical sections of an electronic device and a difference in the resistance of more than 40 Ω essentially means 'a faulty assembly'. The worst-case percentage of faulty assemblies has been calculated by using the outlined algorithm. The empirical probability of a faulty device was 2.9%, obtained on the basis of 100 000 simulation trials.

In this way, a conservative estimate of the likelihood of a faulty assembly can be obtained by sampling only from the distributions of the reliability-critical property which are associated with the maximum possible value of the variance.

12.5 Using Standard Inequalities to Obtain a Tight Upper Bound for the Uncertainty in Mechanical Properties

The use of inequalities to bound uncertainty of mechanical properties is not restricted to the upper bound variance inequality. Consider the equivalent stiffness k_s of n springs in series and the equivalent stiffness k_p of n springs in parallel. The stiffness of the separate springs are $k_1, k_2, \ldots, k_n$. The equivalent stiffness of n springs in series is given by the well-known relationship:

$$k_s = \frac{1}{\frac{1}{k_1} + \frac{1}{k_2} + \ldots + \frac{1}{k_n}} \tag{12.19}$$

and for n springs in parallel, by the relationship:

$$k_p = k_1 + k_2 + \ldots + k_n \tag{12.20}$$

Equation (12.19) follows from the fact that the total deflection δ_s for springs in series, loaded by a common force F, is the sum of the deflections of the separate springs.

$$\delta_s = \delta_1 + \delta_2 + \ldots + \delta_n \tag{12.21}$$

Since, the link between the spring deflection δ and the spring constant k is given by $\delta = \frac{F}{k}$, Eq. (12.21) becomes $\frac{F}{k_s} = \frac{F}{k_1} + \frac{F}{k_2} + \ldots + \frac{F}{k_n}$, from which Eq. (12.19) follows directly.

Equation (12.20) follows from the fact that for springs in parallel, all spring deflections are the same $\delta_s = \delta_1 = \delta_2 = \ldots = \delta_n$ and the total force F acting on the springs is a sum of the forces $F_1, F_2, \ldots, F_n$ acting on the separate springs: $F = F_1 + F_2 + \ldots + F_n$. Since, the link between the spring force F, the spring deflection δ, and the spring constant k is given by $F = k\delta$, it follows that $k\delta_s = k_1\delta_s + k_2\delta_s + \ldots + k_n\delta_s$ from which Eq. (12.20) follows.

The numerical simulation experimentation with random spring constants indicated that the ratio k_p / k_s of the equivalent spring constant (stiffness) k_p of springs in parallel and the equivalent spring constant (stiffness) k_s of springs in series does not fall below n^2 ($k_p / k_s \geq n^2$).

This led to the conjecture that *irrespective of the spring constants of n springs, the equivalent spring constant of a series arrangement is at least n^2 times smaller than the equivalent spring constant of the parallel arrangement*:

$$n^2 k_s \leq k_p \tag{12.22}$$

To prove inequality (12.22), it is sufficient to prove that

$$(k_1 + k_2 + \ldots + k_n)\left(\frac{1}{k_1} + \frac{1}{k_2} + \ldots + \frac{1}{k_n}\right) \geq n^2 \tag{12.23}$$

Clearly, for two springs only

$$(k_1 + k_2)\left(\frac{1}{k_1} + \frac{1}{k_2}\right) \geq 2^2 \tag{12.24}$$

Indeed, expanding the left-hand side of (12.24) gives

$$(k_1 + k_2)\left(\frac{1}{k_1} + \frac{1}{k_2}\right) = 1 + \frac{k_1}{k_2} + \frac{k_2}{k_1} + 1 \tag{12.25}$$

For any positive numbers k_1 and k_2

$$\frac{k_1}{k_2} + \frac{k_2}{k_1} \geq 2 \tag{12.26}$$

is fulfilled and this follows from the standard *Arithmetic Mean-Geometric Mean (AM-GM) inequality* (Kazarinoff 1961; Steele 2004):

$$k_1 + k_2 \geq 2\sqrt{k_1 k_2} \tag{12.27}$$

which follows directly from the inequality $(\sqrt{k_1} - \sqrt{k_2})^2 \geq 0$.

Indeed, squaring both sides of inequality (12.27) (which are positive numbers) and dividing by $k_1 k_2$ yields inequality (12.26), directly.

Expanding the left-hand side of (12.23) gives n^2 terms of the form $\frac{k_i}{k_j}$, $1 \leq i, j \leq n$. From these n^2 terms, n terms are of the type $i = j$ and $\frac{k_i}{k_j} = 1$. The rest of the $n^2 - n$ terms can be paired in the sums $\frac{k_i}{k_j} + \frac{k_j}{k_i}$. According to what has been proved earlier, for each of the paired sums, $\frac{k_i}{k_j} + \frac{k_j}{k_i} \geq 2$ holds. The number of these sums is $(n^2 - n)/2$ and the left-hand side of (12.23) becomes

$$(k_1 + k_2 + \ldots + k_n)\left(\frac{1}{k_1} + \frac{1}{k_2} + \ldots + \frac{1}{k_n}\right) \geq n + \frac{n^2 - n}{2} \times 2 = n^2$$

which proves (12.23) and (12.22).

The conclusion is that irrespective of uncertainty related to the constants of the n springs, the effective spring constant in the series arrangement is always at least n^2 times smaller than the effective spring constant of the parallel arrangement. The upper bound $k_s \leq k_p / n^2$ is tight and equality is attained when all spring constants are equal $k_1 = k_2 = \ldots = k_n = k$. This is another example of a physical property derived from pure mathematical reasoning (see Section 11.1.2.1 for more examples). The physical property must be true because, if not, the underlying mathematical inequality must be wrong, which is impossible.

References

Allen, A.O. (1990). *Probability, Statistics and Queuing Theory with Computer Science Applications*, 2e. New York: Academic Press.

Altshuller, G.S. (1984). *Creativity as an Exact Science: The Theory of the Solution of Inventive Problems*. New York: Gordon and Breach Science Publishing.

Altshuller, G.S. (1996). *And Suddenly the Inventor Appeared, TRIZ, the Theory of Inventive Problem Solving*, Translation from Russian by Lev Shulyak. Worcester, MA: Technical Innovation Center.

Altshuller, G.S. (1999). *The Innovation Algorithm, TRIZ, Systematic Innovation and Technical Creativity*. Worcester, MA: Technical Innovation Center, Inc.

Andrews, J.D. and Moss, T.R. (2002). *Reliability and Risk Assessment*. London: Professional Engineering Publishing.

Ang, A.H.S. and Tang, W.H. (1975). *Probability Concepts in Engineering Planning and Design, Vol. 1, Basic Principles*. New York: Wiley.

Archard, J.F. (1953). Contact and rubbing of flat surface. *Journal of Applied Physics* 24 (8): 981–988.

Ashby, M.F. and Jones, D.R.H. (2002). *Materials and Design: The Art and Science of Material Selection in Product Design*. Butterworth Heinemann.

Aven, T. (2003). *Foundations of Risk Analysis: A Knowledge and Decision-Oriented Perspective*. Wiley.

Aven, T. (2016). Risk assessment and risk management: review of recent advances on their foundation. *European Journal of Operational Research* 253: 1–13.

Awrejcewicz, J., Lewandowski, D., and Olejnik, P. (2017). *Dynamics of Mechatronics Systems Modeling, Simulation, Control, Optimization and Experimental Investigations*. World Scientific.

Baermann, M. (1971). Magnetic worm drive. US Patent 3,814,962.

Barlow, R.E. and Proschan, F. (1965). *Mathematical Theory of Reliability*. New York: Wiley.

Barlow, R.E. and Proschan, F. (1975). *Statistical Theory of Reliability and Life Testing*. New York: Rinehart and Winston, Inc.

Bazovsky, I. (1961). *Reliability Theory and Practice*. Englewood Cliffs: Prentice-Hall, Inc.

Bedford, T. and Cooke, R. (2001). *Probabilistic Risk Analysis, Foundations and Methods*. Cambridge: Cambridge University Press.

Bellman, R. (1957). *Dynamic Programming*. Princeton: Princeton University Press.

Bergman, B. (1985). On the variability of the fracture stress of brittle materials. *Journal of Materials Science Letters* 4: 1143–1146.

Methods for Reliability Improvement and Risk Reduction, First Edition. Michael Todinov.

Bergman, B., de Mare, J., Loren, S., and Svensson, T. (ed.) (2009). *Robust Design Methodology for Reliability*. Chichester: Wiley.

Bessis, J. (2002). *Risk Management in Banking*, 2e. Chichester: Wiley.

Bhat, U.N. (2008). *An Introduction to Queuing Theory: Modelling and Analysis in Applications*. Boston, MA: Birkhauser.

Billinton, R. and Allan, R.N. (1992). *Reliability Evaluation of Engineering Systems*, 2e. New York: Plenum Press.

Bird, G.C. and Saynor, D. (1984). The effect of peening shot size on the performance of carbon-steel springs. *Journal of Mechanical Working Technology* 10 (2): 175–185.

Blischke, W.R. and Murthy, D.N. (2000). *Reliability: Modelling, Prediction, and Optimisation*. New York: Wiley.

Booker, J.D., Raines, M., and Swift, K.G. (2001). *Designing Capable and Reliable Products*. Oxford: Butterworth-Heinemann.

Burrell, N.K. (1985). Controlled shot peening of automotive components. *SAE Transactions*, Section 3 94: 850365.

Calin, O. and Udriste, K. (2014). *Geometric Modelling in Probability and Statistics*. Switzerland: Springer.

Campbel, J. (2015). *Complete Casting Handbook, Metal Casting Processes, Metallurgy, Techniques and Design*, 2e. Amsterdam: Butterwoth-Heinemann.

Carter, A.D.S. (1986). *Mechanical Reliability*. London: Macmillan Education Ltd.

Carter, A.D.S. (1997). *Mechanical Reliability and Design*. London: Macmillan Press Ltd.

Childs, P.R.N. (2001). *Practical Temperature Measurement*. Oxford: Butterworth-Heinemann.

Clausing, D. (2004). Operating window – an engineering measure for robustness. *Technometrics* 46 (1): 25–29.

Coffman, E.G., Mallows, C.L., and Poonen, B. (1994). Parking arcs on the circle with applications to one-dimensional communication networks. *The Annals of Applied Probability* 4 (4): 1098–1111.

Coffman, E.G., Flatto, L., Jelenkovic, P., and Poonen, B. (1998). Packing random intervals on-line. *Algorithmica* 22 (4): 448–476.

Collins, J.A. (2003). *Mechanical Design of Machine Elements and Machines*. New York: Wiley.

Cooper, R. (1981). *Introduction to Queuing Theory*, 2e. Amsterdam: Elsevier.

Costache, A., Glejbol, K., Sivebak, I.M., and Berggreen, C. (2016). Improved friction joint with self-locking grips. *Journal of Offshore Mechanics and Arctic Engineering* 138 (5): 051401.

Cox, L.A. (2008). What's wrong with risk matrices? *Risk Analysis* 28 (2): 497–512.

Crouhy, M., Galai, D., and Mark, R. (2006). *The Essentials of Risk Management*. McGraw-Hill.

Cullen, W.D. (1990). *The Public Enquiry into the Piper Alpha Disaster*. London: HMSO.

Dasgupta, A. and Pecht, M. (1991). Material failure mechanisms and damage models. *IEEE Transactions on Reliability* 40 (5): 531–536.

Davis, T.P. (2006). Science, engineering, and statistics. *Applied Stochastic Models in Business and Industry* 22 (5–6): 401–430.

DeGroot, M. (1989). *Probability and Statistics*. Reading: Addison-Wesley.

Denardo, E.V. (2002). *The Science of Decision Making*. New York: Wiley.

Dhillon, B.S. and Singh, C. (1981). *Engineering Reliability: New Techniques and Applications*. New York: Wiley.

Dowling, N.E. (1999). *Mechanical Behaviour of Materials*. Upper Saddle River: Prentice Hall.

Ebeling, C.E. (1997). *An Introduction to Reliability and Maintainability Engineering*. New York: McGraw-Hill.

Eder, W.E. and Hosnedl, S. (2008). *Design Engineering*. Boca Raton, FL: CRC Press.

Epstein, R.A. (2009). *The Theory of Gambling and Statistical Logic*. Amsterdam: Elsevier.

Erhard, G. (2006). *Designing with Plastics*. Munich: Hanser.

Everitt, B.S. and Hand, D.J. (1981). *Finite Mixture Distributions*. London: Chapman and Hall.

Feng, X., Ouyang, M., Liu, X. et al. (2017). Thermal runaway mechanism of lithium ion battery for electric vehicles: a review. *Energy Storage Materials* doi: 10.1016/j.ensm.2017.05.013.

Fowlkes, W.Y. and Creveling, C.M. (1995). *Engineering Methods for Robust Product Design: Using Taguchi Methods in Technology and Product Development*. Reading: Addison-Wesley.

French, M. (1999). *Conceptual Design for Engineers*, 3e. London: Springer-Verlag London Ltd.

Fu, Y., Ge, E., Su, H. et al. (2015). Cold expansion technology of connection holes in aircraft structures: a review and prospect. *Chinese Journal of Aeronautics* 28 (4): 961–973.

Gadd, K. (2011). *Triz for Engineers: Enabling Inventive Problem Solving*. Wiley.

Gere, J.M. and Timoshenko, S.P. (1999). *Mechanics of Materials*. Cheltenham: Stanley Thornes (Publishers) Ltd.

Giambene, G. (2005). *Queuing Theory and Telecommunications, Networks and Applications*. New York: Springer.

Gildersleeve, M.J. (1991). Relationship between decarburisation and fatigue strength of through hardened and carburising steels. *Materials Science and Technology* 7: 307–310.

Gnedenko, B.V. and Kovalenko, I.N. (1989). *Introduction to Queuing Theory*, 2e. Berlin: Birkhauser.

Gross, D. and Harris, C.M. (1985). *Fundamentals of Queuing Theory*, 2e. New York: Wiley.

Haimes, Y.Y. (2009). On the definition of resilience in systems. *Risk Analysis* 29 (4): 498–501.

Harry, M.J. and Lawson, J.R. (1992). *Six Sigma Producibility Analysis and Process Characterisation*. Reading: Addison-Wesley.

Haugen, E.B. (1980). *Probabilistic Mechanical Design*. New York: Wiley.

Hearn, E.J. (1985). *Mechanics of Materials*, 2e. Butterworth.

Hedlund, F.H., Selig, R.S., and Kragh, E.K. (2016). Large steel tank fails and rockets to height of 30 meters rupture disc installed incorrectly. *Safety and Health at Work* 7: 130–137.

Heitmann, W.E., Oakwood, T.G., and Krauss, G. (1996). Continuous heat treatment of automotive suspension spring steels. *Fundamentals and Applications of Microalloying Forging Steels: Proceedings of a Symposium Sponsored by the Ferrous Metallurgy Committee of TMS*.

Henley, E.J. and Kumamoto, H. (1981). *Reliability Engineering and Risk Assessment*. Englewood Cliffs: Prentice-Hall, Inc.

Heron, R.A. (1998). *System Quantity/Quality Assessment – The Quasi-steady State Monitoring of Inputs and Outputs'in Handbook of Condition Monitoring: Techniques and Methodology*, 1e (ed. A. Davis). Chapman and Hall.

Hobbs, G.K. (2000). *Accelerated Reliability Engineering, HALT and HASS*. Chichester: Wiley.

Hollangel, E. (2016). *Barriers and Accident Prevention*. Abingdon: Routlege.

Hosseini, S., Barker, K., and Ramirez-Marquez, J.E. (2016). A review of definitions and measures of system resilience. *Reliability Engineering and System Safety* 145: 47–61.

Hoyland, A. and Rausand, M. (1994). *System Reliability Theory*. New York: Wiley.

HSE (1992). *The Tolerability of Risk from Nuclear Power Stations*. London: Health and Safety Executive.

Huffer, F.W. and Shepp, L.A. (1987). On the probability of covering the circle by random arcs. *Journal of Applied Probability* 24 (2): 422–429.

International Electrotechnical Commission (IEC). (1991). International vocabulary, Chapter 191: Dependability and quality of service, IEC 50 (191).

Juran, J.M. and Gryna, F.M. (1988). *Juran's Quality Control Handbook*, 4e. New York: McGraw-Hill.

Justicz, J., Scheinerman, E.R., and Winkler, P.M. (1990). Random intervals. *The American Mathematical Monthly* 97 (10): 881–889.

Kalashnikov, V. (2010). *Mathematical Methods in Queuing Theory*. Dordrecht, The Netherlands: Kluwer Academic Publishers.

Kalpakjian, S. and Schmid, S. (2013). *Manufacturing Engineering & Technology*, 7e. Prentice Hall.

Kaplan, S. and Garrick, B.J. (1981). On the quantitative definition of risk. *Risk Analysis* 1 (1): 11–27.

Kaplan, S., Visnepolschi, S., Zlotin, B., and Zusman, A. (1999). *New Tools for Failure and Risk Analysis: Anticipatory Failure Determination*. Detroit: Ideation International Inc.

Kazarinoff, N.D. (1961). *Analytic Inequalities*. New York: Dover Publications, Inc.

Kelly, P.M. and Francis-Rose, L.R. (2002). The martensitic transformation in ceramics—its role in transformation toughening. *Progress in Materials Science* 47 (5): 463–557.

Khintchine, A.Y. (1969). *Mathematical Methods in the Theory of Queueing*. London: Griffin.

Kuo, W., Prasad, V.R., Tillman, F.A., and Hwang, C.L. (2001). *Optimal Reliability Design*. New York: Cambridge University Press.

Lamport, L. (1978). Time, clocks, and the ordering of events in a distributed system (PDF). *Communications of the ACM* 21 (7): 558–565. doi: 10.1145/359545.359563.

Lebkowski, A., Tobiasz, M., Dziedzicki, K., et al. (2005). Areas of dangerous approach to the ship defined with the aid of a domain. *Proceedings of the ESREL 2005 Conference*, Gdyinia-Sopot-Gdansk.

Leveson, N. (2011). *Engineering a Safer World: Systems Thinking Applied to Safety*. Cambridge, MA: MIT Press.

Lewis, E.E. (1996). *Introduction to Reliability Engineering*. New York: Wiley.

Lindroos, M., Apostol, M., Heino, V. et al. (2015). The deformation, strain hardening, and wear behaviour of chromium-alloyed hadfield steel in abrasive and impact conditions. *Tribology Letters* 57 (3): 24.

Mahamood, R.M., Akinlabi, E.T., Shukla, M., and Pityana, S. (2012). Functionally graded material: an overview. *Proceedings of the World Congress on Engineering*, vol. III, pp. 4–6, London.

Mämmelä, O., Hiltunen, J., Suomalainen, J., et al. (2016). Towards micro-segmentation in 5G network security. *Proceedings of the EuCNC 2016 Network Management, QoS and Security Workshop*.

Matthews, C. (1998). *Case Studies in Engineering Design*. Arnold.

Mattson, E. (1989). *Basic Corrosion Technology for Scientists and Engineers*. New York: Wiley (Halsted Press).

Meeker, W.Q. and Escobar, L.A. (1998). *Statistical Methods for Reliability Data*. New York: Wiley.

Melchers, R.E. (2001). On the ALARP approach to risk management. *Reliability Engineering and System Safety* 71 (2): 201–208.

Meriam, J.L., Kraige, L.G., and Bolton, J.N. (2016). *Engineering Mechanics: Statics*, 8e. Wiley.

Miller, K.J. (1993). Materials science perspective of metal fatigue resistance. *Materials Science and Technology* 9: 453–462.

MIL-STD-1629A (1977). *US Department of Defence Procedure for Performing a Failure Mode and Effects Analysis*. Washington, DC: US Department of Defence.

Miner, M.A. (1945). Cumulative damage in fatigue. *Journal of Applied Mechanics* 12: 159–164.

Montgomery, D.C., Runger, G.C., and Hubele, N.F. (2001). *Engineering Statistics*, 2e. New York: Wiley.

Moore, P. (1983). *The Business of Risk*. Cambridge: Cambridge University Press.

Nielsen, T.D. and Jaffray, J.Y. (2006). Dynamic decision making without expected utility: an operational approach. *European Journal of Operational Research* 169: 226–246.

Nelson, W. (2004). *Accelerated Testing, Statistical Models, Test Plans and Data Analysis*. New York: Wiley.

Niku-Lari, A. (1981). *Shot-Peening, in the First International Conference on Shot Peening*, Paris, 14–17 September, 1–27. Oxford: Pergamon Press.

O'Connor, P.D.T. (2002). *Practical Reliability Engineering*, 4e. New York: Wiley.

Oh, J.H., Kim, K., and Jeong, J.S. (2015). A study on the risk analysis based on the trajectory of fishing vessels in the VTS area. *International Journal of e-Navigation and Maritime Economy* 2: 38–46.

Ohring, M. (1995). *Engineering Materials Science*. San Diego: Academic Press, Inc.

Orloff, M. (2006). *Inventive Thinking through TRIZ*, 2e. Springer.

Orloff, M.A. (2012). *Modern TRIZ a Practical Course with EASyTRIZ Technology, a Practical Guide*, 2e. Springer.

Pahl, G., Beitz, W., Feldhusen, J., and Grote, K.H. (2007). *Engineering Design*. Berlin: Springer.

Paley, S.J. (2010). *The Art of Invention*. New York: Prometheus Books.

Paris, P.C. and Erdogan, F. (1963). A critical analysis of crack propagation laws. *Journal of Basic Engineering* 85: 528–534.

Paris, P.C., Gomez, M.P., and Anderson, W.P. (1961). A rational analytic theory of fatigue. *The Trend in Engineering* 13: 9–14.

Pecht, M. (1996). Why the traditional reliability prediction models do not work – is there an alternative. *Electronic Cooling* 2 (1): 10–12.

Pecht, M., Dasgupta, A., Barker, D., and Leonard, C.T. (1990). The reliability physics approach to failure prediction modelling. *Quality and Reliability Engineering International* 6 (4): 267–273.

Phadke, M.S. (1989). *Quality Engineering Using Robust Design*. Englewood Cliffs, NJ: Prentice-Hall.

Pickford, J. (ed.) (2001). *Mastering Risk, Vol. 1: Concepts*. Harlow: Pearson Education Ltd.

Pierre, D.A. (1986). *Optimisation Theory with Applications*. Mineola: Courier Dover Publications.

Ramakumar, R. (1993). *Engineering Reliability, Fundamentals and Applications*. Englewood Cliffs: Prentice Hall.

Rantanen, K. and Domb, E. (2008). *Simplified TRIZ*, 2e. Auerbach Publications.

Reade, C. (1989). *Elements of Functional Programming*. Boston, MA: Addison-Wesley Longman.

Reason, J.T. (1997). *Managing the Risks of Organizational Accidents*. Burlington, VT: Ashgate.

Richardson, J.W. and Outlaw, J.L. (2008). Ranking risky alternatives: innovations in subjective utility analysis, Risk Analysis VI. *Proceedings of the 6th International Conference on Computer Simulation Risk Analysis and Hazard Mitigation*, Cephalonia, Greece (5–7 May 2008), pp. 213–224.

Rosenthal, J.S. (2006). *A First Look at Rigorous Probability Theory*, 2e. Singapore: World Scientific Publishing.

Ross, S.M. (1997). *Simulation*, 2e. San Diego: Harcourt Academic Press.

Ross, S. (1999). Adding risks: Samuelson's fallacy of large numbers revisited. *Journal of Financial and Quantitative Analysis* 34 (3): 323–339.

Samuel, A. and Weir, J. (1999). *Introduction to Engineering Design: Modelling, Synthesis and Problem Solving Strategies*. London: Elsevier.

Samuelson, P.A. (1963). Risk and uncertainty: a fallacy of large numbers. *Scientia* 98: 108–113.

Savransky, S.D. (2000). *Introduction to TRIZ Methodology of Inventive Problem Solving*. CRC Press LLC.

Sedgewick, R. (1992). *Algorithms in C++*. Addison-Wesley Publishing Company.

Shepp, L.A. (1972). Covering the circle with random arcs. *Israel Journal of Mathematics* 11: 328–345.

Smith, D.J. (2001). *Reliability, Maintainability and Risk*. Butterworth.

Solomon, H. (1978). *Geometric Probability*. Philadelphia, PA: SIAM.

Starmer, C. (2000). Developments in non-expected utility theory: the hunt for a descriptive theory of choice under risk. *Journal of Economic Literature* 38: 332–382.

Steele, J.M. (2004). *The Cauchy-Schwarz Master Class: An Introduction to the Art of Mathematical Inequalities*. New York: Cambridge University Press.

Stevens, W.L. (1939). Solution to a geometrical problem in probability. *Annals of Eugenics* 9 (4): 315–320.

Stroock, D.W. (2011). *Probability Theory: An Analytical View*, 2e. Cambridge, UK: Cambridge University Press.

Sundararajan, C.(.R.). (1991). *Guide to Reliability Engineering: Data Analysis, Applications, Implementations and Management*. New York: Van Nostrand Reinhold.

Sunstein, C.R. (2002). *Risk and Reason: Safety, Law, and the Environment*. Cambridge University Press.

Sutton, I.S. (1992). *Process Reliability and Risk Management*. New York: Van Nostrand Reinhold.

Svenson, O. (1991). The accident evolution and barrier function (AEB) model applied to incident analysis in the processing industries. *Risk Analysis* 11 (3): 499–507.

Teall, J.L. and Hasan, I. (2002). *Quantitative Methods for Finance and Investment*. Blackwell Publishing.

Teicher, H. (1963). Identifiability of finite mixtures. *The Annals of Mathematical Statistics* 34: 1265–1269.

Terninko, J., Zusman, A., and Zlotin, B. (1998). *Systematic Innovation: An Introduction to TRIZ*. CRC Press LLC.

Thompson, G. (1999). *Improving Maintainability and Reliability through Design*. London: Professional Engineering Publishing Ltd.

Thompson, W.A. (1988). *Point Process Models with Applications to Safety and Reliability*. London: Chapman & Hall.

Titterington, D.M., Smith, A.F.M., and Makov, U.E. (1985). *Statistical Analysis of Finite Mixture Distributions*. Chichester: Wiley.

Todinov, M.T. (1999). Maximum principal tensile stress and fatigue crack origin for compression springs. *International Journal of Mechanical Sciences* 41: 357–370.

Todinov, M.T. (2000). Residual stresses at the surface of automotive suspension springs. *Journal of Materials Science* 35: 3313–3320.

Todinov, M.T. (2001). Necessary and sufficient condition for additivity in the sense of the Palmgren-Miner rule. *Computational Materials Science* 21: 101–110.

Todinov, M.T. (2002a). Distribution mixtures from sampling of inhomogeneous microstructures: variance and probability bounds of the properties. *Nuclear Engineering and Design* 214: 195–204.

Todinov, M.T. (2002b). Distribution of properties from sampling inhomogeneous materials by line transects. *Probabilistic Engineering Mechanics* 17: 131–141.

Todinov, M.T. (2003). Modelling consequences from failure and material properties by distribution mixtures. *Nuclear Engineering and Design* 224: 233–244.

Todinov, M.T. (2004a). A new reliability measure based on specified minimum distances before the locations of random variables in a finite interval. *Reliability Engineering and System Safety* 86: 95–103.

Todinov, M.T. (2004b). Reliability analysis and setting reliability requirements based on the cost of failure. *International Journal of Reliability, Quality and Safety Engineering* 11 (3): 273–299.

Todinov, M.T. (2004c). Uncertainty and risk associated with the Charpy impact energy of multi-run welds. *Nuclear Engineering and Design* 231: 27–38.

Todinov, M.T. (2005). Limiting the probability of failure for components containing flaws. *Computational Materials Science* 32: 156–166.

Todinov, M.T. (2006a). Equations and a fast algorithm for determining the probability of failure initiated by flaws. *International Journal of Solids and Structures* 43: 5182–5195.

Todinov, M.T. (2006b). Reliability analysis based on the losses from failures. *Risk Analysis* 26 (2): 311–335.

Todinov, M.T. (2006c). Reliability analysis of complex systems based on the losses from failures. *International Journal of Reliability, Quality and Safety Engineering* 13 (2): 1–22.

Todinov, M.T. (2007). *Risk-Based Reliability Analysis and Generic Principles for Risk Reduction*. Amsterdam: Elsevier.

Todinov, M.T. (2009a). A comparative method for improving the reliability of components. *Nuclear Engineering and Design* 239: 214–220.

Todinov, M.T. (2009b). Is Weibull distribution the correct model for predicting probability of failure initiated by non-interacting flaws. *International Journal of Solids and Structures* 46: 887–901.

Todinov, M.T. (2009c). Robust design using upper bound variance theorem. *International Journal of Performability Engineering* 5 (4): 446–462.

Todinov, M.T. (2010). The cumulative stress hazard density as an alternative of the Weibull model. *International Journal of Solids and Structures* 47: 3286–3296.

Todinov, M.T. (2011a). Analysis and optimization of repairable flow networks with complex topology. *IEEE Transactions on Reliability* 60 (1): 111–124.

Todinov, M.T. (2011b). Virtual accelerated life testing of complex systems. In: *Intelligent Decision Systems in Large-Scale Distributed Environment* (ed. P. Bouvry, H. Gonzalez-Vélez and J. Kolodziej), 293–314. Berlin: Springer.

Todinov, M.T. (2013a). *Flow Networks.* Amsterdam: Elsevier.

Todinov, M.T. (2013b). The dual network theorem for static flow networks and its application for maximising the throughput flow. *Artificial Intelligence Research* 2 (1): 81–106.

Todinov, M.T. (2013c). New models for optimal reduction of technical risks. *Engineering Optimization* 45 (6): 719–743.

Todinov, M.T. (2014). Optimal allocation of limited resources among discrete risk-reduction options. *Artificial Intelligence Research* 3 (4): 15–27.

Todinov, M.T. (2015). Reducing risk through segmentation, permutations, time and space exposure, inverse states and separation. *International Journal of Risk and Contingency Management* 4 (3): 1–21.

Todinov, M.T. (2016a). *Reliability and Risk Models: Setting Reliability Requirements*, 2e. Wiley.

Todinov, M. (2016b). Stochastic pruning and its application for fast estimation of the expected total output of complex systems. *Electronic Notes in Theoretical Computer Science* 327: 109–123.

Todinov, M.T. (2017a). Mechanisms for improving reliability and reducing risk by stochastic and deterministic separation. *Journal of Risk Research*, published online doi: 10.1080/13669877.2017.1382561.

Todinov, M.T. (2017b). Reducing risk through inversion and self-strengthening. *International Journal of Risk and Contingency Management* 6 (1): 14–42.

Todinov, M.T. (2017c). Reducing risk by segmentation. *International Journal of Risk and Contingency Management* 6 (3): 27–46.

Todinov, M.T. (2017d). Reliability and risk controlled by the simultaneous presence of random events on a time interval. *ASCE-ASME Journal of Risk and Uncertainty in Engineering Systems, Part B: 4(2), 021003, Mechanical Engineering*, published online, ASME doi: 10.1115/1.4037519.

Todinov, M.T. and Weli, E. (2013). Optimal risk reduction in the railway industry, by using dynamic programming. *WASET, International Journal of Mechanical, Aerospace, Industrial and Mechatronics Engineering* 79 (7): 415–419.

Todinov, M.T., Novovic, M., Bowen, P., and Knott, J.F. (2000). Modelling the impact energy in the ductile/brittle transition region of C-Mn multi-run welds. *Materials Science & Engineering A* A287: 116–124.

Tonelli, L. (1909). Sull'integrazione per parti. *Atti della Accademia Nazionale dei Lincei (5)* 18 (2): 246–253.

Trivedi, K.S. (2002). *Probability and Statistics with Reliability, Queuing and Computer Science Applications*, 2e. Chichester: Wiley.

Tunno, D. and Larsen, S, (2011a). Retrofittable cable mechanical fuse. US Patent 20,110,027,007 A1.

Turner, B. (1978). *Man-Made Disasters*. London, UK: Wykeham Science Press.

Uicker, J.J. Jr., Pennock, G.R., and Shigley, J.E. (2003). *Theory of Machines and Mechanisms*, 3e. New York: Oxford University Press.

Venkatesh, S.S. (2013). *The Theory of Probability: Explorations and Applications*. Cambridge, UK: Cambridge University Press.

Vose, D. (2000). *Risk Analysis, a Quantitative Guide*, 2e. New York: Wiley.

Vugrin, E.D., Warren, D.E., Ehlen, M.A., and Camphouse, R.C. (2010). A framework for assessing the resilience of infrastructures and economic systems. In: *Sustainable Infrastructure Systems: Simulations, Imaging, and Intelligent Engineering*. New York: Springer-Verlag.

Vugrin, E.D., Warren, D.E., and Ehlen, M.A. (2011). A resilience assessment framework for infrastructure and economic systems: quantitative and qualitative resilience analysis of petrochemical supply chains to a hurricane. *Process Safety Progress* 30 (3): 280–290.

Wald, A. (1950). *Statistical Decision Functions*. New York: Wiley.

Weibull, W. (1951). A statistical distribution of wide applicability. *Journal of Applied Mechanics* 18: 293–297.

Weiss, N.A. (2006). *A Course in Probability*. Boston, MA: Pearson/AddisonWesley.

Wirth, N. (1976). *Algorithms + Data Structures = Programs*. Englewood Cliffs: Prentice-Hall.

Zhang, Y.B. (2012). The long-range monitoring system of water level based on GPRS network. In: *Advances in Control and Communication* (ed. D. Zeng). Springer.

Zhang, L., Konyukhov, A., Mok, E., and Choo, H. (2013). Increasing the road safety of e-bike: design of protecting shell based on stability criteria during severe road accidents. *Proceedings of the 2013 International Conference on Electrical, Control and Automation Engineering (ECAE'13)*, pp. 485–489.

Index

u

v

w

y